INTERNATIONAL SERIES IN
NATURAL PHILOSOPHY
General Editor: D. ter HAAR

VOLUME 111

Solid State Nuclear Track Detection

Principles, Methods and Applications

Related Pergamon Titles of Interest

Books

BENTON *et al.* (eds.)
Nuclear Track Registration

CLAYTON
Nuclear Geophysics

ESPINOSA *et al.* (eds.)
Solid State Nuclear Track Detectors 12

FOWLER & CLAPHAM (eds.)
Solid State Nuclear Track Detectors 11

FRANÇOIS *et al.* (eds.)
Solid State Nuclear Track Detectors 10

GRANZER *et al.* (eds.)
Solid State Nuclear Track Detectors 9

LUNAR & PLANETARY INSTITUTE
Proceedings of the Twelfth Lunar and Planetary Science Conference

MCLAUGHLIN
Trends in Radiation Dosimetry

MILLER *et al.* (eds.)
Fission-Track Dating [1985]

TOMMASINO *et al.* (eds.)
Solid State Nuclear Track Detectors 13 [1986]

WAGNER *et al.* (eds.)
Fission-Track Dating [1981]

WAGNER *et al.* (eds.)
Thermoluminescence and Electron-Spin-Resonance Dating [1985]

YORK & FARQUHAR
The Earth's Age & Geochronology, 2nd edition.

Journals

Geochim. Cosmochim. Acta

Health Physics

International Journal of Applied Radiation & Isotopes

International Journal of Nuclear Medicine and Biology

Nuclear Tracks and Radiation Measurements

Radiation Physics and Chemistry

Full details of all Pergamon publications and a free specimen copy of any
Pergamon journal are available on request from your nearest Pergamon office.

Solid State Nuclear Track Detection

Principles, Methods and Applications

by

S. A. DURRANI
University of Birmingham, UK

and

R. K. BULL
AERE, Harwell, UK

PERGAMON PRESS

OXFORD · NEW YORK · BEIJING · FRANKFURT
SÃO PAULO · SYDNEY · TOKYO · TORONTO

U.K.	Pergamon Press, Headington Hill Hall, Oxford OX3 0BW, England
U.S.A.	Pergamon Press, Maxwell House, Fairview Park, Elmsford, New York 10523, U.S.A.
PEOPLE'S REPUBLIC OF CHINA	Pergamon Press, Qianmen Hotel, Beijing, People's Republic of China
FEDERAL REPUBLIC OF GERMANY	Pergamon Press, Hammerweg 6, D-6242 Kronberg, Federal Republic of Germany
BRAZIL	Pergamon Editora, Rua Eça de Queiros, 346, CEP 04011, São Paulo, Brazil
AUSTRALIA	Pergamon Press Australia, P.O. Box 544, Potts Point, N.S.W. 2011, Australia
JAPAN	Pergamon Press, 8th Floor, Matsuoka Central Building, 1-7-1 Nishishinjuku, Shinjuku-ku, Tokyo 160, Japan
CANADA	Pergamon Press Canada, Suite 104, 150 Consumers Road, Willowdale, Ontario M2J 1P9, Canada

First edition 1987

Library of Congress Cataloging in Publication Data
Durrani, S. A.
Solid state nuclear track detection.
(International series in natural philosophy, v. 111)
Includes index.
1. Nuclear track detectors.   I. Bull, R. K.
II. Title.   III. Series.
QC787.N83D87 1985   539.7'7   84-26349

British Library Cataloguing in Publication Data
Durrani, S. A.
Solid state nuclear track detection: principles, methods and applications.—(International series in natural philosophy; v. 111)
1. Particle tracks (Nuclear physics)
2. Solids—Effect of radiation on
I. Title   II. Bull, R. K.   III. Series
539.7'7   QC793.3.T67

ISBN 0-08-020605-0

Cover design

In the lower foreground is a five-prong event produced by the interaction of 16.7 MeV/nucleon ^{238}U ions with natural uranium, and recorded in a mica SSNT detector in 2π-geometry. The irradiation was made on the UNILAC accelerator at GSI, Darmstadt, Fed. Rep. Germany, and analysed by E. U. Khan (Marburg and Gomal universities). The event, believed to be "triple sequential fission", has been reconstructed from ~13 stepwise photomicrographs; the longest prong is ~80 μm (photograph, courtesy P. Vater). The event is superimposed on a picture of electrochemically etched (ECE) spots produced in the CR-39 polymeric detector by ~2.5 MeV neutrons from the University of Birmingham Dynamitron accelerator (photograph, courtesy S.A.R. Al-Najjar).

Printed in Great Britain by A. Wheaton & Co. Ltd., Exeter

Preface

This book has been long in gestation. The senior author (SAD) first proposed the book to Pergamon Press, Oxford, as long as 1973–74, and a contract was drawn up soon thereafter. Before much progress could be made, however, the book by Fleischer, Price and Walker*, the pioneers of the Solid State Nuclear Track Detection (SSNTD) technique, came on the market in 1975. It was a topical and exciting book, and masterfully covered all or most of what was known of the subject and its applications up to around mid-1973. The need for a second book on the subject was thus obviated for some years to come, and the author-designate decided to bide his time. He, in the meantime, became fully engrossed in the launching and nurturing of the journal *Nuclear Track Detection* (now called *Nuclear Tracks and Radiation Measurements*), of which he was the Founder-Editor, and with whose Editorial Board some of the foremost names of the SSNTD discipline (including Robert Fleischer himself) are now associated. This journal was launched with the personal encouragement of Mr Robert Maxwell, Chairman of Pergamon Press, in March 1977, and was a direct product (or by-product) of the contract for the book.

With the passage of time it was felt, around the beginning of the present decade, that enough new material had appeared in the fast-expanding subject area of SSNTD in the meantime to warrant the publication of a new book to bring the reader and research-worker up to date. The original author-designate thereupon invited the second author (RKB), who was then working as a Research Fellow in the Department of Physics, University of Birmingham, and had joined the Editorial Board of *Nuclear Tracks* as its Book Reviews Editor, to join him in the task of writing the book.

The first draft of the book was completed in the winter of 1981–82. It was then very carefully and meticulously read by our colleague in the Department, Dr P. F. Green, who made very many constructive suggestions, for which we are greatly indebted to him. The revised draft was completed and handed over to Pergamon Press, Oxford, at the beginning of 1983. The drawing of figures, collection of illustrations, filling-in of the gaps (in reference details, etc.) took us to the middle of 1983. Thereupon, unfortunately, followed some delays and problems as to the mode of production of the book, which were

---

* *Nuclear Tracks in Solids: Principles and Applications* by R. L. Fleischer, P. B. Price & R. M. Walker. University of California Press, Berkeley (1975).

finally resolved by the autumn of 1983. The final literature-survey was completed in mid-1984, when roughly 20% new references were added (taking the total to over 600). These intercalated references are obvious from their designation (such as 17a, b, etc.). A limited amount of new material was also inserted into the Ms in the summer of 1984 in an effort to update it. The book proofs were ready by mid-February 1985, and were checked in the spring of 1985, when a few paragraphs were added, especially in Chapter 9. The publication of the book was originally aimed to coincide with the 13th International Conference on Solid State Nuclear Track Detectors, held at Rome from 23 to 27 September, 1985.*

This book is much shorter, and more modest in its aims, than the above-mentioned maiden book on the subject by Fleischer, Price, and Walker. The readership it aims at are primarily the newcomers to the SSNTD field of research, who wish to acquire a basic knowledge of the techniques of the discipline and to gain a general view of the present status of the subject; but of course we hope that the more experienced or specialist nuclear-track worker will also find something of interest and value in the book. The readership aimed at also includes the non-specialist professional who wants to apply the SSNTD methods to his (or her) particular field and is looking for a concise introductory text. The book will, moreover, it is hoped, prove of value to graduate students working for their Master's degree in applied physics, nuclear chemistry, or isotope geology, as well as to postgraduate and postdoctoral research workers in related disciplines (e.g. nuclear science and engineering, radiation physics and radiological protection, medical physics, geochronology, radiation dosimetry, etc.). Finally, some parts of the book should also prove helpful in setting up standard undergraduate experiments, and it may be used as supplementary reading for final-year honours undergraduates.

It may be taken as axiomatic that any book will reflect the limitations as well as the strengths of its author or authors. No authors can claim to be omniscient or infallible—and we certainly make no such claim. In a fast-expanding and -developing field of research it is particularly difficult to keep abreast of all the latest developments. The book can, of necessity, only represent those publications and additions to knowledge of which we ourselves became aware, or to which our attention was drawn by colleagues.

---

* *Note added in proof.* The first proof-reading during February–June 1985 unfortunately produced so many changes (but, we hope, improvements) in the text and other printed matter that most of the book had to be reset by the press. The revised proofs were received only in January 1986. This further delay is greatly regretted. Apropos of the Rome conference, the reader may be interested to know that an International Nuclear Tracks Society, originally proposed during the 12th International Conference at Acapulco in 1983, was formally constituted and launched at the Rome conference in September 1985.—*The authors.*

One has then to be selective, if the book is to remain within reasonable bounds of length and coherence. As such, we apologize to any research workers whose contributions to knowledge have not been properly represented in this book for the reasons just stated, or which appeared after our final literature survey had been completed.

It will be seen that the book emphasizes the techniques and applications more than it does the basic mechanisms. This is partly because of the readership at which we have aimed, and partly, or perhaps fundamentally, because the basic solid-state mechanisms have not yet been fully worked out. There is pressing need for further theoretical and experimental work to illuminate the underlying principles of track formation and revelation in dielectric solids. At the recent international SSNTD conferences the sessions on fundamental mechanisms have been notable for their brevity. At the end of the book (Chapter 9) we have outlined what we see as the most promising trends in nuclear track detection and its applications, and have also highlighted some of the successful developments to date.

Throughout the book, we have liberally drawn upon the compendia of knowledge in the form of the Proceedings of the international series of conferences on SSNTD—and in particular the four most recent ones (numbers 9, 10, 11 and 12) held, respectively, at Munich (1976), Lyon (1979), Bristol (1981) and Acapulco (1983)*: all of which have been published as the Supplements (or special issues) of the journal *Nuclear Tracks and Radiation Measurements* (as now entitled). Other particularly useful sources of material have, besides that journal, been *Nuclear Instruments and Methods, Radiation Effects, Health Physics, Earth and Planetary Science Letters, Journal of Geophysical Research, The International Journal of Applied Radiation and Isotopes,* and *Geochimica et Cosmochimica Acta,* besides the more general journals such as *Nature, Science,* and the *Physical Review.*

It is a pleasure for us to thank a number of colleagues, both here and abroad, as well as several students of this Department, who have helped in various ways in removing some of the deficiencies of the book and by making constructive suggestions. Of these, Dr Paul Green has already been mentioned. Among the others, the following merit special mention and our sincere thanks: Professor J. H. Fremlin, Professor W. E. Burcham, Dr G. Somogyi, Dr J. Pálfalvi, J. A. B. Gibson (who pointed out a number of errors in the proofs of the book), Dr K. G. Harrison, Dr A. G. Ramli, Dr S. A. R. Al-Najjar, Dr K. Randle, Dr J. B. A. England, D. Grose, K. James, A. Morsy Ahmed, and A. Abdel-Naby. The help of Matiullah and H. Afarideh with the preparation of the Index is also acknowledged.

---

* These lines were written before the Rome conference (1985), of which the Proceedings will be published as Vol. 12 of *Nuclear Tracks* (1986). We have added a brief Epilogue at the end of Chapter 9 to reflect some of the highlights of the Rome conference.

We also wish to thank a number of authors who readily gave us their permission to reproduce figures from their publications (which is also duly acknowledged at the end of the relevant figure captions). This list includes: R. L. Fleischer, P. B. Price and R. M. Walker; E. C. H. Silk and R. S. Barnes; J. Csikai and A. Szalay; J. Ballam; D. A. Tidman; R. Katz and E. J. Kobetich M. Maurette; D. V. Morgan and D. Van Vliet; E. Dartyge; P. H. Fowler; U. Haack; J. Tripier and M. Debeauvais; W. Enge; R. V. Griffith; M. A. Gomaa; E. V. Benton; L. Tommasino; J. R. Harvey and A. R. Weeks; G. Somogyi; A. G. Ramli; A. J. W. Gleadow and A. J. Hurford; G. Bigazzi; G. Poupeau; J. G. Wilson; K. Thiel; P. Vater; and R. Spohr and C. Riedel; besides several of our colleagues who were co-authors of our joint publications. We are very grateful to R. Brandt and his colleagues (P. Vater, E. U. Khan and others) for allowing us to use the multi-prong fission event, observed in their laboratory, as part of the cover design of our book. We also wish to offer particular thanks to E. V. Benton and R. P. Henke for giving us permission to reproduce their computer code which appears as Appendix 1 of this book.

The manuscript was typed, at various stages, by the following ladies, and we greatly appreciate their skill and patience; Eileen Shinn, Susan Yeomans, Erica Gaize and Pauline Tevlin.

One of us (RKB) wishes to thank the Atomic Energy Research Establishment, Harwell, for permission to publish this book—although, for the sake of record, it ought to be stated that the writing of the manuscript had essentially been completed while he was still a member of the Department of Physics, University of Birmingham.

In the end, we would like to express our appreciation and thanks to various members of Pergamon Press, Oxford, for their help with the production of the book. In particular, we wish to acknowledge the unfailing courtesy and the great forbearance shown to us by Mr P. A. Henn, Senior Publishing Manager (Physical Sciences) at Pergamon Press, during the long months that we kept him waiting while we chopped and changed the manuscript and introduced latest references or subject matter. Similarly, we greatly appreciate the help given by Mr Philip Halsey and his efficient staff at the Drawing Office of the Press for drawing our figures and falling in with all our finical attempts at perfecting them. We also wish to thank Catherine Shephard, latterly the Senior Acquisitions Editor for Pergamon Books Ltd, who is now in charge of the project, and who was also responsible for it at an earlier stage.

It is only right, before we finish these lines, if we say that during the several years that we have been, off and on, occupied in writing this book and bringing it to fruition, often burning the midnight oil, our families have shown great indulgence and fortitude in putting up with all these "unfamilial" preoccupations. Producing a book, it would appear, is a more extended, and worse, business than giving birth! We must, and do, say a big THANK YOU to our families for their loyalty and understanding.

Last, but not least, our readers. We hope that they find that some, at least, of our efforts have been worth the trouble taken.

University of Birmingham   SAEED A. DURRANI
AERE, Harwell   RICHARD K. BULL
24 June, 1985
(and 12 March, 1986)

Acknowledgement for the reproduction of figures

The authors whose figures have been reproduced in this book from published literature (sometimes with modifications) have been thanked individually in the Preface. In addition, the present authors as well as Pergamon Press wish to thank the following publishers for their permission to reproduce the figures as noted below. Complete references to the authors and the journal or book concerned are to be found at the end of the relevant chapter, the reference citation being given at the end of each figure caption.

Taylor and Francis Ltd (Figs. 1.1, 1.4, 3.4 and 8.8).
American Institute of Physics and the American Physical Society (Fig. 2.3).
University of California Press (Fig. 3.1b).
Elsevier Science Publishers (Figs. 3.3, 4.3a, 5.2 and 8.11).
North-Holland Publishing Co. (Figs. 4.3b, 6.4a,b, 6.5a,b, and 6.9).
Gordon and Breach Science Publishers (Fig. 4.9a,b).
Macmillan Journals Ltd (Fig. 6.7).
American Geophysical Union (Figs. 7.9 and 8.10).

Contents

1 Introduction to Nuclear Track Detectors   **1**

  1.1 Cloud, Bubble and Spark Chambers   4
     (a) The cloud chamber   4
     (b) The bubble chamber   5
     (c) The spark chamber   6
  1.2 Nuclear Emulsions   8
  1.3 Silver Halide Crystals   8
  1.4 Etchable Solid State Nuclear Track Detectors (SSNTDs)   10

2 Interactions of Charged Particles with Matter   **13**

  2.1 Nuclear Collision Losses   14
  2.2 Electronic Energy Losses   15
  2.3 Direct Production of Atomic Displacements   19
  2.4 Secondary Electrons   19
  2.5 Range–Energy Relations   20

3 The Nature of Charged-Particle Tracks and Some Possible Track Formation Mechanisms in Insulating Solids   **23**

  3.1 Radiation Damage in Solids   23
     (a) The Seitz model   25
     (b) The Varley model   25
     (c) The Pooley mechanism   25
  3.2 Track-storing Materials   26
  3.3 Track-forming Particles: Criteria for Track Formation   27
     (a) Total rate of energy loss, dE/dx 29
     (b) Primary ionization, J 30
     (c) Restricted energy loss (REL)   31
     (d) Secondary-electron energy loss   31
     (e) Radius-restricted energy loss (RREL)   32
     (f) Lineal event-density (LED)   32
  3.4 Experimental Studies on the Size and Structure of Latent-Damage Trails   33
     3.4.1 Electron microscopy   33
     3.4.2 Low-angle X-ray scattering   34
     3.4.3 Thermal annealing of tracks   37
     3.4.4 The radial extent of the etchable damage   38
     3.4.5 Other experimental evidence for track structure   39
  3.5 Critical Appraisal of Track Formation Models   39
     3.5.1 The thermal-spike model   41
     3.5.2 The ion-explosion spike model   42

4 Track Etching: Methodology and Geometry   **48**

  4.1 Track Etching Recipes   49
  4.2 Track Etching Geometry   51

4.2.1   Constant track etching velocity V_T 52
4.2.2   Determination of track parameters R and V_T 61
4.2.3   Etching efficiencies: internal and external track sources   64
        (i) Internal tracks   64
        (ii) External source of tracks   70
        4.2.3.1   Prolonged-etching factor $f(t)$ 72
4.2.4   Track etching geometry with varying V_T 76
4.2.5   Track etching geometry in anisotropic solids   83
4.3   Some Special Techniques for Track Parameter Measurements   87
4.4   Environmental Effects on Track Etching   89

5   Thermal Fading of Latent Damage Trails   **96**

5.1   The Nature of the Annealing Process   96
5.2   The Effects of Pre-annealing on the Etched Tracks   98
5.3   Typical Annealing Temperatures for Fission Tracks in Various Materials 101
5.4   Closing Temperatures   103
5.5   Annealing Correction Methods   110
5.6   Track Seasoning   111

6   The Use of Dielectric Track Recorders in Particle Identification   **114**

6.1   Calibration   115
      6.1.1   The L-R plot   115
6.2   Charge Assignment   124
6.3   Low-energy Particles   125
6.4   Charge and Mass Resolution   126
6.5   Some Applications of Particle Identification Techniques   127
      6.5.1   Cosmic ray physics   128
      6.5.2   Nuclear physics   131
6.6   The Ancient Cosmic Rays   136

7   Radiation Dosimetry and SSNTD Instrumentation   **144**

7.1   Neutron Dosimetry   145
      7.1.1   Thermal neutrons   145
      7.1.2   Fast and intermediate-energy neutrons   148
7.2   Alpha Particle Dosimetry and Radon Measurements   161
7.3   Charged Particles other than Alphas   167
      7.3.1   Dosimetry of high-LET radiations in space   167
      7.3.2   Microdosimetry of negative-pion beams   168
7.4   SSNTD Instrumentation: Automatic Evaluation and Methods of
      Track Image Enhancement   169
      7.4.1   The spark counter   169
      7.4.2   Other electrical-breakdown devices   173
      7.4.3   Scintillator-filled etch pit counting   173
      7.4.4   Electrochemical etching (ECE)   176
      7.4.5   Automatic and semi-automatic image-analysis systems   184
      7.4.6   Other methods of measurement   192

8   Fission Track Dating   **199**

8.1   Radioactive Dating   199
8.2   The Fission Track Age Equation   200

8.3		Practical Steps in Obtaining a Fission Track Age	203
	8.3.1	The population method	206
	8.3.2	The external-detector method	207
8.4		The Interpretation of Fission Track Ages	209
8.5		Neutron Dosimetry, Fission Decay Constant of ^{238}U, and Age Standards	212
	8.5.1	Neutron fluence measurements	212
	8.5.2	The fission decay constant λ_{f8}	214
	8.5.3	Age standards	216
8.6		Annealing Corrections	218
	8.6.1	The track size correction method	218
	8.6.2	The plateau correction method	220
8.7		Fission Track Dating of Lunar Samples and Meteorites	223
	8.7.1	Heavy cosmic-ray primaries	223
	8.7.2	Cosmic-ray-induced fission	228
	8.7.3	Spallation recoil tracks	230
8.8		^{244}Pu Fission Tracks in Very Ancient Samples	230
	8.8.1	^{244}Pu fission track age equation	231
	8.8.2	The "contact" track density method	232
8.9		Fission Track Dating in Archaeology	236
8.10		Errors in Fission Track Dating	239

9	**Further Applications of Track Detectors and Some Directions for the Future**		**245**
9.1		Applications to Nuclear Physics	245
	9.1.1	Fission phenomena and related studies	245
	9.1.2	Other nuclear reactions	249
9.2		Elemental Distributions and Biological Applications	250
	9.2.1	Elemental mapping	250
	9.2.2	Biological applications	253
		(a) Inhalation of α-active aerosols	254
		(b) Lead in teeth	254
		(c) Filtration of malignant cells by microporous films	255
		(d) Measurement of α-emitters in the environment	256
9.3		Extraterrestrial Samples	257
	9.3.1	Lunar sample studies	257
	9.3.2	Meteorites	260
9.4		Track Detectors in Teaching	260
9.5		Future Developments in Etched Track Techniques and Their Applications	262
9.6		**Epilogue**	266

Appendix 1.	A program to calculate the range and energy-loss rate of charged particles in stopping media	**275**

Subject Index	**285**

Titles in the Series	**301**

CHAPTER 1

Introduction to Nuclear Track Detectors

"During the last 15 years or so the method generally known as 'solid state nuclear track detection' (SSNTD) has grown to such an extent", proclaimed the first editorial[1] written to launch the journal *Nuclear Track Detection* (later renamed *Nuclear Tracks and Radiation Measurements*), in 1977, "that now there is hardly a branch of science and technology where it does not have actual or potential applications. Fields where well-established applications of this technique already exist", it went on to say, "include: fission and nuclear physics; space physics; the study of meteoritic and lunar samples; cosmic rays; particle accelerators and reactors; metallurgy, geology and archaeology; medicine and biology; and many others."

One can only add that, with the passage of time, the above claim has become more than ever true. This is borne out by publications not only in *Nuclear Tracks and Radiation Measurements* itself, but also in many other international research journals covering a diversity of fields. The genuinely multidisciplinary nature of this technique has been highlighted by the periodic gatherings at the international series of conferences on Solid State Nuclear Track Detectors (SSNTDs) held at Clermont-Ferrand (1969),* Barcelona (1970), Bucharest (1972), Munich (1976), Lyon (1979), Bristol (1981), and Acapulco (1983). At these conferences, geologists have rubbed shoulders with nuclear physicists, archaeologists with reactor physicists, and space scientists with biologists.

The story began in 1958 when D. A. Young,[2] working at the Atomic Energy Research Establishment at Harwell in England, discovered that LiF crystals, held in contact with a uranium foil and irradiated with thermal neutrons, revealed a number of etch pits after treatment with a chemical reagent. The number of these pits showed complete correspondence with the estimated number of fission fragments which would have recoiled into the crystal from the uranium foil. It seemed that each pit was formed around the site of solid-state damage produced by the fission fragments.

In 1959 Silk and Barnes,[3] also working at AERE, Harwell, reported direct observations of these damaged regions in mica—which appeared as hair-like

---

* The meeting at Clermont-Ferrand, unlike the others mentioned here, was in fact an International Topical Conference on Nuclear Track Registration in Insulating Solids and Applications. The 13th International Conference on SSNTDs took place at Rome in September 1985.

tracks—using transmission electron microscopy. The abstract of Silk and Barnes's classical paper of 1959 demonstrates their clear insight into the nature of the phenomenon observed by them and is a model of concise and lucid reporting of observation and interpretation. It is reproduced here in full:

Thin sheets, removed by cleavage from the surface layers of mica exposed to fission fragments from uranium, have been examined in transmission in the electron microscope. Fission fragment tracks which are less than 300 Å in diameter and greater than four microns in length have been seen, superimposed upon a general background introduced by associated neutron bombardment. The nature of the tracks is discussed. The technique can be used to study radiation induced atomic displacements, and the nature and behaviour of the particles themselves, with higher resolution than has previously been possible.

Some of the tracks observed by Silk and Barnes are shown in Fig. 1.1.

During the early 1960s the team of R. L. Fleischer, P. B. Price and R. M. Walker, working at the General Electric Research Laboratories at Schenectady, New York, pioneered the extensive development of this method. In particular, they extended the etching technique of Young (whose work was unknown to them at that time) to mica and eventually to a variety of other materials such as glasses, plastics and various mineral crystals. (For an excellent review of the early developments of the technique see the paper by Fleischer *et al.*[4])

Early studies showed that these etchable tracks were:

(1) produced only by heavily ionizing particles (e.g. α-particles in the case of plastics, and fission fragments in the case of crystals);
(2) produced only in electrical insulators or poor semiconductors;
(3) stable even when subjected to light or to high doses of X-rays, β-particles, ultraviolet radiation, etc.

The durability, the simplicity and the markedly specific nature of the response of these detectors led to their rapid application in a wide variety of fields.

While a considerable body of information has now accumulated on the nature of unetched tracks from techniques such as electron microscopy[5] and small-angle X-ray scattering,[6] it is actually this etching process—which essentially "fixes" the damage trail and allows observation to be made under a standard optical research microscope—that has led to the present widespread application of solid state nuclear track detectors mentioned at the beginning of this section. It is, thus, largely with the etched tracks that this book will be concerned.

Before proceeding to an outline of the basic properties of SSNTDs it is perhaps instructive to place these properties in context by discussing, briefly, some of the other types of nuclear detectors which allow an image of a particle trajectory to be formed. This list is not intended to be exhaustive, and readers requiring a fuller treatment of these devices are referred to the many texts on experimental nuclear physics.

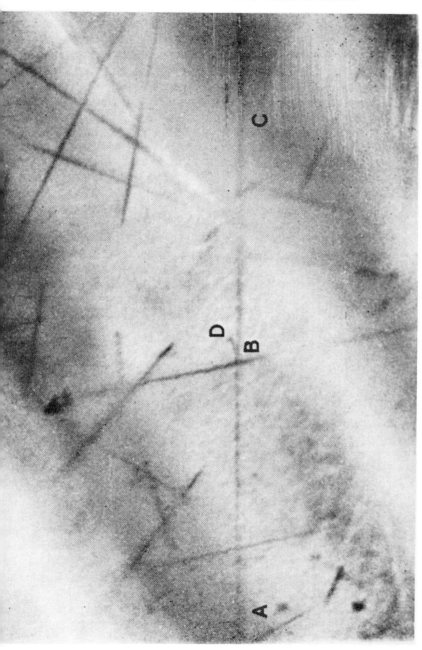

FIG. 1.1   A transmission electron microscope (TEM) photograph of fission tracks in mica (from Silk and Barnes[3]). These tracks, which were produced in thin sheets of muscovite mica exposed to reactor neutrons while in contact with thin layers of natural uranium, were correctly interpreted by Silk and Barnes as being arrays of defects produced by the passage of fission fragments. They noted that "these dark tracks were generally straight, less than 300 Å in diameter" and sometimes over 4 µm long (when lying in the plane of the mica sheet). The photograph also purports to show a colliding fission fragment initially travelling along AB and undergoing a slight deflection at B to proceed along BC; BD being possibly an atom of mica displaced by a Rutherford-type collision.

The detectors are described roughly in the order of increasing similarity to solid state nuclear track detectors.

1.1   Cloud, Bubble and Spark Chambers

(a) The cloud chamber

A body of gas can hold only a limited amount of liquid vapour. The limiting pressure of vapour is known as the "saturated vapour pressure", and is a function of the pressure and temperature of the system. If the conditions governing the system were changed slowly, while maintaining strict thermodynamic equilibrium, then this limiting vapour pressure would not be exceeded. If, however, the temperature of the system is allowed to fall rapidly (e.g. by adiabatic expansion), then the vapour pressure will soon be in excess of its saturation value applicable to the new temperature: in other words the system will become "supersaturated". If the degree of supersaturation is very high, condensation of liquid drops may occur when microscopic fluctuations take place in the density of the vapour. At more moderate supersaturations, droplets can form only on suitable "nuclei", such as dust particles or traces of certain chemical impurities. In particular, droplets may form on ions, such as those produced by the passage of an ionizing particle.

This is the principle of the cloud chamber initially devised by C. T. R. Wilson in 1911 (Fig. 1.2a). By carefully removing all other types of droplet-forming nuclei, the vapour is made to condense specifically along the path of an ionizing particle, rendering it visible under suitable lighting conditions.

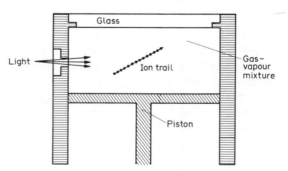

FIG. 1.2a   *Schematic view of the Wilson cloud chamber. On rapidly withdrawing the piston. the temperature of the system fans suddenly and the vapour becomes supersaturated. In such a case, liquid droplets condense from the vapour along the path of an ionizing particle.*

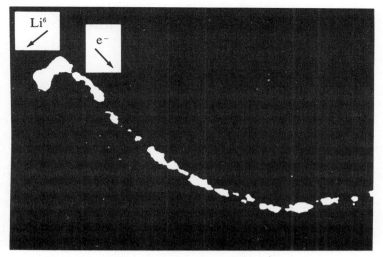

FIG. 1.2b *Cloud chamber photograph of the decay event* ^{6}He →
^{6}Li + β$^-$ + ν̄. *The tracks of the two charged particles, viz. the
electron and the recoiling residual nucleus* ^{6}Li, *can be seen in
the picture. This is believed to be the first experimental evidence
for the emission of the antineutrino. (Figure from reference 6a,
courtesy Prof. J. Csikai.)*

Figure 1.2b shows a typical cloud-chamber photograph of tracks formed by
ionizing particles. A development of the Wilson cloud chamber is the diffusion
cloud chamber, where the working volume remains continuously sensitive
rather than after each successive expansion.

(b) The bubble chamber

In some ways the principle of the bubble chamber, invented in 1952 by D. A.
Glaser, is the reverse of that of the cloud chamber. Instead of forming liquid
drops in a gas, here bubbles of gas are formed in a liquid.

A liquid is maintained under pressure at a temperature somewhat below
its boiling point at that temperature. The pressure is then suddenly lowered
to below the vapour pressure. In the absence of suitable nuclei the liquid,
which is now in a superheated state, will not boil; however, ions can act as
nucleation centres for bubble formation, and these bubbles, on growing to
visible size, can delineate the path of the ionizing particle.

The higher stopping power of the bubble chamber compared with the
cloud chamber allows data on high-energy reactions to be gathered more
rapidly. Bubble chambers, however, are bulky affairs—sometimes running
to thousands of litres in volume—and are very costly. Also, while they have

been enormously valuable in high-energy physics research, their time resolution is poor, and only a few charged particles can usually be handled during each expansion cycle. (Cycling rates up to $\sim 60\,\text{s}^{-1}$ can be achieved.) Figure 1.3 shows tracks as seen in a bubble chamber photograph. Analysing bubble chamber pictures is an enormous task, requiring large human and computing resources. Hybrid systems, involving supplementary detectors, are often used to improve particle identification.

(c) The spark chamber

The limited time resolution of the bubble chamber is overcome in the spark chamber (developed since 1957). Here the triggering takes $\sim 1\ \mu\text{s}$ (though the ensuing dead time of the chamber may be ~ 10 ms). The basic structure of the spark chamber is a bank of thin, parallel, metal plates separated by gaps of a few millimetres, and filled with a noble gas at near-atmospheric pressure. Alternate plates are connected to a high voltage, while the intermediate plates are kept at ground potential. An ionizing particle crossing a gap or a succession of gaps will cause sparks to leap along its trajectory, as in "streamer formation". The high voltage is applied in short pulses actuated by some triggering counters (e.g. large-area fast plastic scintillators or Cherenkov detectors) surrounding the spark chamber; an external magnetic field is also often applied. The spark is accompanied by a sonic shockwave as well as by the emission of electromagnetic radiations. All these phenomena have been exploited in establishing the precise location of a spark. Television cameras etc. have also been incorporated into the system, and sophisticated on-line computer analysis, similar to that used for bubble chamber signals, is employed to handle the complex data.

The spatial resolution of spark chambers can be further improved by replacing the conducting plates by a large number of wires arranged in parallel planes. Here the concept is similar to that employed in the construction of a *multi-wire proportional counter* (MWPC)—which is an array of very thin ($\sim 20\ \mu\text{m}$ diameter) anode wires, with ~ 1 mm spacing, arranged in an accurate plane parallel to, and lying between, two earthed cathodes. In a *drift chamber*, a MWPC is used as the triggering device, separated from the main chamber by a drift space. Greater spatial resolution can thus be achieved by measuring the time taken by the electron avalanche accompanying the trigger event to travel to the nearest anode wire. Here both electrodes of the spark chamber are made up of planes of fine wires.

For further details of the electronically operated detectors described above, the reader is referred to standard texts on experimental nuclear physics, such as that by England.[7]

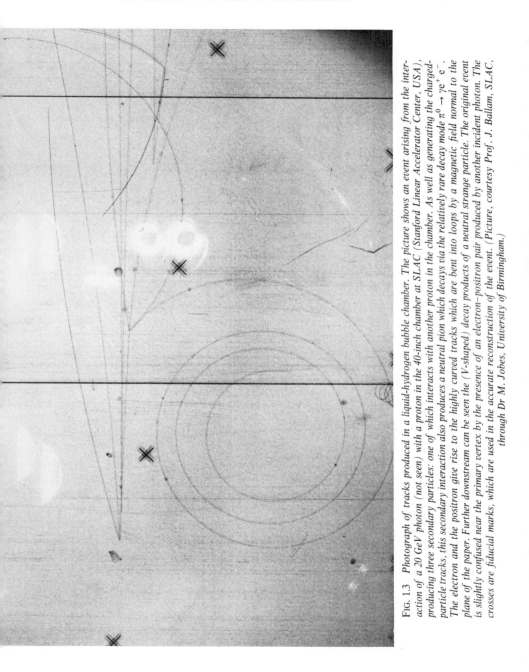

FIG. 1.3   *Photograph of tracks produced in a liquid-hydrogen bubble chamber. The picture shows an event arising from the interaction of a 20 GeV photon (not seen) with a proton in the 40-inch chamber at SLAC (Stanford Linear Accelerator Center, USA), producing three secondary particles: one of which interacts with another proton in the chamber. As well as generating the charged-particle tracks, this secondary interaction also produces a neutral pion which decays via the relatively rare decay mode $\pi^0 \rightarrow \gamma e^+ e^-$. The electron and the positron give rise to the highly curved tracks which are bent into loops by a magnetic field normal to the plane of the paper. Further downstream can be seen the (V-shaped) decay products of a neutral strange particle. The original event is slightly confused near the primary vertex by the presence of an electron–positron pair produced by another incident photon. The crosses are fiducial marks, which are used in the accurate reconstruction of the event. (Picture, courtesy Prof. J. Ballam, SLAC, through Dr M. Jobes, University of Birmingham.)*

1.2 Nuclear Emulsions

In a photographic emulsion the ionization produced by a photon or a charged particle leads to the conversion of some Ag^+ to Ag atoms. These can link together to form complexes which can catalyse the reduction of a whole AgBr grain to metallic Ag under the action of the developer. In the final stage the fixer dissolves away all the undeveloped grains. A particle track is therefore imaged as an array of black Ag grains. Nuclear emulsions have been produced which have high enough sensitivity to record even high-energy electrons.

For accurate work extreme care must be taken with the processing of emulsions to ensure uniform development of the entire emulsion thickness. Shrinkage of the emulsion after processing can lead to problems. Moreover, the emulsions must be carefully stored prior to use, and even afterwards, in order to reduce fogging by cosmic ray particles or by heat. Track images can fade after prolonged storage in air owing to the oxidation of the Ag metal. They must, of course, be kept in complete darkness prior to development, and darkroom conditions must be observed during their treatment.

Despite these limitations, the nuclear emulsion has evolved into a research tool of great importance in nuclear physics.

Particle identification may be made on the basis of detailed studies of the track image. For example, measurements of track width and length, degree of scattering of the charged particle while traversing the emulsion, the grain—or blob—density along its path, and the accompanying δ-ray density may be utilized in determining the charge and the energy of the particle. An event in nuclear emulsion is shown in Fig. 1.4.

1.3 Silver Halide Crystals

These detectors rely on the precipitation of metallic silver specks along the path of a charged particle in an AgCl or AgBr single crystal. The physical processes leading to track formation in these crystals are incompletely understood, but clearly there are similarities to nuclear emulsions.

The first observations of this type were made by Childs and Slifkin[8] in the early 1960s. After irradiation with charged particles, their AgCl crystals were exposed simultaneously to an ultraviolet light and a pulsed electric field for 2 hours. Some of the photoelectrons produced by the u.v. light and held at trapping centres along the path of the particle were able subsequently to combine with interstitial silver ions to form Ag metal. Tiny specks of Ag metal thus showed up as tracks.

In recent years Schopper and co-workers have considerably developed and modified these procedures (for recent descriptions of their work see references 9, 10). Thus, they have irradiated crystals of AgCl with charged particles in the presence of yellow light; these have then been developed by

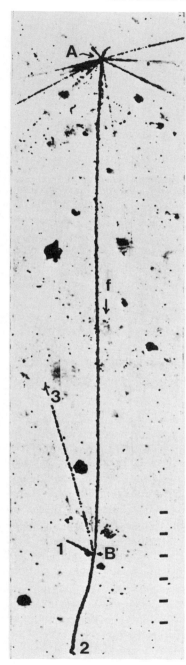

FIG. 1.4 *Photomicrograph of the production and disintegration of a hyperfragment observed in a nuclear emulsion (400 µm thick, Ilford G5) exposed to cosmic rays in a balloon flight at an altitude of ∼29 000 m. The parent star (A) has 16 prongs in addition to the hyperfragment f; estimated energy of the primary, ∼6 GeV. After travelling a distance of 219 µm in the emulsion, the hyperfragment (estimated to have a nuclear charge 2–3e) decays at B, producing a 3-pronged star. The event is almost certainly an example of the production of a hyperfragment containing a Λ^0 hyperon, which seems to have decayed into a proton plus a π^- (possibly the track marked 3), leaving a residual nuclear fragment. (Photograph from Tidman et al.[7a])*

exposure to u.v. light. The sensitivity of the crystals is greatly improved by the addition of Cd^{2+} ions. At a level of 5000 ppm of Cd^{2+} dopant, the crystals have been observed to be sensitive to protons of ~ 10 MeV in energy.[11]

The u.v. exposure provides a supply of photoelectrons which neutralize interstitial silver ions concentrated along the ion path. The rôle of the yellow light is thought to be that of an agent in the redistribution of Ag from small, unstable aggregates to larger, more stable ones.

The addition of Cd^{2+} is believed[9] to result in the formation of a negative space-charge around dislocation lines, which then repel electrons into the regions of track formation rather than acting as sinks for electrons, as occurs in undoped crystals.

AgCl crystals have many advantageous properties, e.g. that they are much more sensitive than most etched-track detectors, they require none of the wet chemical development processes of emulsions, and, since track fading (prior to u.v. development) is rapid in the absence of yellow light, they may be switched on and off selectively: a unique feature among solid track detectors. They are, however, much more complex than the simple dielectric track recorders, and are still in development stages so far as particle identification etc. is concerned. They have, in general, therefore received limited attention and application, so far.

1.4   Etchable Solid State Nuclear Track Detectors (SSNTDs)

Some general properties of these detectors were outlined in the introductory paragraphs of this chapter, and many of their features will be discussed at greater length in the following chapters. Here we will restrict ourselves to a brief description of the track formation and development mechanisms in dielectric materials, and will indicate the range of application of SSNTDs. For an extensive review of the applications of SSNTDs see the excellent book by Fleischer, Price, and Walker.[12]

Heavily ionizing particles passing through insulating media leave narrow (~ 30–100 Å, i.e. ~ 3–10 nm) trails of damage. In crystals this consists of atomic displacements, manifesting themselves as interstitials and vacancies, and surrounded by a region of considerable lattice strain. In plastics, the radiation damage produces broken molecular chains, free radicals, etc.

Certain chemical reagents ("etchants") dissolve or degrade these damaged regions at a much higher rate than the undamaged material. The narrow damage trail is thus gouged out by the etchant, forming a hole. Having reached the boundaries of the damage trail, the etchant continues to enlarge the "burrow-like" hole in all directions at the lower, bulk-etching, rate. This etched track may be enlarged radially until it is visible under an optical microscope.

Track formation is related to the production of dense regions of ionization by a charged particle and, to a first approximation, track formation can be regarded as occurring when the number of ions exceeds a certain threshold value. This threshold varies from one type of material to another. Some plastics are sensitive enough to record slow protons, whereas in most minerals even an argon ion at its maximum ionization rate will be unable to form an etchable track. Even high doses of electrons and photons have relatively little effect on SSNTDs. This is often a very useful property when trying to pick out charged particles in a high field of mixed radiation. At high temperatures, solid state diffusion allows a gradual healing-up (annealing) of unetched tracks to take place, until finally they are rendered unetchable.

The rate of chemical attack on the damage trail is related to the amount of ionization produced by the track-forming ion: a fact which, as will be shown in Chapter 6, can be used to distinguish between tracks formed by different types of particles.

Let us now compare these etchable solid state track detectors with the other systems described earlier in this chapter.

They are clearly the least sensitive of all of the devices described above, this being particularly true of mineral crystals which are insensitive to all but the heaviest ($Z \gtrsim 20$) slow ions. This makes them quite inappropriate for a wide range of applications; but this property leads also to considerable advantages where, for example, very rare heavy particles are sought amid an intense background of lightly ionizing radiations. As with nuclear emulsions and AgCl crystals, they can remain in an "active" state almost indefinitely. Thus meteorite crystals have been able to accumulate tracks of heavy cosmic-ray ions for billions (10^9) of years.

Etched-track detectors are extremely simple to construct compared with bubble, cloud, and spark chambers and even nuclear emulsions. In fact, almost none of the materials used today in etched-track work were specifically designed for particle detection. Some of the plastics which have proved to be most useful were fabricated as electrical insulators (Makrofol*) and for use in the manufacture of a wide range of everyday objects (Lexan†, CR-39‡). The cost of such materials is usually relatively small (especially when available in bulk, e.g. (in 1984) $\sim £40$ sterling per kg for Makrofol purchased as a roll; or a few pounds sterling for a 20 cm × 25 cm sheet of CR-39). Uniquely among all known nuclear track detectors, SSNTDs exist as natural materials (in the case of rock-forming minerals) and have been recording the passage

---

* Manufactured by Bayer AG of Leverkusen, West Germany.

† Registered trade mark of the General Electric Co. of USA.

‡ Manufactured by American Acrylics and Plastics, Stratford, Ct, USA, as well as by Pershore Mouldings Ltd., Pershore, Worcs., England, and other manufacturers in various countries. CR stands for "Columbia Resin". A "super grade" (PM355), with various additives and different casting cycles, is now being made specifically for nuclear track research by Pershore Mouldings, which is considerably more expensive (but is claimed to have a greater uniformity of response).

of ionizing particles (e.g. fission fragments and α-particle recoils from the spontaneous decay of ^{244}Pu, ^{238}U, and ^{232}Th; or heavy cosmic ray particles, especially in meteorites) often for several thousand million years. The detectors are very durable, pose no great handling problems, and are not fogged by exposure to light or affected by moderate degrees of heating. Their simplicity and durability makes them particularly valuable for remote use, such as in high-altitude balloon exposures to cosmic rays, and their robustness enables them to be used in personnel dosimetry. Their minute dimensions can often make them the only practicable systems in congested surroundings. This last mentioned property is possessed also by thermoluminescent dosimeters; but SSNTDs have the extra advantage of retaining their record after readout.

The ability to discriminate between different types of incident particles is comparable, and in some cases superior, to that of nuclear emulsions.

In the rest of this book we will examine these properties and applications in greater detail.

References

1. S. A. Durrani (1977) Editorial. *Nucl. Track Detection* **1**, 1–2.
2. D. A. Young (1958) Etching of radiation damage in lithium fluoride. *Nature* **182**, 375–7.
3. E. C. H. Silk & R. S. Barnes (1959) Examination of fission fragment tracks with an electron microscope. *Phil. Mag.* **4**, 970–2.
4. R. L. Fleischer, P. B. Price & R. M. Walker (1965) Solid-state track detectors: Applications to nuclear science and geophysics. *Ann. Rev. Nucl. Sci.* **15**, 1–28.
5. L. T. Chadderton and I. McC. Torrens (1969) *Fission Damage in Crystals*. Methuen, London.
6. E. Dartyge, M. Lambert & M. Maurette (1976) Structure et enregistrement des traces latentes d'ions Argon et Fer dans l'olivine et le mica muscovite. *J. de Phys.* **37**, 137–41.
6a. J. Csikai & A. Szalay (1959) The effect of neutrino recoil in the beta decay of He^6. *Soviet Phys. JETP* **35**, 749–51.
7. J. B. A. England (1974) *Techniques in Nuclear Structure Physics* (Part 1). Macmillan, London.
7a. D. A. Tidman, G. Davis, A. J. Herz & R. M. Tennent (1953) Delayed disintegration of a heavy nuclear fragment: II. *Phil. Mag.* (Ser. 7) **44**, 350–2.
8. C. B. Childs & L. M. Slifkin (1963) Delineating of tracks of heavy cosmic rays and nuclear processes within large crystals of silver chloride. *Rev. Sci. Instr.* **34**, 101–4.
9. F. Granzer, E. Schopper & T. Wendnagel (1980) Properties and technology of monocrystalline AgCl-detectors. 1. Aspects of solid state physics. In: *Proc. 10th Int. Conf. Solid State Nucl. Track Detectors*, Lyon, and Suppl. 2, *Nucl. Tracks* (eds. H. François *et al.*). Pergamon, Oxford, pp. 47–56.
10. T. Wendnagel, E. Schopper & F. Granzer (1980) Properties and technology of AgCl-detectors. 2. Experiments and technological performance. In: *Proc. 10th Int. Conf. Solid State Nucl. Track Detectors*, Lyon, and Suppl. 2, *Nucl. Tracks* (eds. H. François *et al.*). Pergamon, Oxford, pp. 147–55.
11. G. Haase, E. Schopper & F. Granzer (1978) Solid state nuclear track detectors: Track forming, stabilizing and development processes. In: *Proc. 9th Int. Conf. Solid State Nucl. Track Detectors*, Munich, and Suppl. 1, *Nucl. Tracks* (eds. F. Granzer *et al.*). Pergamon, Oxford, pp. 199–213.
12. R. L. Fleischer, P. B. Price & R. M. Walker (1975) *Nuclear Tracks in Solids: Principles and Applications*. University of California Press, Berkeley.

CHAPTER 2

Interactions of Charged Particles
with Matter

Before proceeding to a discussion of the formation of atomic defects and of the various models which have been proposed to account for the formation of particle tracks, it is useful briefly to consider the manner in which a charged particle communicates energy to the stopping medium.

Charged particles (specifically, protons, deuterons, α-particles, or heavy ions) lose their energy to the stopping material via three principal types of process.

(1) The electrostatic force between the particles and the electrons surrounding the target nuclei can lead to the stripping of these electrons from their orbits, or to raising the electrons to less tightly bound states. These are the processes of ionization and excitation, respectively.
(2) Deceleration of the particle results in the emission of electromagnetic radiation: *bremsstrahlung*. Also, at velocities greater than the phase velocity of light in the stopping medium, the moving charge causes polarization of atoms near to the particle trajectory and a coherent wave front of radiation is formed. This is Cherenkov radiation.
(3) Electrostatic forces can act directly between the moving ion and the target nuclei themselves, and can result in the ejection of target atoms from lattice sites or out of molecular chains.

At the very outset it must be noted that mechanism (2) is not likely to be important for our discussion. Cherenkov radiation occurs only at highly relativistic velocities. At such velocities etchable tracks are formed only in the most sensitive plastics (cellulose nitrate, CR-39, etc.) when irradiated with very heavy ions. As far as bremsstrahlung is concerned, its intensity at a given energy of the incident particle is $\propto (Z_1/M_1)^2 Z_2^2$, where Z_1, M_1 are the charge and mass of the incident particle and Z_2 the charge on the target nucleus. Bremsstrahlung is therefore an important means of energy loss only for light particles such as electrons, where the term $(Z_1/M_1)^2 \simeq 3 \times 10^6$ (when Z is expressed in electronic charge units and M in atomic mass units). For a typical heavy particle, such as an α-particle, $Z_1/M_1 = 0.5$ and $(Z_1/M_1)^2 = 0.25$. The amount of bremsstrahlung radiation is thus less by many orders of magnitude for slowing down atomic nuclei compared with electrons in a given stopping medium.

By restricting our attention to atomic nuclei, which are the only particles of relevance to solid state nuclear track detectors, we need consider only energy-loss mechanisms (1) and (3) above. We may then write the total rate of energy loss in the form

$$\left(-\frac{dE}{dx}\right)_{total} = \left(-\frac{dE}{dx}\right)_{nucl} + \left(-\frac{dE}{dx}\right)_{elec}$$

where the two terms on the right-hand side of the equation refer to nuclear and electronic interactions, respectively.

If a particle of charge Z, energy E, and mass M_1 traverses a stopping medium with N scattering centres, then the number of interactions occurring in a thickness dx of the material in which an energy W is transferred to the scattering centres is given by

$$N\, d\sigma(E, W)\, dx$$

where $d\sigma(E, W)$ is the differential cross-section for transfer of energy W by an incident particle of energy E. The energy transferred in such collisions is therefore given by

$$-dE(W) = NW\, d\sigma(E, W)\, dx$$

The total energy loss dE may be obtained by integrating over all possible values of W between suitable limits W_{min} and W_{max}, so that

$$-\frac{dE}{dx} = N \int_{W_{min}}^{W_{max}} W\, d\sigma(E, W)$$

The scattering problem may be treated classically, provided that the least distance of approach of the moving ion to the scatterer is large compared with the reduced de Broglie wavelength of the incident particle.

The exact form of $d\sigma(E, W)$ depends on the form of the interaction potential between the moving ion and the scattering centre.

2.1 Nuclear Collision Losses

Ions having energies in the MeV region are scattered by the Coulomb potential due to the positively charged nuclei of the target atoms. This is the familiar Rutherford scattering, and the differential cross-section $d\sigma(E, W)$

has the form

$$d\sigma(E, W) = \frac{\pi Z_1^2 Z_2^2 e^4}{(4\pi\varepsilon_0)^2 E} \left(\frac{M_1}{M_2}\right) \frac{dW}{W^2} \tag{2.1}$$

where ε_0 is the permittivity of free space.

As the energy of the incident ion is reduced, the distance of closest approach to the target nucleus increases, with the result that the intervening orbital electrons screen the target nucleus from the incident particle. Also, as the ion velocity approaches that of the orbital electrons, the ion tends to pick up electrons and eventually to become neutralized. A number of empirical relationships have been derived, which give the effective charge Z_{eff} of an ion at velocity v. Heckman *et al.*,[1] for example, give

$$Z_{eff} = Z_1[1 - \exp(-130\beta/Z_1^{2/3})] \tag{2.2}$$

where $\beta = v/c$ and c is the velocity of light. When the ion is completely neutralized, the repulsive force between it and a target atom is due to overlap of the respective electron shells and rises very rapidly at small separation distances. Scattering of this type is known as "hard-sphere" scattering, and the differential cross-section now takes the form

$$d\sigma(E, W) = 4\pi R^2(E) \frac{dW}{W_{max}} \tag{2.3}$$

where $R(E)$ is the effective radius of the hard sphere, and W_{max} is the maximum energy transfer given by

$$W_{max} = \frac{4M_1 M_2}{(M_1 + M_2)^2} E \tag{2.4}$$

Whereas Rutherford scattering favours small values of energy transfer (see Eq. (2.1)), for hard-sphere scattering all values of W up to W_{max} are equally probable.

For ions of energy greater than about 1 MeV, the nuclear collision losses are small compared to electronic energy losses. Nuclear collisions are important only at the end of a heavy-ion track, but will generally be the major mode of energy loss for the target nuclei struck by the primary ion (hard-sphere scattering).

2.2 Electronic Energy Losses

Heavy ions with energies of ~ 1 MeV and upwards (but below the extreme relativistic regime where radiative losses dominate) lose energy primarily through Coulomb interactions with the orbital electrons of the target atoms.

A simple classical derivation of an expression for the energy loss to electrons is as follows.

Consider an ion with charge $Z_1 e$, which passes by an electron at a distance b (see Fig. 2.1). The direction of the force on the electron makes an angle θ with the path of the ion, and the impulse delivered to the electron leads to a momentum transfer p given by

$$p = \int_{-\infty}^{+\infty} F \sin \theta \, dt \tag{2.5}$$

Here F is the Coulomb force, given by $Z_1 e^2 / 4\pi\varepsilon_0 r^2$, where r represents the ion–electron separation, and ε_0 is the permittivity of free space. If the ion velocity is v, then, from elementary kinematics, $r^2(d\theta/dt) = vb$ (the moment of the linear velocity, where $d\theta/dt$ is the angular velocity and b is the perpendicular distance). Hence

$$dt = \frac{r^2}{vb} \cdot d\theta$$

On substituting for F and dt in Eq. 2.5, we get:

$$p = \int_0^\pi \frac{Z_1 e^2 \sin \theta \, d\theta}{4\pi\varepsilon_0 bv} = \frac{Z_1 e^2}{2\pi\varepsilon_0 bv}$$

The kinetic energy W transferred to the electron is therefore given by

$$W = \frac{p^2}{2m_0} = \frac{Z_1^2 e^4}{8\pi^2 \varepsilon_0^2 b^2 m_0 v^2} \tag{2.6}$$

where m_0 is the electron mass. If there are n electrons per unit volume, then the number of electrons in a thickness dx of the absorber, for which the impact parameter lies between b and $b + db$, is $2\pi bn \, db \, dx$, and the energy loss by the ion to these electrons is

$$-dE = 2\pi bn \, db \, dx \, W = 2\pi bn \, db \, dx \, \frac{Z_1^2 e^4}{8\pi^2 \varepsilon_0^2 b^2 m_0 v^2}$$

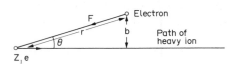

FIG. 2.1 *Schematic diagram showing the interaction between an electron of charge e and a heavy ion of charge $Z_1 e$, with an impact parameter (distance of closest approach) b. At a separation distance r, the Coulomb force between the particles is F, and acts along r at an angle θ with the ion path.*

Upon integration over the allowed range of impact parameters (say, b_{min} to b_{max}) we have

$$-\frac{dE}{dx} = \frac{nZ_1^2 e^4}{4\pi\varepsilon_0^2 m_0 v^2} \cdot \ln\frac{b_{max}}{b_{min}}$$

For large impact parameters, the collision time is long compared with the period of revolution of an electron in its orbit. For such collisions the interaction with the electron is essentially adiabatic. This leads to an upper limit b_{max} for the impact parameter, given by

$$b_{max} = \frac{v}{v}$$

where v is the frequency of the electron in its orbit and v, of course, is the velocity of the approaching ion. If, furthermore, b_{min} is taken as the dimension of the electron as seen by the moving ion, i.e. $\sim$ the de Broglie wavelength $h/(m_0 v)$ for the electron, where h is Planck's constant, then the energy loss becomes

$$-\frac{dE}{dx} = \frac{nZ_1^2 e^4}{4\pi\varepsilon_0^2 m_0 v^2} \ln\frac{m_0 v^2}{hv}$$

If we identify hv with the mean excitation potential $\bar{I}$ for the electrons in the atoms of the stopping medium, then we have:

$$\left(-\frac{dE}{dx}\right)_{elec} = \frac{nZ_1^2 e^4}{4\pi\varepsilon_0^2 m_0 v^2} \ln\frac{m_0 v^2}{\bar{I}} \qquad (2.7)$$

A quantum-mechanical approach leads to the well-known Bethe–Bloch formula[1a,b]:

$$\left(-\frac{dE}{dx}\right)_{elec} = \frac{nZ_{eff}^2 e^4}{4\pi\varepsilon_0^2 m_0 v^2}\left[\ln\frac{2m_0 v^2 W_{max}}{\bar{I}^2(1-\beta^2)} - 2\beta^2 - \delta - U\right] \qquad (2.8)$$

where we have replaced Z_1 by the effective charge Z_{eff} of the ion at velocity v. In the non-relativistic limit this is very similar to Eq. (2.7). Here δ is a correction for polarization of the medium, important only at high energies, and U is a term which takes account of the non-participation of inner electron shells.

By using the simplified classical approach it is possible to derive equations for the number of ions, J, produced per unit path length of the primary charged particle, and their mean energy $\bar{W}$. Thus we have

$$J = \frac{Z_1^2 e^4 n}{8\pi\varepsilon_0^2 m_0 v^2}\left(\frac{1}{\bar{I}} - \frac{1}{W_{max}}\right) \qquad (2.9)$$

and

$$\bar{W} = \frac{\ln\left(\dfrac{W_{max}}{\bar{I}}\right)}{\left(\dfrac{1}{\bar{I}} - \dfrac{1}{W_{max}}\right)} - \bar{I} \tag{2.10}$$

From Eqs. (2.7) and (2.8) it can be seen that the loss rate $(dE/dx)_{elec}$ increases with decreasing velocity roughly as $1/v^2$. As, however, the ion slows down, it gradually loses its positive charge and finally becomes neutralized. The rate of energy loss of the ion, therefore, passes through a maximum. At very low energies the electronic energy loss rate becomes proportional to v (see reference 2). The general form of $(dE/dx)_{elec}$ versus energy is shown in Fig. 2.2.

So far we have looked at the manner in which the primary particle interacts with the stopping medium. We have seen that heavy ions at non-relativistic velocities lose energy largely through the production of ionization, thus leading to the release of large numbers of electrons from the target atoms. To a lesser extent, collisions with atomic nuclei occur, leading to

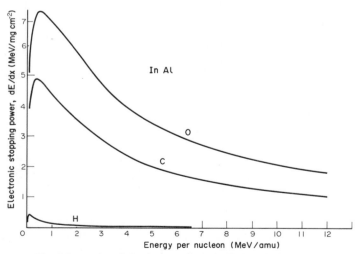

FIG. 2.2   *A plot of the electronic stopping power of H, C and O ions in Al as a function of the energy per nucleon of these ions. The data used in this plot were obtained from the tables of Northcliffe and Schilling.[11]*

the production of displaced atoms. We must now examine the fate of these electrons and displaced atoms.

2.3   Direct Production of Atomic Displacements

The moving ion, as a result of nuclear collisions, produces a number of energetic knock-on atoms. These atoms will produce further displacements via interactions which approximate to the hard-sphere-type collisions at low energies and to the Rutherford-type scattering at higher energies (see §2.1). Several models have been devised to calculate the number of displacements produced by the "primary" displaced atom. A very simple model is one in which the knocked-on atoms are assumed to lose energy in hard-sphere collisions, sharing their energy equally in each collision with a target atom. A sharp displacement-energy threshold E_d is assumed to exist. This model leads to a total number of displacements being equal to $E_R/2E_d$, where E_R is the energy of the primary knock-on atom. This approach almost certainly overestimates the number of displacements produced, but probably yields a correct order-of-magnitude estimate of displacement production. A 1 keV primary recoil might thus produce ~ 20 displacements for $E_d \sim 25$ eV.

More elaborate models have been developed, but a description of them is beyond the scope of this book. The interested reader is referred to one of the many works on radiation damage in solids (see, for example, references 3 and 4).

2.4   Secondary Electrons

The electrons released by the incident ion will have a broad spectrum of kinetic energies, strongly peaked at low energies owing to the nature of the Rutherford scattering cross-section (Eq. 2.1), but extending up to $W_{max} - I_i$ for electrons released from the ith electron shell (I being the ionization potential).

If the incident ion is a fission fragment of energy 100 MeV and $A = 100$ amu, then W_{max} imparted to the electron will be about 2 keV (cf. Eq. (2.4)). A δ-ray of this energy would have a range in Al, for example, of approximately 500 Å (i.e. ~ 50 nm). Such δ-rays will produce a considerable amount of ionization of their own; and this ionization will, to some degree, be spread throughout an appreciable volume of the stopping material along the path of the incident ion.

Kobetich and Katz,[5] have used a formula for the number of δ-rays per unit length of ion path, along with empirical formulae for the range and transmission probabilities of electrons as a function of energy, to calculate the energy deposition by δ-rays as a function of distance from the ion path (see Fig. 2.3). The bulk of the energy carried by δ-rays is deposited within a few tens of ångströms of the ion path. More detailed calculations of energy deposition around the paths of ionizing particles have now been carried out by other authors (Paretzke,[6] Paretzke and Burger,[7] and Fain *et al.*[8]). For fast ions, about half of the energy loss is converted into the kinetic energy of electrons; of the remaining energy loss about half has been used up in the excitation of the electrons. The residual energy ($\sim\frac{1}{4}$ of the original energy loss) is accounted for in overcoming the ionization potential of target atoms (see Paretzke[9]).

2.5 Range–Energy Relations

Because heavy charged particles, over most of their trajectory, lose energy in a large number of small-energy-transfer collisions with electrons, they can

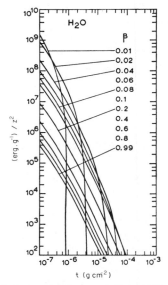

FIG. 2.3 *Calculations by Katz and Kobetich*[5a] *of the spatial distribution of ionization energy deposited in water by δ-rays as a function of distance* t *from the ion trajectory and of ion velocity. To obtain the dose (erg g*$^{-1}$*) deposited at a radial distance* t *(g cm*$^{-2}$*) by an ion of effective charge* z *moving at a velocity βc, the values on the vertical axis must be multiplied by* z^2*. The effective charge of an ion of atomic number* Z*, moving at a speed β relative to that of light (c), is given by:* z = Z[1 − exp(− 125βZ$^{-2/3}$)]. *(N.B. Many authors use 130 instead of 125 for the value of the constant in these calculations.) (Figure from reference 5a.)*

be regarded, to a first approximation, as having well defined ranges. In principle the range R of an ion of energy E can, thus, be computed from the formulae for stopping power $(-dE/dx)$, i.e. by using the relationship

$$R(E) = \int_0^E \left(-\frac{dE}{dx} \right)^{-1} dE \qquad (2.11)$$

Instead of substituting theoretical stopping power equations such as (2.7) and (2.8) into Eq. (2.11), range–energy relations are usually computed on a semi-empirical basis by making extrapolations, guided by theory, from experimental stopping-power data. Two range–energy compilations widely used in particle track work are those of Henke and Benton[10] and Northcliffe and Schilling.[11]

Voluminous literature now exists on the interactions of charged particles with matter. Valuable reviews of this subject are to be found in references 12–15.

References

1. H. H. Heckman, B. L. Perkins, W. G. Simon, F. M. Smith & W. Barkas (1960) Ranges and energy loss processes of heavy ions in emulsion. *Phys. Rev.* **117**, 544–56.
1a. H. A. Bethe (1930) Theory of the passage of rapid corpuscular rays through matter (*in German*). *Ann. Physik* **5**, 325–400.
1b. F. Bloch (1933) Stopping power of atoms with many electrons (*in German*). *Z. Physik* **81**, 363–76.
2. J. Lindhard, M. Scharff & H. E. Schiott (1963) Range concepts and heavy ion ranges. *Kgl. Danske Videnskab. Selskab., Mat. Fys. Medd.* **33** (14), 1–42.
3. G. J. Dienes & G. H. Vineyard (1957) *Radiation Effects in Solids*. Interscience, New York.
4. D. S. Billington & J. H. Crawford Jr. (1961) *Radiation Damage in Solids*. Princeton University Press, Princeton.
5. E. J. Kobetich & R. Katz (1968) Energy deposition by electron beams and δ rays. *Phys. Rev.* **170**, 391–6.
5a. R. Katz & E. J. Kobetich (1968) Formation of etchable tracks in dielectrics. *Phys. Rev.* **170**, 401–5.
6. H. G. Paretzke (1974) Comparison of track structure calculations with experimental results. In: *Proc. 4th Symp. on Microdosimetry* (Verbania Pallanza, Italy, 1973) (J. Booz, H. G. Ebert, R. Eickel and A. Waker, eds.). EUR 5122 d-e-f, pp. 141–68. Commission of the European Communities, Luxembourg.
7. H. G. Paretzke & G. Burger (1970) Spatial distribution of deposited energy along the path of heavy charged particles. In: *Proc. 2nd Symp. on Microdosimetry* (Stresa, Italy, 1969) (H. G. Ebert, ed.). EUR 4452 d-f-e, pp. 615–27. Commission of the European Communities, Brussels.
8. J. Fain, M. Monnin & M. Montret (1974) Spatial energy distribution around heavy ion paths. *Radiat. Res.* **57**, 379–89.
9. H. G. Paretzke (1977) On primary damage and secondary electron damage in heavy ion tracks in plastics. *Radiat. Effects* **34**, 3–8.
10. R. P. Henke & E. V. Benton (1966) Range-energy and range-energy loss tables. *US Naval Radiological Defense Laboratory, San Francisco, Report* TR-1102.
11. L. C. Northcliffe & R. F. Schilling (1970) Range and stopping power for heavy ions. *Nucl. Data Tables, A7* 233–63.

12. L. C. Northcliffe (1963) Passage of heavy ions through matter, *Ann. Rev. Nucl. Sci.* **13,** 67–102.
13. U. Fano (1963) Penetration of protons, alpha particles and mesons. *Ann. Rev. Nucl. Sci.* **13,** 1–66.
14. R. M. Sternheimer (1961) Interaction of radiation with matter. In: *Methods of Experimental Physics* (L. Marton, ed.), vol. 5, part A, pp. 1–89. Academic Press, New York.
15. S. P. Ahlen (1980) Theoretical and experimental aspects of the energy loss of relativistic heavily ionizing particles. *Revs. Mod. Phys.* **52,** 121–73.

CHAPTER 3

The Nature of Charged-Particle Tracks and Some Possible Track Formation Mechanisms in Insulating Solids

In the previous chapter some features of the processes by which ionizing particles communicate energy to the stopping medium were described. These processes can be considered as the first of three stages in the response of a medium to irradiation by ionizing particles.[1,2] This first, *physical*, stage is followed by a *physico-chemical* stage, in which the initially generated primary products (i.e. ions; excited atoms and molecules; free electrons) rapidly undergo secondary reactions (dissociation of some excited molecules, etc.), until the system reaches thermodynamic equilibrium. This is followed by the *chemical* stage, in which ions and free radicals react with each other and with other atoms and molecules to yield the final products of the irradiation.

Before proceeding to a discussion of experimental evidence on the nature of the track-forming process and to a review of track formation models, it is useful to examine some of the general features of the damage produced in solids as a result of irradiation.

3.1 Radiation Damage in Solids

The type of damage produced by irradiation of solids depends not only on the nature of the ionizing radiation but also on the nature of the solid itself. There are considerable differences in the extent and type of damage produced in the two major classes of track-storing solids, viz. inorganic crystals and glasses, and synthetic organic polymers. These differences seem to be reflected in the hiatus which exists between the sensitivities for track production of these two classes of material.

Let us consider plastics first. Here ionizing radiations directly produce ionized and excited molecules, and electrons. Some excited molecules may de-excite through the emission of radiation or through non-radiative transitions. Excitation energy can also be transferred from one molecule to another. Electrons are trapped at various sites, or can combine with molecules to form negative ions, or recombine with positive ions yielding excited molecules. Ions may participate in charge-transfer reactions.

Both ions and excited molecules may acquire considerable vibrational energy and undergo bond rupture to form a complex array of stable molecules, free radicals, ionized molecules, and radical ions. Further reactions between these ions, radicals, and molecules will then take place.

The net effect on the plastic will be the production of many broken molecular chains, leading to a reduction in the average molecular weight of the substance. (Radiation can, in fact, also initiate crosslinking and thus produce an increase in the molecular weight of the substance; but this process is probably not relevant to track formation.) Fleischer *et al.*[3] have reported that the rate of chemical attack on a plastic increases as the average molecular weight decreases. A number of authors have also shown that the rate of reaction of a given etchant with a plastic increases as a function of the dose of radiation absorbed by the plastic (see, for example, Benton[4]).

Environmental factors play an important rôle in determining the magnitude of the effects of radiation on plastics. For example, the presence of oxygen either before, during, or after irradiation tends to increase the final chemical reactivity generated by the irradiation.[5] It is thought that oxygen tends to combine with ions and radicals, thus preventing their subsequent recombination. Crawford *et al.*,[6] Benton and Henke,[7] DeSorbo and Humphrey,[8] and DeSorbo[9] have discussed effects such as track length increase and enhancement in track revelation in irradiated plastics brought about by exposure to u.v. light in the presence of air, oxygen, etc., photo-oxidation phenomena, the "aging" of film prior to etching, and exposure to high-energy electrons and X-rays. (Such environmental factors are much less important in the irradiation of inorganic crystals.) Benton[5] has reviewed some of the radiation effects produced in plastics which are of importance to track formation. For a more general discussion of radiation effects in polymers, see references 10 and 11.

As will be seen in §3.2, tracks are not observed in metals or good semiconductors. Therefore radiation effects in these materials will not be described here.

Taking inorganic insulating crystals next: the effect of radiation upon them will also be to produce ionization and excitation of atoms or molecules. Electrons are raised across the forbidden energy band. Some of these may return to the valence band via luminescence centres with the emission of radiation; while others, after diffusing through the crystal, will either be trapped at the sites of various imperfections or will return, via non-radiative transitions, to positive ions. Low-energy heavy ions will produce numerous atomic displacements directly through elastic collisions. Electron irradiation can also cause direct atomic displacements. An electron of kinetic energy E

and rest mass m_0 can transfer a maximum energy W_{max} to an atom of mass M given by

$$W_{max} = \frac{2(E + m_0c^2)E}{Mc^2} \qquad (3.1)$$

which reduces to Eq. (2.4) in the non-relativistic limit. An electron of energy 1 MeV can transfer up to ~ 100 eV to a silicon atom for example, whereas only ~ 25–30 eV is required to produce displaced atoms. However, it has been known for some time[12] that the number of atomic displacements produced by electrons can exceed the levels expected simply from direct collisions between electrons and atomic nuclei. Some of the models put forward to account for displacement production by electrons are described below. These models have been primarily constructed to explain defect formation in alkali halides.

(a) The Seitz model

Seitz[13] proposed that irradiation of alkali halides leads to the formation of excitons (bound electron–hole pairs) which move through the crystal, giving up their energy at crystal defects (such as jogs* on dislocation lines) and resulting in the formation of lattice vacancies.

(b) The Varley model

According to the model proposed by Varley,[12] multiple ionization of Cl^- in an alkali chloride crystal by incident radiation results in the formation of Cl^+. This ion then feels strong repulsive forces from its nearest neighbours, e.g. Na^+ ions. If the Cl^+ has a lifetime against electron recapture greater than a typical lattice vibration time ($\sim 10^{-13}$ s), then displacement of the Cl^+ can occur. Some workers, for example Chadderton and Torrens,[14] have questioned whether this mechanism can in fact lead to defect formation. Nevertheless it has been incorporated, in modified form, into one track formation model (see §3.5.2).

(c) The Pooley mechanism

In the mechanism proposed by Pooley,[15] an electron–hole pair is formed by the ionizing radiation, and the hole is captured between two halogen ions which form a halogen molecule. In certain conditions non-radiative

---

* In its simplest form, a jog is a one-atom step in a dislocation line.

recombination of the electron and the hole occurs, and the energy release results in the formation of a vacancy and an interstitial.

Whatever the exact mechanism, the general conclusion is that irradiation of insulating crystals results in the formation, by various means, of lattice vacancies and interstitials, or complex aggregates of such defects. These defects can give rise to the coloration of crystals: for example the centre in alkali halides (the F centre) consists of an electron trapped at a negative-ion vacancy.

3.2 Track-storing Materials

We now commence a review of some experimental data of relevance to the problem of track formation. We wish to consider the types of material capable of storing tracks, and to examine the properties of the incident particle which will determine whether or not an etchable track is formed by it in a given medium.

Etched tracks have now been observed in a large number of materials. These materials are generally polymers, inorganic glasses, mineral crystals, and some poor semiconductors. The most important generalization that can be made about these substances is that they are all dielectric solids, i.e. poor conductors of electricity.

Fleischer, Price, and Walker[16] quote a value of 2000 Ω cm as the limiting resistivity below which tracks are not observed.

A considerable body of evidence also exists on the observation by transmission electron microscopy (TEM) of latent tracks in crystals.[17,14] Some caution must be exercised in comparing these results to those for etchable tracks. With the TEM, essentially two classes of tracks are observed: (1) tracks which are formed throughout the body of the crystal and which seem to be composed of a linear array of defects, and (2) surface tracks formed by the physical ejection of material from within the crystal. It seems that etchable tracks of the type under discussion here are most closely related to class 1.

Morgan and Chadderton[18] report a series of experiments in which crystals of the semiconducting molybdenite family, having very similar mechanical and thermal properties but widely varying electrical resistivity (from 10^2 Ω cm to 2.6×10^{-4} Ω cm), were irradiated with fission fragments and examined with a TEM for latent tracks. Also crystals of WTe_2, in which the resistivity was varied from 10^{-3} Ω cm to 1.7×10^{-3} Ω cm by doping with niobium impurity, were studied in a similar way. In both cases a minimum resistivity for the observation of tracks, viz $\sim 10^{-3}$ Ω cm, seemed to exist. The reason for very large differences in the limiting-resistivity values for track storage found by Fleischer *et al.*[16] and by Morgan and Chadderton[18] is not yet apparent, but the serious discrepancy does call into question the concept of resistivity threshold as the criterion for track storage.

Evidence of a different kind has been obtained by Sigrist and Balzer.[19] They find that the threshold in dE/dx for ionizing particles, above which tracks can be etched, depends on the thermal conductivity of the irradiated crystal (the crystals with the highest thermal conductivity yielding the highest threshold, $(dE/dx)_c$). This interesting result may have important implications for models of track formation—although it should be noted that spinel, which according to Sigrist and Balzer revealed no etchable tracks (thus conforming to their expectation from its high thermal conductivity), has in fact been shown to exhibit etched fission tracks.[20]

It is important that a successful track formation model should correctly predict the types of materials which will and will not reveal tracks. We shall have more to say on these points when the specific track formation models are discussed in §3.5.

3.3 Track-forming Particles: Criteria for Track Formation

The validity of various models of track formation may be judged by critical appraisal of parameters such as charge, mass, and energy of such incident particles as are able to form etchable tracks. A wide variety of heavy ions are now available from a number of accelerators, with energies ranging from keV up to hundreds of MeV per nucleon.

Various authors have suggested that track formation should be related to a number of different parameters, such as total energy loss rate, primary ionization, restricted energy loss, etc. One then sets up a track formation criterion, which takes the form of a statement that tracks are formed in a medium when, and only when, the chosen parameter exceeds some critical value, whatever the bombarding particle. These track formation criteria can be tested by irradiating a given solid with a number of ions at various energies and recording those cases for which etchable tracks are formed on diagrams such as Figs. 3.1a–c, on which the various parameters are plotted as a function of ion velocity for the different ions. Filled circles correspond to those cases for which tracks were observed, and open circles to those where no tracks were observed. For a valid track formation parameter, X, it should be possible to draw a unique threshold value, X_c, separating the region of track etchability from that corresponding to non-etchability.

In recent years this view of track formation criteria has been modified somewhat in that less emphasis is now placed on threshold values of X. This is because experimental data[21,22] have accrued to show that the track etching rate V_T (the linear rate at which the chemical reagent travels along the tracks) is a smoothly varying function of incident-particle parameters. It is now usual to seek a parameter X such that $V_T(X)$ always has the same value (for a given combination of detector and etching conditions) at a given value of

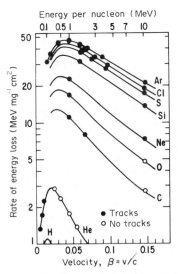

FIG. 3.1a *Curves of linear rate of energy loss dE/dx as a function of β (velocity relative to that of light) and energy per nucleon for various ions. Data for track etchability of accelerated ions in Lexan polycarbonate are marked on these curves: solid symbols represent ions and energy regions for which tracks were observed; open symbols represent failure of track revelation. It is not possible to draw a horizontal line to separate solid from open symbols; and so it is concluded that dE/dx is not a suitable parameter for predicting the etchability of tracks. (Figure from Fleischer.[58])*

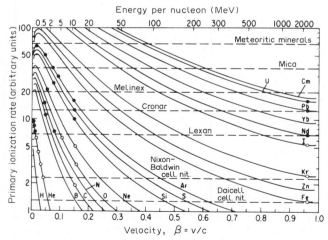

FIG. 3.1b *Curves of primary ionization rate J as a function of relative velocity β and energy per nucleon for a number of ions. Track-etchability data for a number of accelerated ions in Lexan are marked on the curves (with solid and open symbols with the same meaning as in Fig. 3.1a); but, unlike that figure, a horizontal line can now be fitted to separate the registration and non-registration regions. The registration thresholds for a number of other plastic and mineral detectors are also shown by dashed lines in the figure. Note that the registration threshold of the highly sensitive plastic CR-39 lies below the x-axis of the figure. (After Fleischer, Price, and Walker.[16])*

28

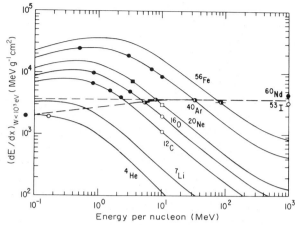

FIG. 3.1c *Restricted energy loss (REL) data for track-etchability in Lexan polycarbonate, taking the upper limit of the energy (W), imparted by different ions to electrons, as 1000 eV ($= W_0$). Since a given ion moving faster in a medium yields a larger fraction of delta rays possessing higher energies, therefore $(dE/dx)_{W < W_0}$ will represent a smaller fraction of the total energy loss at higher, than at lower, energies of the ion. Solid symbols represent etchability in Lexan, and open symbols the lack of it. Once again (as in Fig. 3.1a), no horizontal line quite fits the data. (Figure from Fleischer.[58])*

X, regardless of the combination of ionic charge and velocity that may lead to the X value in question.

Such correlations of track etch velocity with a particle parameter, in addition to placing constraints on track formation mechanisms, are of practical value in that once the function $V_T(X)$ is determined, the etching rate of any ion at any energy may be calculated. The importance of this fact will become more apparent when particle identification is discussed in Chapter 6.

(a) Total rate of energy loss, dE/dx

The earliest, and perhaps the most natural, explanation of track formation was that it depended on the total amount of energy deposited per unit path length by the incident ion. This criterion was proposed by Fleischer *et al.*[23] The proposition was that tracks are formed when dE/dx exceeds some critical value $(dE/dx)_c$. Data from heavy-ion irradiation experiments were plotted on calculated dE/dx curves (see Fig. 3.1a) and a threshold level drawn which, it was hoped, would allow a clear demarcation of data points representing the revelation and non-revelation of etchable tracks. This approach became untenable when results of more extensive irradiations, particularly in the high-energy region, became available and was rejected by Fleischer *et al.*[24] in favour of the primary-ionization criterion described below.

(b) Primary ionization, J

In this model, the formation of etchable tracks is related to the number of primary ionizations produced close to the ion path. A relationship for the primary ionization, J, based on the work of Bethe[25] is usually employed:

$$J = C' \frac{Z_{\text{eff}}^2}{\beta^2} \left[\ln \left(\frac{\beta^2}{1 - \beta^2} \right) - \beta^2 - \delta + K \right] \qquad (3.2)$$

where Z_{eff} and β are, respectively, the effective charge and the velocity of the ion relative to the velocity of light, c. The terms K and C' are constants for a given stopping medium, and δ is a relativistic correction term which is related to the polarization of the stopping medium. It has not yet proved possible to calculate K from first principles or to make an absolute determination of J for most stopping media. K is frequently used as a fitting parameter and J is calculated in arbitrary units. This procedure does not affect correlations, since the fitting of the data depends on the shape of the J versus β curves, rather than on the absolute values of J.

It was found that primary ionization did indeed give a good fit to the results of heavy-ion irradiation experiments and that a consistent value of primary-ionization threshold, J_c, could be obtained for a given stopping medium (see Fig. 3.1b).

The reason for the success of the primary-ionization criterion in comparison to dE/dx is that the higher-energy δ-rays carry off an appreciable energy outside the central track region (~ 50–100 Å), and this energy is unlikely to play a part in track formation. In the formula for the total rate of energy loss, the energy of these δ-rays makes a significant contribution to dE/dx; whereas in the primary-ionization formula such high-energy events are given the same weighting factor as low-energy δ-rays and thus assume comparatively less importance.

The primary-ionization criterion is open to the criticism that it takes no account of the ionization produced by even the low-energy δ-rays[16], and that the ionization potentials implied by the values of K used in obtaining the best fit to experimental data are physically unrealistic.[26]

With regard to the second of these criticisms it must be said that Fleischer *et al.*[24] regard Eq. (3.2) as phenomenologically useful rather than of deep theoretical significance. Moreover, since in crystals the radial extent of un-etched tracks is only a few tens of angstroms, it seems reasonable that primary processes should predominate over secondary ones as the basis for track formation—although this may not be true for polymers, or generally in the case of low-energy ions (< 1 MeV/nucleon).

The primary-ionization criterion has proved highly successful in practice, and has survived to the present day—except that the concept of a threshold J is regarded nowadays as being less decisive. Thus, for both plastics and minerals[21,27] it has been found that the track etch velocity V_T is a continuous function of J. Typically, $V_T \propto J^\alpha$, where a value of $\alpha \simeq 2$ has been found to hold for some plastics. Nevertheless, V_T versus J curves generally rise very steeply, as will be seen in Chapter 6; and, particularly for minerals, the concept of a threshold remains a reasonable approximation to reality.

(c) Restricted energy loss (REL)

This criterion, proposed by Benton,[28] makes allowance for the energy removed from the track core by high-energy δ-rays. This is done by imposing a cutoff ceiling, W_0, on the energy of delta rays, such that only that part of the energy loss is assumed to be relevant to track formation where the energy carried by the δ-ray electrons is less than W_0. Restricted energy loss can be calculated from the following expression:

$$\left(\frac{dE}{dx}\right)_{W < W_0} = \frac{n_0 e^4}{8\pi\varepsilon_0^2 m_0 c^2} \cdot \frac{Z_{\text{eff}}^2}{\beta^2}\left(\ln\frac{W_{\max}W_0}{\bar{I}^2} - \beta^2 - \delta - U\right) \quad (3.3)$$

where all the symbols are as defined earlier in this chapter and in Chapter 2. For the purposes of fitting this equation to experimental data, W_0 is variously taken as 200, 350, or 1000 eV. This quantity, $(dE/dx)_{W < W_0}$, varies with ion energy E in a manner rather similar to primary ionization. It gives a good fit to most experimental data (see Fig. 3.1c). Siegmon et al.[29] have found that it fits track etch rate data for Fe isotopes in plastics equally as well as does primary ionization. It may be noted, however, that Ahlen[30] has recently shown that some cosmic ray track data are less well fitted by REL than by primary ionization.

The choice of W_0 as a fitting parameter introduces a certain arbitrariness into REL calculations—a feature that REL shares with the primary-ionization criterion. Also, even electrons of energy ~ 350 eV will deposit some of their energy far from the track core; so the physical basis of W_0 is not clearly understood.

(d) Secondary-electron energy loss

The approach of Katz and his co-workers[31] is diametrically opposed to that of Fleischer, Price, and Walker (criteria (a) and (b) above). The Katz hypothesis maintains that energy deposited by δ-rays, rather than the primary-ionization events themselves, is the crucial factor for the formation of an etchable track.

Kobetich and Katz[32] have calculated the amount of energy deposited in the medium as a function of distance from the ion path for different ion-velocities. Some of their spatial dose distributions have been shown in Fig. 2.3 of Chapter 2. Katz and Kobetich[31] take the energy deposition *at* about 20 Å from the ion path as the critical parameter for track formation. Katz and Kobetich[31] show that, by plotting track etchability data on curves of "20 Å dose" versus ion energy, self-consistent dose thresholds can be drawn which can separate the etchable and non-etchable regions. These thresholds are about 3.5×10^5 Gy ($= 3.5 \times 10^7$ rad) for muscovite mica and 2.5×10^4 Gy for cellulose nitrate.

Physically, it seems unreasonable that, on this model, primary-ionization events should be completely neglected. Also, Fleischer *et al.*[16] note that the secondary-electron energy loss criterion predicts a minimum detectable charge of 70 for relativistic ions in Lexan plastic, whereas experimentally it is found that relativistic ions of as low as $Z = 57 \pm 2$ leave etchable tracks in it. Some other shortcomings of this model have recently been discussed by Ahlen.[30]

(e) Radius-restricted energy loss (RREL)

Paretzke[26] has discussed the concept of radius-restricted energy loss L_r. This differs from the normal REL criterion of Benton[28] in that it includes the energy deposited in *all* events occurring *within* a radius r of the particle trajectory; whereas REL neglects those ionization events close to the track which are due to electrons carrying off an energy greater than W_0 and yet takes account of energy deposited outside the track core by electrons of energy $W < W_0$. Thus the L_r criterion gives central importance to the region near the track, whereas the REL approach gives primary importance to the limiting energy W_0. Paretzke[26] gives as an approximate representation of L_r the relation:

$$L_r = L_\infty - \frac{aZ_{\text{eff}}^2}{\beta^2}\left[\ln\frac{R}{r} - \left(1 - \frac{r}{R}\right)\right] \tag{3.4}$$

where L_∞ is the total rate of energy loss, dE/dx; R is the maximum track width; and a is a constant for a given medium. The radial distance r from the particle trajectory is the adjustable parameter in this model.

(f) Lineal event-density (LED)

Paretzke[26] has also described a parameter which he calls the lineal event-density (LED). This is the *number* of primary and secondary ionization and excitation events within a distance r of track core (cf. L_r, which considers the *energy* deposited within r). This quantity would probably be rather difficult to calculate, but should lead to a more accurate representation of the track formation process.

The overall conclusion to be reached from studies of track formation by ions of different energies is that tracks are generally formed by ions at energies for which electronic interactions are the dominant mode of energy loss. Etchable damage is usually not produced at the very end of the ion path where nuclear interactions become important.[14,33,34] This result places an important constraint on acceptable models of track formation.

3.4   Experimental Studies on the Size and Structure of Latent-Damage Trails

We now consider some of the experiments which have yielded information on the structure and radial extent of unetched tracks.

3.4.1   Electron microscopy

Unetched tracks are not visible under an optical microscope. Direct viewing of damage trails was, in fact, achieved[35] with a transmission electron microscope (TEM) very early on in the history of SSNTD, as described in Chapter 1.

Observations of tracks and other defects in crystals often make use of diffraction contrast. The principles of this technique are illustrated in Fig. 3.2.

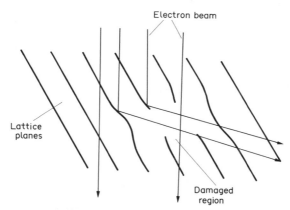

FIG. 3.2   *Principle of "diffraction contrast". Undamaged lattice planes shown here are not at the correct angle to the electron beam for the Bragg reflection to take place. Around the track, however, planes are distorted, and some are exactly in the right orientation with respect to the beam to produce diffraction in the latter. If the viewing through the objective of a transmission electron microscope (TEM) is so organized that all else is obscured except the diffracted rays, this is called "dark field" micrography: tracks will appear as bright images against a dark background. If, on the other hand, the opposite prevails, i.e. all diffracted rays are removed, this will result in "bright field" images, with dark tracks observed against a light background.*

The electron beam in a TEM will generally pass undeflected through the normal regions of the crystal lattice. However, if for some reason any of the lattice planes are bent or tilted, the electron beam may strike the plane at exactly the right angle for Bragg reflection to take place. If so, the beam will, in this region, be strongly diffracted. If an objective aperture is placed so as to block out such diffracted rays, then the image of the region of lattice distortion will appear as a dark area on a bright background. Since such lattice distortions seem to occur around trails of strong, interconnected radiation damage, these may be conveniently imaged by this method (see Fig. 3.3). It is important to realize, however, that—as has been emphasized by Fleischer *et al.*[16]—the image is produced by diffraction effects due to the distorted planes around the track, and so the width of the image represents only an upper limit to the width of the region of primary atomic damage. Typically, these upper limits to the track width are found to be ∼ 100 Å (10 nm).

Another method of visualization of tracks makes use of Moiré fringes.[17,14] If two thin crystals are placed on top of each other—with their lattice planes either laterally displaced with respect to one another or making a small angle to each other—then a pattern of Moiré interference fringes is produced (they are respectively termed parallel and rotational Moiré fringes). The interference fringes obtained represent what is essentially a magnified image of one or both of the crystal lattices. Figure 3.4 shows the Moiré patterns produced in fission-fragment-irradiated crystals of molybdenum trioxide. These images exhibit voids extending over several lattice planes, surrounded by regions of lattice strain produced by planes which have relaxed inwards towards the void.

The evidence from electron microscopy thus points to severe strain and disruption of the lattice in the region of the track, extending over up to ∼ 100 Å (10 nm).

3.4.2   Low-angle X-ray scattering

X-ray diffraction is a common method of investigating crystal structure. Defects in the crystal will distort the X-ray diffraction pattern.

In recent years, small-angle X-ray scattering has been used to probe the nature of the crystal defects which comprise particle tracks.[26,39] Here individual tracks are not imaged: track densities of $\sim 10^{10}$ cm^{-2} are employed; and the information obtained is statistical—being averaged over many millions of individual tracks.

In brief, the results of these experiments show that a track consists of atomic defects. A linear chain of extended defects, each typically tens of angstroms across, is seen to be surrounded by a "soup" of point defects—presumably single interstitials and vacancies. Track etchability seems to be related primarily to the abundance of these extended defects, which overlap

FIG. 3.3 *Dark field micrograph of a crystalline grain from an Apollo-12 lunar fines sample (12070). The fine hairline-like features are the "latent" (unetched) tracks, caused mostly by solar-flare particles of intermediate and VH atomic masses and of relatively low energies incident on the lunar surface. Very high densities (up to ~ 10^{11} cm^{-2}) of these latent tracks (typically, ≤10 nm in diameter) were found in small grains (~2 μm). Such tracks—exhibited by ~90% of the crystalline grains from the lunar fines—proved to be unetchable in a mixed-acid etchant (HF + H_2SO_4), and were easily annealable. They were difficult to see under "bright field" viewing, and generally appeared only on "dark field" micrographs (see caption to Fig. 3.2 for an explanation of the principle of diffraction contrast), according to the authors (Borg et al., Earth Planet. Sci. Lett.* **8** *(1970) 379–86). (Photograph, courtesy M. Maurette, CNR, Orsay.)*

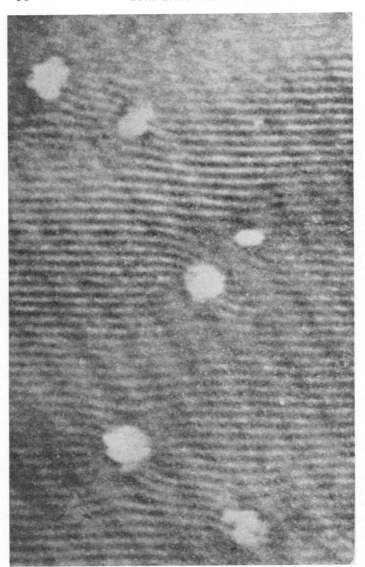

Fig. 3.4 *Transmission electron micrograph showing curvature of Moiré interference fringes near fission-track holes in a crystal of molybdenum trioxide. The lattice planes adjacent to the holes relax into the void left by the fission fragment. (Figure from Morgan and Van Vliet.[17])*

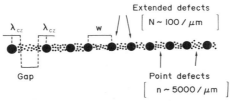

FIG. 3.5 *Schematic representation of latent-damage trails, as deduced from small-angle X-ray scattering analysis of single crystals exposed to high fluences of artificially-accelerated heavy ions.[36-39] The track consists of extended defects (large, filled circles) surrounded by more numerous point defects (small dots). The extended defects are tens of ångströms (several nanometres) across, but the size is dependent on the charge of the track-forming ion. The etching rate of these extended defects is very high, whereas regions containing only point defects are relatively weakly etchable. The effective diameter of the core-zone of high etchability surrounding each extended defect is designated by λ_{cz}. W is the distance between the centres of successive extended defects; it becomes progressively smaller as dE/dx increases. Towards the end of a heavy-ion track, where the rate of energy loss due to electronic interactions is high, extended defects become very crowded and will eventually overlap (not shown). The extended defects are much more resistant to thermal annealing than the point defects. (Figure from Dartyge et al.[39])*

in the regions of most intense damage. The extended defects are produced only when the bombarding ions have been slowed to the domain where their linear rate of energy loss is high. A schematic diagram of track structure as deduced from these studies is shown in Fig. 3.5.

3.4.3 Thermal annealing of tracks

Annealing is a process by which, at elevated temperatures, the solid-state damage is wholly or partially repaired. It is found that the "activation energies" associated with the repair of damage trails are several electron-volts. These energy values are typical of those involved in atomic diffusion, and thus provide further evidence of the atomic nature of the defects. Annealing is discussed at greater length in Chapter 5. Edmonds and Durrani[40] and Edmonds[41] made an extensive study of radiation damage in polycarbonate plastic, using thermoluminescence (TL), electron spin resonance (ESR), and nuclear track detection techniques. They related the most prominent peak in the TL glow curve for the polycarbonate with the recombination of broken molecular fragments. The thermal decay characteristics of this peak were found to match very closely the thermal annealing data for fission tracks in polycarbonate.

3.4.4 *The radial extent of the etchable damage*

From the point of view of the study of etchable tracks the most relevant "track-size" is the radial extent over which preferential etching occurs. The measurement of this region has been carried out in a series of elegant experiments by Bean, Doyle and Entine[42] and by DeSorbo.[9] The principle of their approach is as follows. As the track is etched out, the width of the etched track will at first increase very rapidly, since the damaged region, with its concentration of more highly reactive species, is quickly dissolved. Once the core region has been removed, the rate of increase of the width (i.e. of the diameter) is reduced to just twice that at which the undamaged material is being dissolved. Figure 3.6 shows a schematic plot of etched track radius r versus etch time t. If measurements of etched track radius are performed with an optical microscope, it is found that a plot of r versus t passes through the origin within experimental errors. This is because the maximum attainable resolution of an optical microscope is only ~ 1000 Å (0.1 μm) at best. Provided that a high-resolution measurement of track width can be made, however, it should, in principle, be possible to deduce r_0, the radius of the reactive zone. In the apparatus described by Bean *et al.*,[42] a thin sheet of track-recording material (for example, mica) is used as the dividing wall in an electrolytic cell which contains a suitable track etchant as the electrolyte. The appearance of etched holes in the foil results in a flow of electric current across the cell. The a.c. conductance of the cell is measured, and from this the radius of the etched holes may be deduced. The results of experiments of this type have indicated a radius of about 30 Å for the etchable region of fission tracks in mica[12] and about 80 Å in polycarbonate.[16,9]

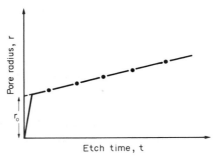

FIG. 3.6 *A schematic diagram to show the principle of determining the radius of the etchable region of a track by making precise measurements of etched-track radius* r *as a function of etching time* t. *The track grows rapidly to a radius* r$_0$ *which is obtained by extrapolating the* r *versus* t *data back to zero etch time. In practice, a perfectly linear plot, such as the one shown above, is not obtained. For a full discussion of these measurements see references 9, 16, 42.*

3.4.5 *Other experimental evidence for track structure*

Chambaudet *et al.*[43] have studied irradiated polymers, using electron spin resonance spectroscopy, and have concluded that the existence of heavy-ion tracks might be correlated with the formation of carbon-like radicals similar to those produced in polymer pyrolysis.

Evidence on track formation processes has also come to hand from heavy-ion sputtering experiments.[44,45] This approach can give information on the mechanisms by which atoms are ejected from the crystal lattice, and will be discussed in §3.5.2.

3.5 Critical Appraisal of Track Formation Models

From the foregoing discussion it is clear that heavy ions traversing an insulating medium produce narrow regions (a few tens of ångströms across) of atomic defects—lattice vacancies, interstitials, etc.—or broken molecular chains. These concentrations of damaged material will be chemically more reactive than the surrounding, undamaged matter, and will therefore give rise to the phenomenon of etchable tracks. Moreover, tracks are formed in materials of low electrical and thermal conductivity by ions of such velocities for which the dominant mode of energy loss is via electronic interactions. These are the essential facts which must be incorporated in any successful model of the track formation process.

At first sight one plausible mechanism would seem to be the direct formation of displacements by the incident ion as it slows down and undergoes collisions with lattice atoms. These displaced atoms will themselves produce further displacements, as discussed in §2.3. Brinkman[46,47] has calculated that, in copper for example, a primary knock-on atom of energy less than about 10 keV would produce at least one secondary displacement for every interatomic spacing traversed. Thus the end of the path of the knock-on atom would be surrounded by a cluster of secondary displacements. Such a mechanism should be equally effective in both insulators and metals, and should also operate most readily when the incident ion is slowed down to energies of ≪1 MeV. Neither of these facts is compatible with our knowledge of etchable tracks—which are not observed in metals, and which do not occur over the last few microns of the ion path in minerals. It is possible that the direct-displacement mechanism may be important for the formation of tracks due to the recoil of heavy atoms after α-particle emission.[48,49] Atomic collisions cannot, however, be of general importance for the formation of tracks, a point emphasized by Fleischer.[50]

According to some models, no "special" mechanism is required for the formation of heavy-ion tracks. Since it is known that large doses of electrons and γ-rays can increase the bulk-etching rate of various materials,[4] hence it is supposed that on these models track formation is an inevitable consequence of the passage of heavy ions through these materials—for such ions will deposit a large amount of energy, via δ-rays, around the track core. Also, since the initially produced complement of reactive species is concentrated within a narrow region around the ion path, it is possible that a host of such secondary reactions will occur which would rarely take place if energy deposition were distributed over a large volume.

The problem of track formation then reduces to the question as to which fraction of the energy deposited by the ion is responsible for the formation of a track. Katz and Kobetich[31] suggest that energy deposition by δ-rays is more important than the primary events themselves; and, in particular, that the energy deposited by electrons at about 20 Å from the ion path is the crucial factor in determining whether or not a track is formed. For a given material, if the variation of chemical reactivity with electron (or gamma) dose is known, it should be possible, by folding-in this response function with the calculated electron dose around an ion-path, to predict the track-etching rate produced by that ion. Katz (see reference 51 for a recent review of his work) interprets the dose-response function of the detector in terms of target theory, according to which one or more "hits" on a sensitive volume (the "target") of the detector are required in order for the target to be activated. A hit in this sense means that a secondary electron must pass through the target. In this formalism the formation of etchable tracks is a many-hit process, in which more than one electron must pass through a sensitive target in order for activation to occur. Since the probability of such multiple hits will rise rapidly as the spatial density of secondary electrons increases, it is expected that the track etchability will rise rapidly with increasing electron-energy deposition around the ion path.

As stated in subsection (d), §3.3, Katz and Kobetich[31] have estimated that the critical dose at 20 Å from the ion path, above which tracks were formed, should be about 2.5×10^4 Gy ($= 2.5 \times 10^6$ rad) for cellulose nitrate and 3.5×10^5 Gy for muscovite mica. According to this model electron or gamma doses of such magnitudes should produce changes in the chemical reactivity of these materials. In cellulose nitrate, doses of the order of 10^4–10^5 Gy will indeed produce an increase in the etching rate of the plastic;[4] but in mica, doses very much larger than 3.5×10^5 Gy are known to have been absorbed without increasing the bulk etching rate.[16] Hashemi-Nezhad et al.[52] have found that a dose of $\sim 10^9$ Gy of 1 MeV electrons is required to increase the bulk etching rate of biotite mica. Thus, while secondary-electron energy deposition may play a rôle in track formation in plastics, it is probably not important in the case of minerals. Also, as was mentioned in §3.3, track etching rates in plastics have been found to correlate less well with secondary-electron energy deposition than with primary ionization or restricted energy loss.

Benton[5,24] has proposed that track formation is due to the radiolytic action of low-energy δ-rays and excitations around the ion path, and that the parameter which determines track formation is restricted energy loss (see §3.3). The model is principally directed to the problem of track formation in plastics rather than in crystals. The correlation between track etchability and REL is good, although some discrepancies have been reported (see Ahlen[30]).

The aforementioned models appear to describe track formation in polymers reasonably well. However, in inorganic crystals it seems that appeal must be made to some more specific mechanism for track formation, which is triggered only by the passage of very heavily ionizing particles through a stopping medium. Some mechanisms of this kind are considered below.

3.5.1 The thermal-spike model

The concept of the thermal spike was introduced some years ago[53] as a mechanism by which energetic particles could produce considerable disruption of a crystal lattice. In this model the passage of an energetic particle is assumed to produce intense heating of a localized region of the lattice. This region is therefore raised to a high temperature, from which it cools rapidly via heat conduction. As a result of this heating episode various atomic processes are activated, and damage to the lattice is produced. This idea has been extended to the case of fission-fragment tracks in crystals by Bonfiglioli *et al.*[54] and lately by Chadderton and co-workers.[14,55,56]

If a particle deposits an energy Q per unit path length at time $t = 0$, then the temperature T as a function of t and of radial distance r from the axis of the ion path is given, from considerations of the classical laws of heat conduction, by:[57]

$$T(r, t) = T_0 + \frac{Q}{4\pi cd} \frac{1}{Dt} \bar{e}^{r^2/4Dt} \tag{3.5}$$

where T_0 is the initial temperature of the lattice; c is the heat capacity of the medium; d is its density; and D is related to the thermal conductivity σ of the medium through the relationship:

$$D = \sigma/cd$$

Simple calculations of this type indicate that a fission fragment could raise the temperature of a narrow cylindrical region of lattice by many thousands of degrees Kelvin for a short time.

Chadderton *et al.*[56] and Chadderton and Torrens[14] have presented more detailed considerations of the manner in which the electronic excitation produced by the fission fragment is transferred to the lattice atoms. They find that, in metals, energy is lost by δ-rays primarily via electron–electron collisions down to very low electron energies—because the energy loss per collision is greater, and the relaxation time is shorter, than for electron–phonon interactions. Thus the excitation is spread throughout a large volume by the electrons before significant energy transfer to the lattice occurs; and therefore the peak temperature to which the lattice is raised in a metal is low.

In insulators, on the other hand, electrons can interact readily with polar and acoustic modes of lattice vibration; and, indeed, once the δ-ray energy falls below that equal to the band-gap width—i.e. several electron-volts— release by the δ-rays of further electrons into the crystal's conduction band becomes impossible. Therefore electron–phonon collisions are expected to be the predominant energy loss process in insulators, and the excitation is communicated to the lattice more efficiently in such materials.

In this way, the inability of metals to show etchable tracks is explained by virtue of the fact that the thermal spike quickly becomes too broad and diffuse in the metallic lattice, whereas in insulators a narrow, intense spike is produced, leading to sufficiently severe localized radiation damage capable of producing etchable tracks.

It seems clear that high-energy, long-range δ-rays will not contribute significantly to the formation of a narrow, intense thermal spike, and so a detailed formulation of the thermal spike model would predict a correlation between track etchability and some restricted energy loss parameter such as REL or RREL (see §3.3).

3.5.2 *The ion-explosion spike model*

In 1965, Fleischer, Price, and Walker[3] proposed a semi-quantitative model for track formation which is outlined below.

The passage of a heavily-ionizing particle leaves in its wake a narrow region containing a high concentration of positive ions. Provided that the time for electron-positive-ion recombination is long compared with the lattice vibration time ($\sim 10^{-13}$ s), mutual repulsion can drive these ions into interstitial positions. Subsequent processes include neutralization of the positive ions and relaxation of the surrounding lattice into the disrupted region. Lattice strains are set up around the track core (see Fig. 3.7). This model has similarities to the Varley mechanism for defect formation in alkali halides[12] described in §3.1. It leads to some quantitative criteria for track formation (see references 16, 7, and 58 for a more detailed discussion of these criteria).

(1) For tracks to form, the "electrostatic stress" must be greater than the mechanical strength of the material. If two atoms have received an *average* ionization of n unit charges e each, and are separated by a distance a_0 (the interatomic spacing), then the force between them is

$$\frac{n^2 e^2}{4\pi\varepsilon_0\varepsilon a_0^2}$$

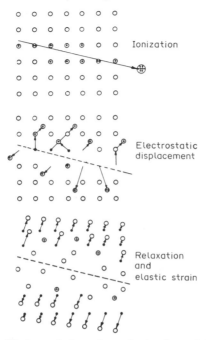

FIG. 3.7 *The ion explosion spike mechanism for track formation in inorganic solids. The orginal ionization left by the passage of a charged particle (top) is* **unstable, and** *ejects ions into the solid, creating vacancies and interstitials (middle). Later, the stressed region relaxes elastically (bottom), straining the undamaged matrix. It is this region of extended damage that is eventually enlarged still further by the etchant to reveal the latent-damage trails as microscopically observable etched tracks. (Figure from Fleischer et al.[16])*

where ε is the dielectric constant for the material, ε_0 is the permittivity of free space, and e is the electronic charge. The force per unit area—the electrostatic stress—is then

$$\frac{n^2 e^2}{4\pi\varepsilon_0 \varepsilon a_0^4}$$

Fleischer et al.[3] took, as a working approximation, a value of $Y/10$, where Y is Young's modulus, for the mechanical strength of the material. The criterion for track formation may thus be expressed as:

$$n^2 > \frac{4\pi\varepsilon_0 \varepsilon a_0^4 Y}{10 e^2} \tag{3.6}$$

The quantity on the right-hand side of the above inequality has been termed by Fleischer et al.[3] the "stress ratio" of the material. This stress ratio shows a fairly good correlation with the sensitivity of track-record-ing materials. For polymers the ratio is ~ 0.01 (i.e. $n \gtrsim 0.1$; so that the inequality (3.6) is readily satisfied by comparatively lightly ionizing par-ticles), whereas for inorganic crystals it is ~ 1.

(2) For tracks to be atomically continuous, there must be at least one ion-ization event per atomic plane. Fleischer *et al.*[3] showed that, roughly speaking, this criterion is satisfied over the regions of incident-particle trajectory where tracks are formed.

(3) If the positive-ion core is to survive long enough (i.e. for $\gtrsim 10^{-13}$ s) for a track to form, then the free-electron density n_f must be low. Fleischer *et al.* showed that, quantitatively, this implies

$$n_f < \frac{en}{\pi a_0 \mu_n kTt}$$

where, besides the symbols defined for (1) above, μ_n is the electron mobil-ity, T the absolute temperature, k Boltzmann's constant, and t diffusion time for electrons, typically 10^{-13} s. This condition is satisfied by insula-tors but not by metals, as indeed should be the case *a fortiori*.

(4) The positive-ion core contains a high concentration of holes. The hole mobility should not be too high, otherwise the core would be neutralized before repulsion could take place. Good semiconductors are correctly predicted, on this basis, as not being able to register tracks.

The ion-explosion spike model, therefore, correctly predicts the observed gradation of sensitivity values among track-recording materials and accounts for the inability of metals and good semiconductors to show etchable tracks. It is suggestive of a quantity, such as primary ionization, which should provide a good criterion for track etchability; and this criterion is in agreement with most experimental data. As we have seen, plastics show enhanced chemical reactivity after electron and gamma bombardment;[4] and it may be that direct scission of polymer chains by ionizing particles and δ-rays plays some rôle in track formation. Some recent work[60] has shown that track formation in polymers correlates with the G value for molecular chain scission (i.e. the number of chain scissions per 100 eV of deposited energy), and it seems that this process, resulting from the decay of excited electrons, may be primarily responsible for track formation in polymers.[61] In any event, for crystalline solids the ion-explosion spike seems to provide an adequate model for track formation.

Chadderton and Torrens[14] have carried out a computer simulation of an explosion spike produced by a fission fragment in a two-dimensional KCl crystal, and find that no permanent damage results from such a spike. How-ever, it is not clear whether a two-dimensional model is a sufficiently realistic basis for tackling a phenomenon as complex as track formation.

Some recent work on heavy-ion sputtering of dielectrics[45,46] has provided additional evidence in favour of the ion-explosion mechanism. Griffith *et al.*[45] found that when UF_4 targets were bombarded with ions possessing energies near to those corresponding to the peak of the electronic stopping power, the yield of sputtered uranium atoms was much larger than could be explained by standard sputtering theory. According to simple calculations by Seiberling *et al.*[59] the time required for transfer of energy from the electrons to the lattice atoms ($\sim 10^{-10}$ s) is much longer than the time needed for the energy to be removed by heat conduction ($\sim 10^{-12}$ s). They therefore suggest that an ion-explosion spike is responsible for the ejection of atoms. The ions are repelled from each other and transfer their energy to other lattice atoms, and a condition approaching local thermodynamic equilibrium is eventually attained. Using this thermalized ion-explosion model, Seiberling *et al.*[59] are able to predict a velocity distribution for the sputtered uranium atoms which agrees well with experimental data.

To conclude: the ion-explosion model seems to be favoured by most workers in the field of solid-state nuclear track detectors at present, at least as a mechanism for track formation in inorganic materials. Clearly, however, much further experimental and theoretical work on the formation of particle tracks in dielectric materials remains to be done.

References

1. R. L. Platzman (1958) In: *Radiation Biology and Medicine*, W. Claus (ed.). Addison-Wesley, Reading, Mass., Ch. 2.
2. R. L. Platzman (1966) Energy spectrum of primary activations in the action of ionizing radiation. In: *Proc. Third Int. Congress of Rad. Research*, G. Silini (ed.). North Holland, Amsterdam, pp. 20–42.
3. R. L. Fleischer, P. B. Price & R. M. Walker (1965) The ion explosion spike mechanism for formation of charged particle tracks in solids. *J. Appl. Phys.* **36**, 3645–52.
4. E. V. Benton (1968) A study of charged particle tracks in cellulose nitrate. *US Naval Radiological Defense Lab., San Francisco, Report* NRDL-TR-68-14.
5. E. V. Benton (1970) On latent track formation in organic nuclear charged particle track detectors. *Radiat. Effects* **2**, 273–80.
6. W. T. Crawford, W. D. DeSorbo & J. S. Humphrey, Jr. (1968) Enhancement of track etching rates in charged particle-irradiated plastics by a photo-oxidation effect. *Nature* **220**, 1313–4.
7. E. V. Benton & R. P. Henke (1969) Sensitivity enhancement of Lexan nuclear track detector. *Nucl. Instrum. Meth.* **70**, 183–4.
8. W. DeSorbo & J. S. Humphrey (1970) Studies of environmental effects upon track etching rates in charged particle irradiated polycarbonate film. *Radiat. Effects* **3**, 281–2.
9. Warren DeSorbo (1979) Ultraviolet effects and aging effects on etching characteristics of fission tracks in polycarbonate film. *Nucl. Tracks* **3**, 13–32.
10. F. A. Bovey (1958) *The Effects of Ionizing Radiations on Natural and Synthetic High Polymers*. Interscience, New York.
11. A. Charlesby (1960) *Atomic Radiation and Polymers*. Pergamon, London.
12. J. H. O. Varley (1954) A mechanism for the displacement of ions in an ionic lattice. *Nature* **174**, 886–7.

13. F. Seitz (1954) Color centers in alkali-halide crystals, II. *Revs. Mod. Phys.* **26**, 7–94.
14. L. T. Chadderton & I. McC. Torrens (1969) *Fission Damage in Crystals*. Methuen, London.
15. D. Pooley (1966) F-centre production in alkali halides by electron–hole recombination and a subsequent (110) replacement sequence: A discussion of the electron–hole recombination. *Proc. Phys. Soc.* **87**, 245–55.
16. R. L. Fleischer, P. B. Price & R. M. Walker (1975) *Nuclear Tracks in Solids: Principles and Applications*. University of California Press, Berkeley.
17. D. V. Morgan & D. Van Vliet (1970) Charged particle tracks in solids. *Contemp. Phys.* **11**, 173–93.
18. D. V. Morgan & L. T. Chadderton (1968) Fission fragment tracks in semiconducting layer structures. *Phil. Mag.* **17**, 1135–43.
19. A. Sigrist & R. Balzer (1978) Investigations on the formation of tracks in crystals. In: *Proc. 9th Int. Conf. Solid State Nucl. Track Detectors,* Munich, and Suppl. 1, *Nucl. Tracks*. Pergamon, Oxford, pp. 387–91.
20. J. Shirck (1974) Fission tracks in a white inclusion of the Allende chondrite. *Earth Planet. Sci. Lett.* **23**, 308–12.
21. P. B. Price, D. Lal, A. S. Tamhane & V. P. Perelygin (1973) Characteristics of tracks of ions of $14 \leq Z \leq 36$ in common rock silicates. *Earth Planet. Sci. Lett.* **19**, 377–95.
22. G. Somogyi, K. Grabisch, R. Scherzer & W. Enge (1976) Revision of the concept of registration threshold in plastic track detectors. *Nucl. Instrum. Meth.* **134**, 129–41.
23. R. L. Fleischer, P. B. Price, R. M. Walker & E. L. Hubbard (1967) Criterion for registration in various solid state nuclear track detectors. *Phys. Rev.* **133**, 1443–9.
24. R. L. Fleischer, P. B. Price, R. M. Walker & E. L. Hubbard (1967) Criterion for registration in dielectric track detectors. *Phys. Rev.* **156**, 353–5.
25. H. A. Bethe (1930) Theory of the passage of rapid corpuscular rays through matter. *Ann. Physik.* **5**, 325–400.
26. H. G. Paretzke (1977) On primary damage and secondary electron damage in heavy ion tracks in plastics. *Radiat. Effects* **34**, 3–8.
27. P. B. Price & R. L. Fleischer (1971) Identification of energetic heavy nuclei with solid dielectric track detectors: Applications to astrophysical and planetary studies. *Ann. Rev. Nucl. Sci.* **21**, 295–334.
28. E. V. Benton (1967) Charged particle tracks in polymers. No. 4: Criterion for track registration. *US Naval Radiological Defense Laboratory, San Francisco, Report* USNRDL-TR-67-80.
29. G. Siegmon, H. J. Kohnen, K.-P. Bartholomä & W. Enge (1978) The dependence of the mass-identification scale on different track formation models. In: *Proc. 9th Int. Conf. Solid State Nucl. Track Detectors*, Munich, and Suppl. 1, *Nucl. Tracks*. Pergamon, Oxford, 137–43.
30. S. P. Ahlen (1980) Theoretical and experimental aspects of the energy loss of relativistic heavily ionizing particles. *Revs. Mod. Phys.* **52**, 121–73.
31. R. Katz & E. J. Kobetich (1968) Formation of etchable tracks in dielectrics. *Phys. Rev.* **170**, 401–5.
32. E. J. Kobetich & R. Katz (1968) Energy deposition by electron beams and δ rays. *Phys. Rev.* **170**, 391–6.
33. P. B. Price, R. L. Fleischer & C. D. Moak (1968) On the identification of very heavy cosmic ray tracks in meteorites. *Phys. Rev.* **167**, 277–82.
34. M. Maurette (1966) Study of the registration of fission fragment tracks in certain substances. *J. de Phys.* **27**, 505–12.
35. E. C. H. Silk & R. S. Barnes (1959) Examination of fission fragment tracks with an electron microscope, *Phil. Mag.* **4**, 970–1.
36. E. Dartyge & M. Lambert (1974) Formation de défauts dans les échantillons de mica muscovite irradiés par des ions de grand energie. *Radiat. Effects* **21**, 71–9.
37. E. Dartyge, M. Lambert & M. Maurette (1976) Structure et enregistrement des traces latentes d'ions Argon et Fer dans l'olivine et le mica muscovite. *J. de Phys.* **37**, 137–41.
38. E. Dartyge, J. P. Duraud & Y. Langevin (1977) Thermal annealing of iron tracks in muscovite, labradorite and olivine. *Radiat. Effects* **34**, 77–9.

39. E. Dartyge, J. P. Duraud, Y. Langevin & M. Maurette (1978) A new method for investigating the past activity of ancient solar flare cosmic rays over a time scale of a few billion years. In: *Proc. Lunar Planet. Sci. Conf. 9th*, Pergamon, New York, pp. 2375–98.
40. E. A. Edmonds & S. A. Durrani (1979) Relationships between thermoluminescence, radiation-induced electron spin resonance and track etchability of Lexan polycarbonate. *Nucl. Tracks* 3, 3–11.
41. E. A. Edmonds (1978) Radiation effects in the polycarbonate of bisphenol-A. Thermoluminescence, electron spin resonance and charged particle track studies. PhD thesis, Physics Department, University of Birmingham.
42. C. P. Bean, M. V. Doyle & G. Entine (1970) Etching of submicron pores in irradiated mical. *J. Appl. Phys.* 41, 1454–9.
43. A. Chambaudet, A. Bernas & J. Roncin (1977) On the formation of heavy ion latent tracks in polymeric detectors. *Radiat. Effects* 34, 57–9.
44. L. E. Seiberling, J. E. Griffith & T. A. Tombrello (1980) Enhanced sputtering of dielectric materials and its relationship to track registration. In: *Lunar and Planet. Sci. XI* (Abstracts), Lunar and Planetary Institute, Houston, pp. 1021–3.
45. J. E. Griffith, T. A. Weller, L. E. Seiberling & T. A. Tombrello (1980) Sputtering of uranium tetrafluoride in the electronic stopping region. *Radiat. Effects* 51, 223–31.
46. J. A. Brinkman (1954) On the nature of radiation damage in metals. *J. Appl. Phys.* 25, 961–70.
47. J. A. Brinkman (1956) Production of atomic displacements by high energy particles. *Am. J. Phys.* 24, 246–67.
48. W. H. Huang & R. M. Walker (1967) Fossil alpha-particle recoil tracks: a new method of age determination. *Science* 155, 1103–6.
49. S. R. Hashemi-Nezhad & S. A. Durrani (1981) Registration of alpha recoil tracks in mica: the prospects for alpha-recoil dating method. *Nucl. Tracks* 5, 189–205.
50. R. L. Fleischer (1976) Fission tracks are not from atomic collisions. *Radiat. Effects* 28, 113–4.
51. R. Katz (1978) Track structure theory in radiobiology and in radiation detection. *Nucl. Track Detection* 2, 1–28.
52. S. R. Hashemi-Nezhad, R. K. Bull & S. A. Durrani (1981) Electron and alpha-particle damage in biotite mica: Implications for track formation mechanisms. In: *Proc. 11th Int. Conf. Solid State Nuclear Track Detectors*, Bristol, & Suppl. 3, *Nucl. Tracks*. Pergamon, Oxford, pp. 23–6.
53. F. Seitz & J. S. Koehler (1956) Displacement of atoms during irradiation. *Solid State Phys.* 2, 305–448.
54. G. Bonfiglioli, A. Ferro & A. Nojoni (1961) Electron microscope investigation on the nature of tracks of fission products in mica. *J. Appl. Phys.* 32, 2499–2503.
55. L. T. Chadderton & H. M. Montagu-Pollock (1963) Fission fragment damage to crystal lattices. Heat sensitive crystals. *Proc. Roy. Soc.* A274, 239–52.
56. L. T. Chadderton, D. V. Morgan, I. McC. Torrens & D. Van Vliet (1966) On the electron microscopy of fission fragment damage. *Phil. Mag.* 13, 185–95.
57. G. J. Dienes & G. H. Vineyard (1957) *Radiation Effects in Solids*. Interscience, New York.
58. R. L. Fleischer (1980) Nuclear track production in solids. In: *Progress in Materials Science* (Chalmers Anniversary Volume), pp. 97–123. Pergamon, New York.
59. L. E. Seiberling, J. E. Griffith & T. A. Tombrello (1980) A thermalized ion explosion model for high energy sputtering and track registration. *Radiat. Effects* 52, 201–10.
60. D. O'Sullivan, P. B. Price, K. Kinoshita & C. G. Willson (1982) Predicting radiation sensitivity of polymers. *J. Electrochem. Soc.* 129, 811–3.
61. P. B. Price (1982) Applications of nuclear track-recording solids to high-energy phenomena. *Phil. Mag.* 45, 331–46.

CHAPTER 4

Track Etching: Methodology and Geometry

Chemical etching is the most widely used method of "fixing" and enlarging the image of the latent damage trail in a solid state track detector. Essentially, etching takes place via rapid dissolution of the disordered region of the track core which exists in a state of higher free energy (see Chapter 3) than the undamaged bulk material. The linear rate of chemical attack along the track is termed the track etching velocity V_T. The surrounding undamaged material is attacked at a rate V_B, the bulk etching velocity. The bulk etching rate is generally constant for a given material and for a given etchant applied under a specific set of etching conditions, although in crystals it will often depend on the crystallographic orientation, and in some polymers it may vary with depth below the original surface.* Track etching rates will depend, in addition to the above mentioned factors, on the amount of damage located in the track core region (and hence on the properties of the track forming particle), and will usually vary along an individual track.

The chemistry of track etching has been little studied, although Paretzke et al.[1] have examined the reaction between bisphenol-A polycarbonate (e.g. Lexan**) and NaOH. The main reaction here seems to be:

Thus the bisphenol A anion is released from the polymer chain. More recently, the highly sensitive track recording plastic CR-39, produced from

---

* Both V_T and V_B may also depend on the pre- and post-irradiation treatment of the material.
** Lexan is the registered trade mark of the General Electric Co. of USA.

diethylene-glycol bis(allylcarbonate), has been studied by Gruhn *et al.*[2] They find that attack by the hydroxide ion results in the hydrolysis of the carbonate ester bonds and the release of poly-allylalcohol (PAA) from the polymer network. The reaction is

$$\{CH_2CHCH_2O\overset{\overset{\displaystyle O}{\|}}{C}OCH_2CH_2OCH_2CH_2O\overset{\overset{\displaystyle O}{\|}}{C}OCH_2CHCH_2\} + 4OH^-$$

$$\rightarrow 2\{CH_2\overset{\overset{\displaystyle CH_2OH}{|}}{CH}\} + HOCH_2CH_2OCH_2CH_2OH + 2CO_3^{2-}$$

In addition to the polymeric etch product PAA, 2,2'-oxydiethanol is also formed in the above reaction.

The most intensive studies of track etching have concentrated, however, on track shape geometry and on the effects of environmental conditions on track etching (e.g. effects of temperature, concentration of etchant and of etch products, etc.). These factors are considered in subsequent sections of this chapter.

4.1 Track Etching Recipes

Before moving on to a discussion of track-etching geometry, we will first mention, briefly, some of the etchants in common use.

For *plastics*, the most frequently used etchant is the aqueous solution of NaOH with concentrations typically within the range 1 to 12 M;* the temperatures usually employed are in the range 40–70°C. In some cases, ethyl alcohol is added to the etchant; this has the effect, generally, of reducing the threshold value of primary ionization at which tracks become etchable,[3] although this treatment also renders the plastic more brittle. Alcohol has, however, been found to *reduce* the sensitivity of CR-39[3a] for track revelation. In the case of Lexan, an increase in the etching sensitivity has also

---

* *Etchant concentration.* In expressing the etchant concentration a number of conventions have been followed by nuclear track workers (e.g. weight or volume %; weight to weight or weight to volume ratio; molarity; normality; molality, etc.). In our view, it is best to stick to molarity. Molarity represents the number of moles of a substance per litre of solution (mole being the molecular weight in grams). Thus a 1 M aqueous solution of NaOH (molecular weight, $23 + 16 + 1 = 40$ g) contains 40 g of NaOH per litre of solution.

In simple solutions (e.g. NaOH in water), normality and molarity are often identical (normality being defined as the number of equivalent weights of solute, in grams, per litre of solution). Molality represents the number of moles of a substance per kg of solvent (and, in the case of an aqueous solution of NaOH, is practically equivalent to molarity). However, in complicated reactions and solutions normality can be ambiguous, and is best avoided.

been observed when high concentrations of etch products accumulate in the NaOH etchant.[4]

Glasses are almost invariably etched in aqueous HF solutions (ranging downwards from the undiluted 48 vol.% HF) at room temperature. Some *mineral crystals*, such as quartz, mica, or certain pyroxenes, also respond to this etchant; although the bulk etch rates are frequently highly directional.[5,6] A more usual etchant for silicate crystals is concentrated aqueous NaOH solution[7] (with concentrations varying from 10 M to greater than 40 M), boiling under reflux.

Etch times can vary from a few seconds (e.g. 48% HF on soda–lime glass) to many hours (for example, cosmic ray tracks in Lexan may require etching

TABLE 4.1   *Some useful etchants for nuclear track detectors**

Material	Reference(s)	Etchant
Polycarbonate plastics	41	Aqueous NaOH solution; typically 1–12 M. Temperature: 40–70 °C.
	3	Alternatively, 15 g KOH + 45 g H_2O + 40 g C_2H_5OH. 70°C.
Cellulose nitrate plastics	42	NaOH; 1–12 M. Temperature: 40–70 °C
CR-39 plastic (allyl diglycol carbonate)	23; 43	NaOH, KOH solutions; 1–12 M. Temperature: 40–70 °C.
Orthopyroxenes and clinopyroxenes	7	6 g NaOH + 4 g H_2O. Boiling, under reflux.
Mica	44; 45	48% HF. Temperature: 20–25°C.**
Glasses	46	1–48% HF. Temperature: 20–25 °C.
Feldspars	7	1 g NaOH + 2 g H_2O. Boiling, under reflux.
Apatite, Whitlockite[†]	47	0.1–5% HNO_3. Temperature: 20–25 °C.
Zircon	48	11.5 g KOH + 8 g NaOH (eutectic). Temperature: 200–220°C.
Olivine	49	1 ml H_3PO_4 + 1 g oxalic acid + 40 g disodium salt of EDTA + 100 g H_2O; NaOH added to bring pH to 8.0 (the "WN solution"). Boiling, under reflux.
Sphene	50	$1HF:2HNO_3:3HCl:6H_2O$. Temperature: 20°C.

* Etching times will vary according to the exact etching conditions (temperature and concentration of etchant) and the nature of the track-forming particle. In most cases they are a few hours (but of the order of a few seconds or minutes for some glasses or micas etched in 48% HF). They should be determined by trial and error for each detector type.

** Note that muscovite needs ~20–30 min, but biotite only a few min, of etching.

[†] In current mineralogical usage whitlockite is termed "merrillite".

for 96 hours in 6.25 M NaOH at 40 °C.[8]) For work of high accuracy, long etch times are preferable. Also, fine control of the etching temperature (to within ± 0.1 °C) is often necessary.

If reproducible results are to be obtained, it is important that the quality of the etchant should be carefully controlled. After prolonged use high concentrations of etch products build up in the etchant, and it is therefore necessary that fresh batches of etchant are prepared frequently. A layer of etch products can accumulate on the surface of the detector. In the case of Lexan etched with NaOH, the sodium salt of the bisphenol-A anion is precipitated.[1,2] Some workers (see for example Enge *et al.*[9]) use constant stirring during the etching process to prevent etch product buildup. Khan[10] has pointed out that a series of etching intervals interrupted by cleaning steps can prevent the accumulation of etch-product layers.

Table 4.1 lists some of the more commonly used etchants. A longer, but by no means exhaustive, list of etching recipes is given in the book by Fleischer, Price, and Walker.[11]

4.2 Track Etching Geometry

Nuclear particles incident on a track detector may be characterized by their charge Z (or effective charge Z_{eff}), mass M, and energy E (or relative velocity, $\beta = v/c$). Differences in these particle parameters manifest themselves as changes in the track length R and the track etching rate V_T (as well as in the variation of V_T with position along the track). The most easily measurable parameters of an etched track are the etched-cone length L_e (or the length S projected onto the detector surface) and the major and minor axes D, d of the etch pit opening (see Figs 4.1 and 4.4). It is the purpose of the study of etched-track geometry to relate these latter parameters to the track and bulk etch rates V_T and V_B and to the track length R (we shall use R in the treatment that follows to denote the length of the damage trail in the detector, whereas L_e will refer to the length of the conical etched-out track), so that quantities of interest such as Z, M, or β may ultimately be deduced. The track etching geometry also determines quantities such as the efficiency of a given detector—which we define here as the ratio of the number of observed etched tracks to the number of latent damage trails crossing a unit area of the "original" surface of the detector (i.e. where the etching first starts).

In the remainder of this section we indicate some of the general features of track etching geometry and present some of the simpler derivations. Interested readers are referred to more extensive treatments by Fleischer, Price, and Woods,[12] Henke and Benton,[13] Somogyi and Szalay,[14] and Ali and Durrani.[15]

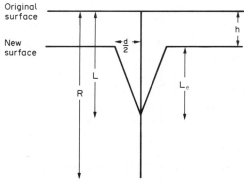

FIG. 4.1 *Some parameters used to describe the geometry of etched tracks.* R, *full length of the latent (unetched) track;* L, *the length of track attacked by the etchant up to a given moment;* L_e, *the observed length of the etched track (or of the etch-cone);* h, *the thickness of surface removed by etching (= $V_B t$, where t is the duration of etching);* d, *the diameter of the etch-pit opening. The track shown has a track-etching to bulk-etching velocity ratio* V(= V_T/V_B) *of 3; for simplicity, a vertical track has been depicted. The more general case, where the track makes an angle θ with the surface, is shown in Fig. 4.4. In the case of the inclined track, the diameter of the opening is replaced by the major and minor axes,* D *and* d, *of the elliptical opening; the etched cone then subtends a projected length* S *on the detector surface.*

4.2.1   Constant track etching velocity V_T

The calculation of etched-track parameters is comparatively simple when the track etch rate V_T can be taken to be constant along that portion of the latent damage trail which is etched out. This condition will apply in many instances where the ionization rate of the particle is not varying rapidly (such as the case of an energetic cosmic-ray nucleus). Furthermore, the constant V_T model allows a number of features of the track etching process to be established fairly simply.

Initially, we shall make a further simplification by concentrating on a track which is normally incident upon the detector surface. Now, the linear rate of attack down the track is V_T (the radial extent of the enhanced etchability is assumed to be very small compared with the final dimensions of the etched track), so that in an etching time t the etch pit will extend to a distance L from the point of origin, where $L = V_T t$. However, the surface is also being removed at a rate V_B; so that the length of the etch pit is

$$L_e = V_T t - V_B t \qquad (4.1)$$

At each point along the track, the etchant moves outwards at a rate V_B. Any point at a distance y from the beginning of the track is reached by the etchant at a time $t(y) = y/V_T$; and there is a residual time $t - t(y)$ available for the etchant to attack radially outwards from the point at y to a distance $V_B \times (t - t(y))$ in the medium. The three-dimensional pit wall is then formed by the locus of all the spheres of radius $V_B \times (t - t(y))$, where $t(y)$ is the variable. It will be seen from Fig. 4.2 that this leads to the formation of a cone with semi-cone angle δ given by

$$\sin \delta = \frac{V_B t}{L} = \frac{V_B t}{V_T t} = \frac{V_B}{V_T} \tag{4.2}$$

This angle $\delta = \sin^{-1} (V_B/V_T)$ is also known as the "critical angle of etching".[16] From the triangle O'PT in Fig. 4.2 it is apparent that

$$\frac{d/2}{L_e} = \tan \delta \tag{4.3}$$

From Eq. (4.2) it follows that

$$\tan \delta = \frac{V_B}{\sqrt{(V_T^2 - V_B^2)}}$$

so that, from Eq. (4.3),

$$d = \frac{2 V_B L_e}{\sqrt{(V_T^2 - V_B^2)}}$$

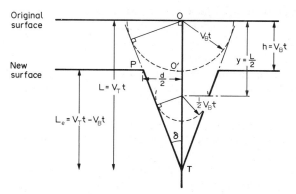

FIG. 4.2 *Construction for the calculation of etched-track parameters for a track of constant* V_T *(with* $V = V_T/V_B = 3$*), lying normally to the detector surface. The semi-cone-angle is denoted by* δ*; the other parameters are as explained in the caption to Fig. 4.1. In the present figure, a point half way along the etched length (with* $y = L/2$*) has been illustrated, from which the etchant is considered to have travelled outward for time* $t/2$ *to the edge of the etched cone.*

Using the value of L_e from Eq. (4.1) we then obtain

$$d = 2V_B \frac{(V_T - V_B)t}{\sqrt{(V_T^2 - V_B^2)}}$$

or

$$d = 2V_B t \sqrt{\frac{V_T - V_B}{V_T + V_B}} \tag{4.4}$$

Some features of track etching have emerged from this simple calculation, which are in fact found to apply in more general cases.

(1) The semi-cone angle $\delta = \sin^{-1}(V_B/V_T)$; in certain materials such as glasses where V_T is not very much greater than V_B, etched tracks with large cone angles are produced (see Fig. 4.3a). In plastics (Fig. 4.3b), and minerals (Fig. 4.3c), long needle-like tracks of much smaller cone angle are usually produced. This is a consequence of the fact that $V_T \gg V_B$ in these cases.
(2) The diameters of the surface openings of etched tracks increase with increasing V_T, reaching a maximum of $2V_B t$ when $V_T \gg V_B$ (see Eq. (4.4)).

Next, let us calculate D, d (the major and minor axes of the opening, respectively) and L_e for a track incident at some arbitrary dip angle θ with the detector surface (Fig. 4.4). The track opening being the intersection of a cone with a plane surface making an angle θ with the axis of the cone, the resulting shape is a conic section: an ellipse, in fact, with major axis D and minor axis d. (For a normally-incident track, of course, $D = d$.)

The length of the etch pit, L_e, as measured from the cone tip to the point where the damage trail crosses the final etched surface of the detector, is now $V_T t - (V_B t/\sin \theta)$. Note that this point of intersection O' no longer divides D into two equal portions. Let us construct perpendiculars from the latent damage trail to the extremities P, P' of the etch pit opening; these have lengths x_1, x_2 (see Fig. 4.4). The points of intersection of these perpendiculars with the track lie at distances Δy_1 and Δy_2 from O'. Now from Fig. 4.4 we see that the etch pit radii r_1 and r_2 (where $D = r_1 + r_2$) are given by

$$r_1 = x_1/\sin \theta$$

and

$$r_2 = x_2/\sin \theta$$

also

$$\Delta y_1 = r_1 \cos \theta$$

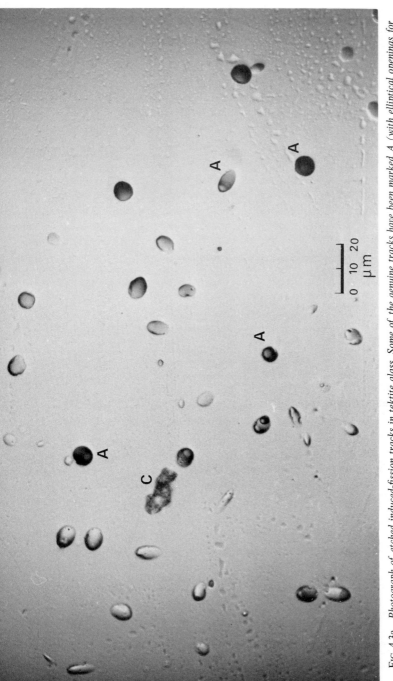

FIG. 4.3a Photograph of etched induced-fission tracks in tektite glass. Some of the genuine tracks have been marked A (with elliptical openings for tracks inclined to the etched surface, and circular ones for tracks normal to it). Also shown are some artefacts (marked C). The etching was done in 40 vol.% HF for ~60 s at room temperature, after a high degree of polishing of the section following reactor irradiation (~10^{15} thermal n cm^{-2}) of the tektite (an indochinite). (Figure from Durrani and Hancock.[16a]) In glasses, where the ratio of the along-the-track and bulk etching velocities, viz. $V = V_T/V_B$, is relatively small (typically ~1.7 to 2.3), the critical angle $\theta_c = \sin^{-1}(1/V)$, which equals half the etched-cone angle, is relatively large (~30°), so that the etched tracks appear like rather "blunt" cones. It may be worth mentioning here that, since the critical angle depends on the charge and the energy of the ionizing particle, tracks in plastics can, under certain circumstance, look like those in glass. The polycarbonate plastic CR-39, now becoming increasingly popular because of its high sensitivity, can, for example, behave like a glass for particles with low values (up to ~50) of Z_{eff}/β (where Z_{eff} is the effective charge and $\beta = v/c$), and exhibit a large critical angle (e.g. $\theta_c \sim 25°$ for ~5 MeV α-particles in CR-39; see Khan et al.[16b]).

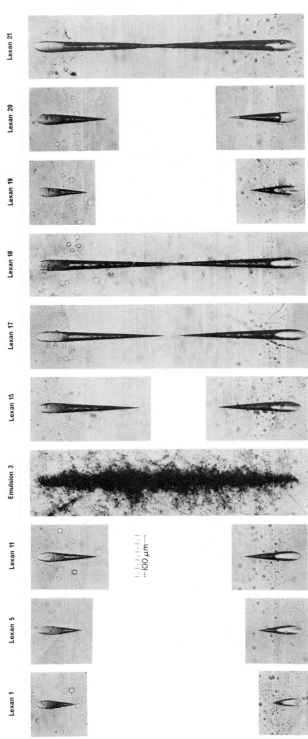

FIG. 4.3b   The track of a cosmic-ray nucleus of charge $Z = 78$, having an energy at entry: 700 MeV/nucleon. Segments of the track obtained in several layers of a Lexan polycarbonate stack, flown in a balloon (Sioux Fall, 1973), are shown, with sheet 1 uppermost. In any given Lexan sheet, the etchant attacks the track from either side: the etched-cone length increases as the cosmic ray particle slows down and its track becomes more etchable. The change in the etch rate between Lexan sheets 18 and 19 is due to a nuclear interaction which is believed to have taken place in the intervening iron plate and to have reduced the charge of the heavy ion. A segment of the track, as developed in a nuclear emulsion (no. 3) also flown along with the plastic stack, is shown as well. (Figure from Fowler et al.[16c]). In plastics, where V is usually relatively large (typically ~15 to 30), the critical angle is small (up to a few degrees of arc), so that the etched track looks like a fine "needle" or a long cone (as here). The critical angle also depends on the charge and the energy of the particle concerned.

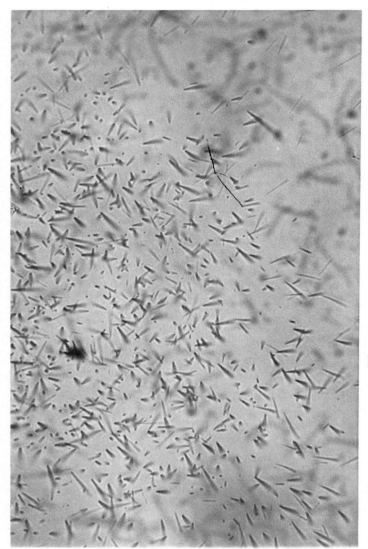

FIG. 4.3c   *Photograph of etched ^{252}Cf fission tracks in muscovite mica. Etching in 20% to 48% HF at room temperature for times of the order of a minute to several tens of minutes can reveal fission tracks in different types of mica. These tracks usually have semi-cone-angles of a few degrees of arc, with polygonal surface-openings (e.g. rhombus-like in muscovite). The bulk etching velocity of the etchant is much faster parallel to the mica cleavage planes than across them.*

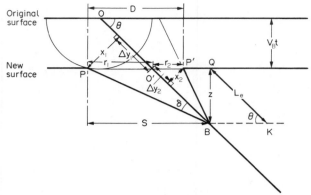

FIG. 4.4 *Construction for the calculation of the major axis of an elliptical track-opening and related etched track parameters. The track shown has a constant V_T (with $V = V_T/V_B = 3$), and lies at an angle $\theta = 45°$ to the detector surface. For an explanation of the parameters, see caption of Fig. 4.1, for the rest, see text. D is the major axis of the elliptical track-opening which, at the etching time t, is divided by the track at O' into unequal parts, r_1 and r_2. δ (which equals the critical angle θ_c) is the semi-angle of the etched cone, whose projection on the detector surface has a length $S = L_e \cos \theta + r_1$. (Here $O'Q = BK = L_e \cos \theta$).*

and

$$\Delta y_2 = r_2 \cos \theta$$

We can also calculate $\tan \delta$, and find that:

$$\tan \delta = \frac{x_2}{V_T t - (V_B t/\sin \theta) - \Delta y_2} = \frac{x_1}{V_T t - (V_B t/\sin \theta) + \Delta y_1}$$

Some rearrangement leads to:

$$r_1 = \frac{\tan \delta [V_T - (V_B/\sin \theta)]t}{\sin \theta - \cos \theta \tan \delta} \tag{4.5}$$

and

$$r_2 = \frac{\tan \delta [V_T - (V_B/\sin \theta)]t}{\sin \theta + \cos \theta \tan \delta} \tag{4.6}$$

On adding r_1 and r_2, and after some further manipulation, we obtain for the major axis of the track opening:

$$D = \frac{2V_B t \sqrt{(V_T^2 - V_B^2)}}{V_T \sin \theta + V_B} \tag{4.7}$$

An alternative expression for Eq. 4.7 is obtained, by putting $V = V_T/V_B$, as

$$D = \frac{2V_B t \sqrt{(V^2 - 1)}}{V \sin \theta + 1} \tag{4.7'}$$

Notice that, unlike the circular case, the latent damage trail does not cross the etched surface at the centre C of the ellipse (see Fig. 4.5). The intersection is displaced from the centre by an amount

$$\Delta r = \frac{V_B t}{\tan \theta (V \sin \theta + 1)} \tag{4.8}$$

where, again, the track etch rate to bulk etch rate ratio V is given by

$$V = V_T/V_B \tag{4.9}$$

In order to calculate the minor axis d of the track opening, consider a right circular section (centred on A) of the etch cone (Fig. 4.5). The plane of this circle passes through the centre C of the ellipse; the circle itself passes through the track opening at the extremities X, X' of the minor axis. The distance of A from the tip B of the cone is $L_e + \Delta r \cos \theta$; this distance L' reduces to

$$L' = \frac{(V^2 - 1) \sin \theta}{V \sin \theta + 1}$$

The radius of the circle is $AK = L' \tan \delta$. In the right-angled triangle ACO', the distance AC is given by $\Delta r \sin \theta$. Hence, from the right-angled triangle ACX, whose side AX equals the radius (AK) of the circle, the semi-minor axis $CX \, (= d/2)$ is obtained as

$$(d/2)^2 = L'^2 \tan^2 \delta - (\Delta r)^2 \sin^2 \theta$$

which leads to an expression for the minor axis of the track opening:

$$d = 2V_B t \sqrt{\frac{V \sin \theta - 1}{V \sin \theta + 1}} \tag{4.10}$$

A further important feature of track etching is revealed by this last equation: when θ assumes a value such that $V \sin \theta = 1$, then d vanishes and for all lower values of θ, d is imaginary. The reason for this is that when $V \sin \theta = 1$, the component of the track etch velocity normal to the surface is equal to the bulk etch velocity, so that the tip of the etch cone never keeps ahead of the advancing surface. The track therefore is never observable. The angle θ_c, defined by $\sin \theta_c = 1/V = V_B/V_T$, is known as the critical angle of

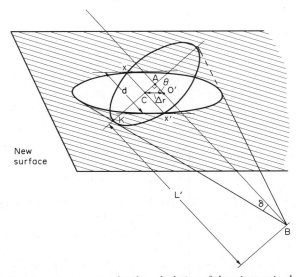

FIG. 4.5 *Construction for the calculation of the minor axis,* d, *of an elliptical etched-track opening lying in the plane of the paper (termed the "new surface"). The track, running along AB, makes an angle θ with the (horizontal) plane of the paper. As the etching has proceeded from the original surface downwards, the point where the track intersects the major axis of the opening has moved further and further away from the centre of the ellipse, C; so that, in the situation depicted, the intersection point is O', a distance Δr away from C. In order to determine the minor axis XX' of the ellipse, proceed as follows. Imagine a vertical plane containing both the track and the major axis running along CO'. Drop a perpendicular from point C onto the track; call the point of intersection A, so that CÂO' = 90°, and CA = Δr sin θ. Next, imagine the etch cone to have a (circular) base at right angles to the central axis BA, the circle (centred on A) passing through points X, K, and X', as shown; so that the radii AK and AX are equal (where AK = AB tan δ; δ being the semi-cone-angle). Then, since the major and minor axes of the ellipse are at right angles to each other, and the track AB is continued in the vertical plane passing through AB and O'C (to which plane the semi-minor axis XC is normal), hence XCA is also a right angle. But CA and AX are known; CX (= ½d) can, therefore, be obtained by Pythagoras' theorem.*

etching,[16] and represents the minimum angle to the surface that a track can make in order to be revealed by etching. It will be seen that $\theta_c = \delta$, the semi-cone-angle (see Eq. (4.2)).

Finally, in the case of the "conical phase" (Fig. 4.6), the projected length S of the track pit PQR' on to the final etched surface is found (see Fig. 4.4) to be

$$S = L_e \cos \theta + r_1 \qquad (4.11)$$

where r_1 is given by Eq. (4.5). When the etchant reaches the end of the latent damage trail, the etched track ceases to be a perfect cone: the cone tip starts to become rounded. The time t_0 at which this stage is reached (still assuming V_T to be constant) is given by

$$t_0 = R/V_T$$

The radius of curvature, σ, of the rounded tip at time t ($> t_0$) is given by

$$\sigma = V_B(t - t_0)$$

and the projected length now becomes

$$S = L_s \cos \theta + r_1 + \sigma \qquad (4.12)$$

where L_s is $R - (V_B t/\sin \theta)$. It may be noted that the formula (4.12) only applies when the dip angle θ is such that the etch cone projects from beneath the track opening. In the cases when this does not occur, the "projected" length is simply the major axis D itself.

At much longer etch times than t_0, the equations (4.7) (or 4.7') and (4.10) for the major and minor axes of the track opening become invalid. The new top surface of the detector has by then descended beyond and below the end of the latent damage trail (this takes a time t greater than $t_1 = (R \sin \theta)/V_B$). The surface opening eventually becomes a section through the (by then) much enlarged "rounded cone tip", and therefore the track opening becomes circular for all dip angles θ greater than θ_c. It is beyond the scope of this chapter to derive the track geometry equations for this situation (which is rarely encountered in practice); but these are extensively documented in papers by Henke and Benton,[13] Somogyi and Szalay,[14] and Ali and Durrani.[15] Figure 4.6 shows the evolution of the etch pit profile with increasing etch time.

4.2.2 *Determination of track parameters R and V_T*

Among the most important parameters of the latent damage trail which are often required to be determined are the length R and the mean value of the track etch velocity V_T. It is relatively easy to measure the track parameters S, D, and d (see §4.2.1) by using a micrometer attachment to the microscope eyepiece or from a photomicrograph/video screen projection. V_B can also be measured from the changes in the thickness of the plastic or the glass sheet. Alternatively, we can irradiate a sheet of plastic with ^{252}Cf fission fragments at normal incidence. Then, from Eq. (4.7') or (4.10) (for $\theta = 90°$),

$$D = d = 2V_B t \sqrt{\left(\frac{V - 1}{V + 1}\right)} \qquad (4.13)$$

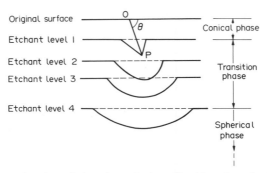

FIG. 4.6   *The evolution of an etch pit profile with prolonged etching. The track shown has* V *(= V_T/V_B) = 2, and makes a dip
angle* θ = 70° *with the detector surface O. The track profile goes
through three phases: conical, transition, and spherical, as the
etching proceeds. The conical phase lasts while the general etchant
descends from the original surface to level 1; this takes time
t = R/V_T, by when the track-etchant has reached P, the end of
the latent track (of length OP = R). Throughout this phase, the
profile of the etch pit remains conical and the surface opening
elliptical. After this, the transition phase sets in. In the transition phase, the bottom of the etch pit begins to get increasingly
rounded, since the etchant now proceeds at the bulk etching velocity V_B in all directions. During the time that the general etchant
descends from level 1 to level 2, the pit profile becomes progressively more rounded at the bottom, but the surface opening still
remains elliptical. During the progress of the general etchant from
level 2 to level 3, the surface opening is partly circular and partly
elliptical, while the profile continues to become more and more
spherical. Finally, when the general etchant has reached level 4,
the whole track profile becomes (and remains ever afterwards)
completely spherical, and the surface opening of the etch pit becomes (and remains thereafter) completely circular. The above
applies to all tracks making dip angles θ greater than θ_c. For
tracks with θ = 90°, of course, the surface opening is circular
throughout. (Figure after Somogyi and Szalay,[14] and Ali and
Durrani.[15])*

where $V = V_T/V_B$. Since for most plastics and etching conditions used the
track etch rate for fission fragments is very much higher than the bulk etch
rate (so that $V \gg 1$), we have

$$D = d \simeq 2V_B t$$

i.e.

$$V_B \simeq D/2t \qquad (4.14)$$

Thus a measurement of the diameter of the normally incident fission-
fragment tracks at a known etching time yields a value for the bulk etch rate.

If the ratio of the diameters of some tracks (having an etch rate ratio V) to those of fission fragments is denoted by x, then we can write (cf. equations (4.13) and (4.14))

$$x = \sqrt{\left(\frac{V-1}{V+1}\right)}$$

and upon rearranging, we get the useful formula

$$V = \frac{1 + x^2}{1 - x^2} \tag{4.13'}$$

Most treatments of track geometry are formulated directly in terms of the thickness of the removed surface layer $h = V_{\mathrm{B}}t$. In Table 4.2, the main track geometry formulae (for constant V_{T}) are summarized, after making this substitution; V_{T} has also been expressed in terms of V, and the appropriate expressions have been used to eliminate L_{e}, L_{s}, r_1, and σ (as defined in §4.2.1).

TABLE 4.2   *Track geometry formulae (for constant V_{T})*

Parameter	Symbol[†]	Formula	Eq. No.[‡]
Major axis of etch pit opening	D	$\dfrac{2h(V^2-1)^{1/2}}{V\sin\theta + 1}$	(4.7a)
Minor axis of etch pit opening	d	$2h\left(\dfrac{V\sin\theta - 1}{V\sin\theta + 1}\right)^{1/2}$	(4.10a)
Projected* length (at $t \le t_0$)	S_0	$h\dfrac{V\sin\theta - 1}{\tan\theta}$ $+ \dfrac{h}{\sin\theta}\cdot\dfrac{V\sin\theta - 1}{(V^2-1)^{1/2}\cdot\sin\theta\,-\cos\theta}$	(4.11a)
Projected* length (at $t > t_0$)	S_1	$\left(R - \dfrac{h}{\sin\theta}\right)\cos\theta$ $+ \dfrac{h}{\sin\theta}\cdot\dfrac{V\sin\theta - 1}{(V^2-1)^{1/2}\cdot\sin\theta\,-\cos\theta}$ $+ h - R/V$	(4.12a)

* Applies only to tracks for which the etch cone projects from beneath the track opening when viewed from above. Time t_0 refers to the time when the etchant reaches the end of the latent damage trail.

† For a discussion of these and other symbols, see text; $h = V_{\mathrm{B}}t$ = thickness of the surface layer removed; $V = V_{\mathrm{T}}/V_{\mathrm{B}}$; R = residual range; t = duration of etching; $t_0 = R/V_{\mathrm{T}}$; θ = dip angle of particle with the detector surface. It is useful to remember (cf. Eq. (4.5)) that: $\tan\delta = (V^2-1)^{-1/2}$, where $\delta \equiv \theta_{\mathrm{c}}$.

‡ The equation numbers correspond to, or are variants of, the formulas given in the text (e.g. for (4.7a) see Eqs. (4.7), (4.7') in the text).

In the general case where the dip angle is unknown, the three equations (see Table 4.2) (4.7a), (4.10a) and (4.11a) are sufficient to establish V_T and θ, provided that the etch time t is less than $t_0 = R/V_T$ (this can be ascertained by observing that for all $t < t_0$ the etch cone has a sharply pointed tip). A further set of measurements on the projected length allows R to be obtained from Eq. (4.12a) (the projected length at $t > t_0$ being denoted by S_1; see Table 4.2).

We emphasize that other parameters could be measured, and other equations used to obtain V_T, R, and θ (see, in particular, Henke and Benton[13] and Somogyi and Szalay[14]). The above treatment simply shows one particular approach to the problem.

4.2.3 Etching efficiencies: internal and external track sources

Before moving on to a discussion of the evolution of the etch pit for the case of varying V_T (which will be considered in the next subsection), we will first describe how simple calculations of the solid angle, circumscribed by the limitation of the critical angle of etching,[16] can be used to determine the efficiencies of detectors for track registration.

Let us calculate those fractions of the latent damage trails crossing a detector surface, which are revealed by etching, for the respective cases of (1) internally produced tracks and (2) externally produced tracks. It should be noted that, although the discussion presented, and the formulas derived, pertain primarily to fission tracks (where each fission event yields *two* fragments emitted isotropically and proceeding in opposite directions), yet the same arguments are applicable to other isotropic sources of track forming particles: with the only difference that the corresponding track densities of singly emitted particles have half the values of the fission track densities.

(i) INTERNAL TRACKS

(a) Thick source

Internally generated tracks can result, for example, from the spontaneous fission of ^{238}U impurities within natural minerals and crystals as well as from the induced fission of ^{235}U (or any other fissile element) within these, or man-made, materials. We want to relate the density of etched tracks on any internal surface (revealed by cutting the sample, and treated as the "top" surface during etching) to the number of fission events per unit volume within the sample.

If we consider the horizontal plane O (Fig. 4.7a), it is seen that the internal tracks will contribute to the surface track density on O only if the fission events take place within a vertical distance R below it, where R is the range

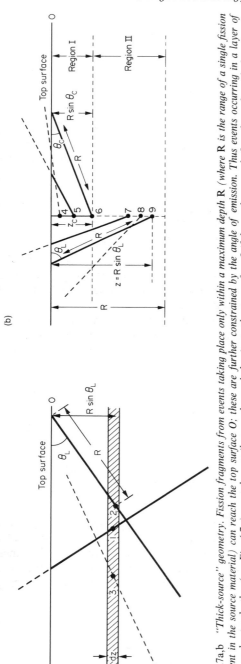

FIG. 4.7a,b *"Thick-source" geometry. Fission fragments from events taking place only within a maximum depth* **R** *(where* **R** *is the range of a single fission fragment in the source material) can reach the top surface* O*; these are further constrained by the angle of emission. Thus events occurring in a layer of thickness* dz *at a depth* z *at the top surface which is greater than* θ_L*, where* $\sin\theta_L = z/\mathbf{R}$*, can only contribute to the surface track density. For instance, events 1, 2 would contribute to the surface track density, but event 3 would not. A further constraint, however, must be placed on latent tracks originating from layers close to the top. The situation is illustrated by Fig. 4.7b, where, for the sake of clarity, we only show the track made by the "forward-hemisphere" fission fragment in each case. Here, we have subdivided the body of the source into two contiguous regions: Region I and Region II. Region I stretches from depths* z = 0 *to* z = z_c = **R** \sin \theta_c, *where* θ_c *is the critical angle of etching. In this region,* θ_c *is the governing factor: if the latent track makes an angle* $>\theta_c$ *(e.g. track 5), it will be revealed by etching; if* $\theta < \theta_c$ *(e.g. for track 4), it will fail to be revealed; track 6 just makes it (with* $\theta = \theta_c$*). Region II extends from depth* z_c *to* z = **R**. *Here the direction of emission, i.e. angle* θ_L *is the governing factor. Thus from depth 9, a track making a minimum angle* $\theta_L = \sin^{-1} z/\mathbf{R}$ *with the surface will be revealed by etching. Latent tracks 7 and 8 represent those that will, and will not, etch out, respectively. Integration over the relevant angular limits yields, in each case, the total fraction of tracks from a given depth that will be etched. A further integration over the whole depth of each region then yields the total etched fission track density from within the thick source (see text for derivation). Adding together the two contributions gives the full etched track density from within the thick source, whether the tracks are produced internally (in an "intrinsic" detector) or externally (on the surface of an external detector in contact with the source), so long as the source involved is thick (i.e. is of thickness* $\geq \mathbf{R}$*, the range of a single fission fragment in the source material). The critical angle* θ_c*, however, pertains to the detector material.*

of a *single* fission fragment. Even within this region, many tracks at low angles of inclination with the horizontal will not reach the surface. From a given depth z ($\leq R$), tracks will only reach the surface if the dip angle θ with the surface is such that

$$z \leq R \sin \theta$$

Neglecting, for the moment, the effect of the critical angle of etching, the above relation means that at a given z only those tracks which have θ greater than θ_L, where $\sin \theta_L = z/R$, will contribute to the track density at surface O; those with an angle $\theta < \theta_L$ will fail to reach the "top" surface. It is seen from Fig. 4.8 that the element of solid angle corresponding to tracks originating from point P in the source plane and proceeding (in

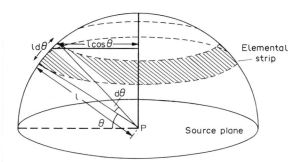

FIG. 4.8   *Solid angle considerations. For fission fragments emanating from a point P in directions that make angles between θ and $\theta + d\theta$ with the source plane, the corresponding solid angle is that subtended at P by the area of the shaded element strip on the surface of a hemisphere of radius l. The area of this strip is $dS = 2\pi l \cos \theta \cdot l d\theta$, and the element of solid angle subtended at P is $d\Omega = 2\pi \cos \theta \, d\theta$. If the detector sits adjacent to the source plane, and if no critical-angle considerations apply, then of course all fission fragments issuing into the forward hemisphere (over a total solid angle $\Omega = 2\pi$) will be registered in the detector (assumed to be larger than R in extent, where R is the range of a single fission fragment in the detector). If, however, only etchable tracks are to be taken into account, then the total solid angle for revealable tracks is*

$$\Omega = \int_{\theta_c}^{\pi/2} 2\pi \cos \theta \, d\theta = 2\pi(1 - \sin \theta_c)$$

The etched tracks produced on an external detector, expressed as a fraction of all latent tracks incident on the detector surface, are therefore $f_{ext} = 1 - \sin \theta_c$. The same formula would apply to a thin "intrinsic" detector; i.e. one on whose surface its own internal tracks can be revealed, and whose thickness is $\ll R$. In such a case $(f_{int})_{thin\ source} = 1 - \sin \theta_c$ as well.

the "forward" hemisphere) in directions contained within the angles θ and $\theta + d\theta$ with the plane (and hence also with the detector surface) is given by $d\Omega = [(2\pi l \cos \theta \cdot l \, d\theta)/l^2] = 2\pi \cos \theta \, d\theta$. Hence the total solid angle contained within a cone with a *semivertical* angle $(\pi/2 - \theta_L)$ is obtained by integrating the above element of solid angle $d\Omega$ over the range θ_L to $\pi/2$ (which is equivalent to integration over the complementary-angle range 0 to $(\pi/2 - \theta_L)$). This yields a total solid angle within the cone:

$$\int_{\theta_L}^{\pi/2} 2\pi \cos \theta \, d\theta = 2\pi(1 - \sin \theta_L)$$

Then, since the total solid angle subtended by the "forward" hemisphere is 2π, the *fraction* of tracks reaching the "top" surface is given by: $1 - \sin \theta_L$. Hence the fraction of tracks from the layer between z and $z + dz$ which reach the surface is:

$$1 - \sin \theta_L = 1 - z/R \qquad (4.15A)$$

If n is the number of fission events per unit volume (remembering that, from each fission event in the layer at depth z, one of the fission fragments always goes "upwards" towards the "top" surface), the total number of tracks originating from a unit area of this layer (of thickness dz at a depth z) is given by:

$$n \, dz(1 - z/R) \qquad (4.15B)$$

For a source layer of sufficiently large lateral dimensions (i.e. large compared to a fission-fragment range) it is readily seen that the number of tracks crossing each unit area of surface (far away from the edges of the surface) due to fission events in the above-mentioned source layer is also $n \, dz$ $(1 - z/R)$. Upon integrating over all z from 0 to R, we have the total number of tracks per unit surface area, $\rho_{2\pi}$, given by

$$\rho_{2\pi} = \int_0^R n\left(1 - \frac{z}{R}\right) dz = \tfrac{1}{2}nR \qquad (4.15)$$

This is the number of tracks crossing a "top" surface in the case of a thick internal-tracks source with 2π geometry. For 4π geometry (when the surface in question was well within the body of the material, i.e. deeper than one range length R from the original top surface of the sample, before it was revealed by cutting), an equal contribution arises from fission events occurring "above" the surface O, and the number of tracks per unit area is therefore $\rho_{4\pi} = nR$.

[It may be noted that, as pointed out at the beginning of this subsection, both this last track density $\rho_{4\pi}$ and the one given by Eq. 4.15 (viz. $\rho_{2\pi}$) are twice the respective values in the case of a thick internal source of, say, α-particles, where only one particle is emitted per decay unlike a fission source, where one fragment is emitted in the forward and one in the backward hemisphere.]

For a finite critical angle of etching, a similar calculation applies; except that for small z values the limiting angle of incidence to the surface is no longer determined by the purely geometric condition $\sin \theta_L = z/R$, but is governed by the critical angle of etching, θ_c, given by $\sin \theta_c = V_B/V_T$, so that no tracks making an angle $\theta < \theta_c$ with the detector surface can be revealed by etching.[16]

It is then helpful to divide the total depth below the top surface, which can contribute etchable tracks, into two regions (see Fig. 4.7b). Region I extends from $z = 0$ to a depth $z = z_c$, such that

$$z_c = R \sin \theta_c$$

in this region, the lower limit θ_c applies to θ. By following arguments similar to those leading up to Eqs. (4.15A,B), the contribution from a layer of thickness dz (at a depth z) in this region, to the etched-track density, is seen to be

$$n \, dz \int_{\theta_c}^{\pi/2} \cos \theta \, d\theta = n \, dz(1 - \sin \theta_c) \tag{4.15C}$$

The whole of Region I (extending from $z = 0$ to $z = z_c = R \sin \theta_c$) thus yields an etchable track density given by:

$$n \int_0^{z_c} (1 - \sin \theta_c) \, dz = n[z - z \sin \theta_c]_0^{R \sin \theta_c}$$

$$= nR \sin \theta_c(1 - \sin \theta_c) \tag{4.15D}$$

The second region, termed Region II, extends from $z = z_c = R \sin \theta_c$ to $z = R$; here the only criterion for etchability is that a fission fragment (i.e. the "upper" limb of a fission event) must be able to reach the top surface: θ is now always $> \theta_c$. As before (cf. Eq. 4.15B), the contribution from a layer dz in this second region is

$$n \, dz \int_{\theta_L}^{\pi/2} \cos \theta \, d\theta = n \, dz(1 - \sin \theta_L)$$

$$= n \, dz(1 - z/R) \tag{4.15E}$$

The total contribution from Region II (remembering that $z_c = R \sin \theta_c$) is thus:

$$n \int_{z_c}^{R} (1 - z/R) \, dz = n\left[z - \frac{z^2}{2R} \right]_{R \sin \theta_c}^{R}$$

$$= nR(\tfrac{1}{2} - \sin \theta_c + \tfrac{1}{2} \sin^2 \theta_c) \tag{4.15F}$$

On adding together the contributions given by Eqs. (4.15D) and (4.15F), the total etched-track density of fission tracks contributed from *below* the "top" surface (i.e. in 2π-geometry), as recorded on a unit area of that surface, is readily seen to be:

$$\rho_{2\pi} = \tfrac{1}{2}nR \cos^2 \theta_c \qquad (4.16)$$

which differs from Eq. (4.15) by the multiplicative factor $\cos^2 \theta_c$.

In the case of 4π geometry, when contributions from both *above and below* the etched surface are taken into account, one obtains from Eq. (4.16), the etched fission-track density as:

$$\rho_{4\pi} = nR \cos^2 \theta_c \qquad (4.17)$$

[For a thick source of α-particles, with only one particle per decay event, one obtains the track-density values $(\rho_{2\pi})_\alpha$ and $(\rho_{4\pi})_\alpha$ which are one-half of the corresponding 2π- and 4π-geometry values for fission tracks.]

The "etching efficiency" of a detector for internally generated tracks in all cases (whether in 2π- or 4π-geometry, and whether for singly emitted particles such as α-particles or doubly-emitted particles per event such as fission fragments) is given by

$$f_{int} = \cos^2 \theta_c$$

where the etching (or registration) efficiency is defined as the proportion of tracks etched out expressed *as a fraction of particles actually incident on the detector surface* (e.g. $\rho_{4\pi}/nR$ for fission fragments in 4π geometry; or $(\rho_{2\pi})_\alpha/\tfrac{1}{4}nR$ for 2π detection geometry in the case of a thick source of α-particles).

(b) Thin source

The above formula ($f_{int} = \cos^2 \theta_c$) is the general one that applies to a *thick* source. If, however, the source is "thin" (such that its thickness $z \ll R$, and implicitly $z < R \sin \theta_c$), then, as mentioned above, considerations pertaining to Region I will apply, namely that θ_c will be the governing factor. In such a case, the fraction of internal tracks reaching the top surface is given by:

$$\frac{\int_{\theta_c}^{\pi/2} 2\pi \cos \theta \, d\theta}{2\pi} = 1 - \sin \theta_c$$

Hence

$$(f_{int})_{\text{thin source}} = 1 - \sin \theta_c \qquad (\text{for } z \ll R)$$

Note that if $R \sin \theta_c$ is not negligible compared to R (e.g. for glass, when θ_c may be $\simeq 30°$ and therefore $R \sin \theta_c \simeq \frac{1}{2}R$), the yields from layers of thickness dz (and, say, of unit area) must be integrated over the range $z = 0$ to, say, z_1, where $z_1 \leq R \sin \theta_c$. It may then be easily seen that whereas the actual number of tracks incident on the top surface is

$$\int_0^{z_1} n(1 - z/R) \, \mathrm{d}z = nz_1(1 - z_1/2R)$$

those actually etched are

$$\int_0^{z_1} n(1 - \sin \theta_c) \, \mathrm{d}z = nz_1(1 - \sin \theta_c)$$

Thus the revealed fraction (expressed *in terms of particles actually reaching the top surface*, and not merely emitted from the volume concerned) is given by

$$(f_{\text{int}})_{\text{intermediate}} = \frac{1 - \sin \theta_c}{1 - \frac{1}{2}(z_1/R)} \qquad \text{(for } 0 \leq z_1 \leq R \sin \theta_c)$$

This leads, in the case of an "intrinsic" glass detector of thickness $z_1 = R \sin \theta_c = \frac{1}{2}R$, say (with $\theta_c = 30°$), to a value of the efficiency for revealing its own internal tracks $= (1 - \sin \theta_c)/(1 - \frac{1}{2} \sin \theta_c) = (\frac{1}{2}/\frac{3}{4}) = \frac{2}{3}$; which should be compared to an efficiency value of 1, had no critical angle been involved. It should be mentioned, however, that such a situation is rarely encountered in reality.

(ii) EXTERNAL SOURCE OF TRACKS

(a) Thick source

In the case of an external source which is "thick" (i.e. is thicker than one particle range R), the calculations are identical to those for 2π geometry in the case of the internal source already treated above, namely when the radiation reaching the top surface O (Fig. 4.7) came only from "below". The density of etched-out fission tracks on the detector surface is, thus, again given by Eq. (4.16) seen earlier, viz.

$$\rho_{2\pi} = \frac{1}{2}nR \cos^2 \theta_c \qquad (4.16)$$

This is also the value applicable to an external detector placed in intimate contact with a source of induced fission, e.g. in reactor irradiation, where R signifies the range of a single fission fragment in the sample (i.e. the source); θ_c, of course, refers to the detector.

(b) Thin source

Consider next a "thin" external source (i.e. one of thickness much smaller than the particle range R). Let us imagine it to be a plane source whose surface dimensions are much larger than R, and which (as before) emits particles isotropically (i.e. with equal probability into all equal solid-angle intervals).

Our chief interest is in measurements with "2π geometry". Hence we shall take the source to be in intimate contact with a detector placed on one side of it. We wish to calculate the fraction of latent damage trails made on the surface of the detector which are actually revealed by etching.

As in the case of internal tracks treated in (i) above, the element of solid angle corresponding to particles emitted in the "forward" hemisphere and lying between angles θ and $\theta + d\theta$ with the surface is (see Fig. 4.8): $d\Omega = 2\pi \cos \theta \, d\theta$; whereas the total solid angle subtended by the hemisphere is 2π. Hence the latent tracks lying between angles θ and $\theta + d\theta$ expressed as a fraction of all particles *actually incident on the top surface of the detector* are given by:

$$f(\theta) \, d\theta = \cos \theta \, d\theta$$

Of these, only those tracks will be revealed by etching which make angles θ with the surface greater than the critical angle of etching θ_c, i.e. those making angles between θ_c and $\pi/2$. The etched fraction is thus given by

$$f = \int_{\theta_c}^{\pi/2} \cos \theta \, d\theta$$
$$= 1 - \sin \theta_c$$

Thus the etching (or registration) efficiency of the detector, defined (as before) as the proportion of tracks etched out expressed as a fraction of particles actually incident on the detector surface (or, in other words, etched tracks expressed as a fraction of the latent tracks) is—by redesignating f as f_{ext}—simply:

$$f_{\text{ext}} = 1 - \sin \theta_c$$

This expression actually remains valid even in the case of 4π-geometry, if the thin plane source is sandwiched between two external detectors; and, in fact, it applies equally to a *point* source, whether in 2π- or 4π-geometry. Moreover, this expression for etching efficiency applies both to singly emitted particles per decay such as α-particles and to doubly-emitted particles per event such as fission fragments.

Note that as $\theta_c \to 0$ (which is approximately true for many crystals and plastics), $f_{\text{int}} \to f_{\text{ext}} \to 1$. For glasses, however, where θ_c may be $30°$ or more, $f_{\text{int}} = \cos^2 \theta_c \lesssim 0.75$; and $f_{\text{ext}} = 1 - \sin \theta_c \lesssim 0.5$.

It should be emphasized that the above treatment of etching efficiency calculations applies only when V_T, and hence $\sin \theta_c$ and $\cos \theta_c$, are constant at all points along the track. When V_T is variable, it is generally not possible to obtain analytical expressions for these efficiencies.

4.2.3.1 PROLONGED-ETCHING FACTOR $f(t)$

A tacit assumption is often made in the case of internal tracks in a thick detector that, as etching proceeds, the number of new tracks revealed from deeper layers is compensated by the loss of a corresponding number of the earlier tracks originating near the initial top surface. In other words, that if a layer of thickness h is removed from the top during etching, the number of etch pits (or surface openings) lost from that layer is—assuming normally incident tracks for simplicity—equal to those added by tracks originating from a layer of thickness h lying between R and $R + h$, where R is the range of the particle concerned in the detector medium. This is, in fact, strictly speaking not true; for many of the early etch pits, once formed, continue to be visible even after long etching times, even though they may progressively lose their sharpness. There is, thus, a monotonic increase in the number of observed etched tracks with etching time.[16d]

In applications where only the ratios of the observed track densities are of importance, e.g. in fission track dating (where the age might be calculated from ρ_s/ρ_i, viz. the ratio of the spontaneous to the induced fission track density in the sample), the duration of etching does not affect the results, provided that the etching conditions are identical in the two cases: for both track densities increase with etching time at the same rate. The effect can, however, be quite significant if the actual track densities are used separately on their own, e.g. in determining the uranium content in a sample by etching its induced-fission tracks, or if, in age determination, the value of ρ_i is measured by an "external detector" method and ρ_s in the sample itself. The effect of prolonged etching becomes highly significant in cases when large thicknesses h are removed during the etching process, e.g. in glass etching where the bulk etching velocity V_B is large in relation to the track etching velocity V_T.

The effect of prolonged etching can be easily seen by considering Fig. 4.7(b) and by examining Eqs. (4.15C) *et seq.* We shall, for simplicity, assume in what follows that any etch pits once formed on the etched surface will continue to be observed as the etching proceeds. We shall also primarily consider only fission tracks, although the same considerations will apply to any other etchable tracks. We have established above in §4.2.3 that the revealed internal tracks can be treated as originating from two regions (cf.

Fig. 4.7b): those from an upper Region I, where the overriding factor for revelation is the critical angle of etching θ_c; and those from the adjoining lower Region II, where revelation depends on whether or not the fission fragment emitted in the "upper hemisphere" can reach the top surface, i.e. on its depth and angle of emission. Region I extends from $z = 0$ to $z = z_c = R \sin \theta_c$; and Region II from $z = z_c$ to $z = R$, where R is the range of a *single* fission fragment in the medium.

Suppose now that a thickness $h = V_B t$ is removed by etching, where V_B is the bulk velocity of etching and t the length of etching time. This has the effect of extending Region I by a thickness h, so that Region I now extends from $z = 0$ to $z = z_c + h$. Some of the tracks in the region z_c to $z_c + h$ which were previously unable to reach the initial top surface O (Fig. 4.7b) can now reach the final etched surface (say O', a distance h below O); they will be revealed provided that they make an angle $\theta > \theta_c$ with the detector surface. The total contribution from this modified (and larger) region, termed Region I', is then given (cf. Eqs. 4.15C,D) by:

$$n \int_0^{z_c + h} (1 - \sin \theta_c)\, dz = n \int_0^{R \sin \theta_c + h} (1 - \sin \theta_c)\, dz$$
$$= nR \sin \theta_c (1 - \sin \theta_c) + nh(1 - \sin \theta_c) \quad (4.15D')$$

The additional contribution from Region I', over and above that from Region I (cf. Eq. 4.15D) is thus:

$$nh(1 - \sin \theta_c)$$

As far as the modified Region II is concerned, viz. Region II', it now extends from $z = z_c + h$ to $z = R + h$. A little consideration will show that the contribution from Region II' is identical with that from the former Region II; for the modification is tantamount to moving the reference frame from the initial etched surface O to the final etched surface O' (a distance h below the former). More explicitly, if depths z from plane O are changed to depths $z'\ (= z - h)$ from plane O', the limits of integration of z' revert to their former values $z_c = R \sin \theta_c$ and R, respectively, so that (cf. Eqs. 4.15E,F):

$$n \int_{R \sin \theta_c}^{R} (1 - z'/R)\, dz' = nR(\tfrac{1}{2} - \sin \theta_c + \tfrac{1}{2} \sin^2 \theta_c) \quad (4.15F')$$

On adding together the contributions from Regions I' and II' (Eqs. 4.15D',F'), we obtain the total fission track density in 2π-geometry resulting from an etching for time t as:

$$\rho_{2\pi}(t) = \tfrac{1}{2}nR \cos^2 \theta_c + nh(1 - \sin \theta_c)$$
$$= \tfrac{1}{2}nR \cos^2 \theta_c + nV_B t(1 - \sin \theta_c) \quad (4.16')$$

This should be compared with Eq. 4.16, from which it differs by the additional, second factor. Similarly, for 4π-geometry—remembering that the *additional* contribution is only made by fission fragments originating from *below* the etched surface (and not by those from above)—the total etched fission track density resulting from prolonged etching for a time t is given by:

$$\rho_{4\pi}(t) = nR \cos^2 \theta_c + nh(1 - \sin \theta_c)$$
$$= nR \cos^2 \theta_c + nV_B t(1 - \sin \theta_c) \qquad (4.17')$$

[Similar expressions are obtained for α-decay events taking place in a thick source of α-particles—if their tracks can be etched—except that the corresponding values are one-half of those for fission events.]

Khan and Durrani[16d] have defined the "prolonged-etching factor" $f(t)$ as follows. If the track densities given by Eqs. (4.16) and (4.17) are taken to be those obtained by extrapolating the growth curves for the track densities $\rho(t)$ to zero etching times, viz. $\rho_{2\pi}(0)$ and $\rho_{4\pi}(0)$, respectively, then

$$f(t) = \rho(t)/\rho(0)$$

Hence, by dividing Eqs. (4.16') and (4.17') by their counterparts, we obtain:

$$f_{2\pi}(t) = 1 + \frac{2V_B t(1 - \sin \theta_c)}{R \cos^2 \theta_c}$$

$$= 1 + \frac{2V_B t}{R(1 + \sin \theta_c)} \qquad (4.16'')$$

and similarly

$$f_{4\pi}(t) = 1 + \frac{V_B t}{R(1 + \sin \theta_c)} \qquad (4.17'')$$

note that

$$f_{2\pi}(t) = 2f_{4\pi}(t) - 1$$

The value of $f(t)$ depends, of course, on V_B (and hence on the medium and the etching conditions), on t, and on the critical angle of etching θ_c for the medium, and can therefore be very large for media such as glass, particularly for long etching times (it is usually less important for crystals for which V_B is small so that the removed layer thickness $h = V_B t$ is generally small). Figures 4.9(a,b) vividly show the large effect that prolonged etching has on the observed track densities in various types of glass etched for up to ~ 180 s in 48% and 40% HF at different temperatures. The values of $f(t)$ are seen to be as high as ~ 4.6 in the case of soda-lime glass under certain etching conditions. The neglect of the prolonged-etching factor could thus introduce serious errors in some applications.

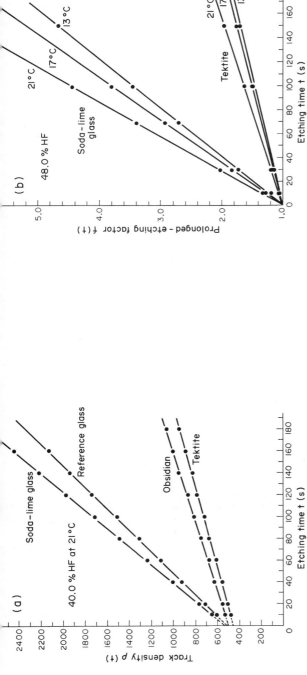

FIG. 4.9a,b *Effect of prolonged etching and determination of the prolonged-etching factor* f(t). *In Fig. 4.9a is shown the growth in the number density* $\rho(t)$ *of induced-fission etch pits in several different types of glass as a function of etching time* t. *The observations were made in* 4π-*geometry by removing a layer* $\approx 20\ \mu m$ *thick at the top after irradiation in a reactor. The black dots represent the experimentally observed values of* $\rho_{4\pi}(t)$ *over arbitrary units of area;* $\rho_{4\pi}(0)$ *is obtained by extrapolation. 40 vol.% HF at 21°C* ($\pm 0.5°C$) *was used as the etchant. The tracks, especially the circular ones, remained recognizable for very long etching times (e.g. even beyond 1100 s of etching in the case of tektite glass). The tektite was an australite; the obsidian, a volcanic glass from Borsod, Hungary; the reference glass (U-2), made by Corning Glass Works with 43 ppm of depleted uranium; and soda-lime glass, a glass slide made by Chance Bros. Figure 4.9b shows the values of the prolonged-etching factor* $f(t) = \rho_{4\pi}(t)/\rho_{4\pi}(0) = 1 + [V_{B}t/R(1 + \sin\theta_{c})]$ *(see Eq. 4.17''), calculated from curves similar to these in (a) but for etching in 48 vol.% HF at the temperatures indicated on the figure. The black dots show the experimentally determined values of* $\rho_{4\pi}(t)/\rho_{4\pi}(0)$, *the solid lines being the least-squares fits. The* f(t) *factor is seen to closely follow a straight line of the form* $f(t) = 1 + mt$ *in each case (with errors of* <1% *in the value of the slope* m), *even when large thicknesses of the detector had been removed by etching. Separate measurements were made of the critical angle of etching* θ_{c} *for each type of glass; and* V_{B} *was measured by etching each piece of glass for up to 300 s under the given etching conditions (e.g. in 48% HF at 21°C, the* V_{B} *values were found to be: tektite* ($\theta_{c} = 25°\ 45'$), 0.127 $\mu m\,s^{-1}$; *obsidian* ($\theta_{c} = 26°\ 00'$): 0.147 $\mu m\,s^{-1}$; *reference glass* ($\theta_{c} = 31°\ 45'$), 0.616 $\mu m\,s^{-1}$; *soda-lime glass* ($\theta_{c} = 35°\ 30'$), 0.804 $\mu m\,s^{-1}$). *From the data given, the values of the fission fragment range R in glass could be calculated (Eq. 4.17''). (Figure and data from Khan and Durrani.*[16d]*)*

It may be noted that the above theoretical values of $\rho(t)$ and hence of $f(t)$ remain valid even when $h = V_{\rm B}t$ is larger than R (for the derivations can be made by considering stepwise or gradual increases in h), provided that the assumption as to the continued retention of earlier etch pits as recognizable tracks remains true. For instance, Khan and Durrani[16d] report the following values for soda-lime glass ($\theta_c = 35°\ 30'$): $V_{\rm B}$ (in 48% HF at 21 °C) = 0.804 μm s^{-1}; $f_{4\pi}(100\ {\rm s}) = 4.435$: the growth curve of $f(t)$ with time remains a straight line $f(t) = 1 + mt$, with $m = 3.435 \times 10^{-2}$ s^{-1}, up to 100 s, during which time the thickness of the surface layer removed $h \simeq 80\ \mu$m—much larger than the range ($\sim 10\ \mu$m) of single fission fragments in glass. These authors[16d] have, in fact, employed the $f(t)$ technique to derive the value of R in glass (from Eqs. (4.16″) and (4.17″); θ_c and $V_{\rm B}$ being known or measured).

4.2.4 Track etching geometry with varying $V_{\rm T}$

The calculation of etch pit profiles as a function of etching time (or of removed surface layer thickness $h = V_{\rm B}t$) for the case of non-constant $V_{\rm T}$ has been described by Somogyi and Szalay,[14] Paretzke et al.,[17] Fleischer et al.[12] and Somogyi et al.[17a]. We will attempt here only to outline the solutions of the problem and to indicate how $V_{\rm T}$ may be determined experimentally when it varies with the residual range.

Figure 4.10 shows the case of $V_{\rm T}$ slowing down towards the end of the track (which is true of fission fragment tracks and also of heavy ions towards the very end of their range). For simplicity, the track is taken to be normal to the top surface. We must calculate the co-ordinates (x_0, y_0) of some arbitrary point $P(x_0, y_0)$ on the wall of the etch pit.

Consideration of Fig. 4.10 leads quite simply to the following useful relations.

$$x_0 = [(y_0 - y')^2 + x_0^2]^{1/2} \sin \zeta \tag{4.18}$$

$$y_0 = y' + [(y_0 - y')^2 + x_0^2]^{1/2} \cos \zeta \tag{4.19}$$

$$\tan \zeta = \frac{x_0}{y_0 - y'} \tag{4.20}$$

where ζ is the angle between the (vertical) axis of the track and the normal to the etch pit wall at $P(x_0, y_0)$.

Now the time taken for the etchant to reach P along the path $OQ'P$ is given by the sum of (1) the time for the etchant to reach Q' at a variable rate $V_{\rm T}(y)$ and (2) the time for it to move from Q' to $P(x_0, y_0)$ at a constant rate $V_{\rm B}$. Thus

$$t = \left[\int_0^{y'} \frac{dy}{V_{\rm T}(y)} \right] + \frac{[(y_0 - y')^2 + x_0^2]^{1/2}}{V_{\rm B}} \tag{4.21}$$

The actual path followed by the etchant is determined by the value of y' for which the time t is a minimum. For this value

$$\frac{dt}{dy'} = 0$$

Upon differentiating Eq. (4.21) with respect to y' and setting the result equal to 0, we obtain

$$y' = y_0 - x_0 \left[\left(\frac{V_T(y')}{V_B} \right)^2 - 1 \right]^{-1/2} \tag{4.22}$$

From Eqs. (4.22) and (4.20) we have:

$$\tan \zeta = \left[\left(\frac{V_T(y')}{V_B} \right)^2 - 1 \right]^{1/2} \tag{4.23}$$

From Eq. (4.23), using the simple trigonometric relation $\sec^2 \theta = \tan^2 \theta + 1$, we get:

$$\cos \zeta = V_B/V_T(y') \tag{4.24}$$

which immediately gives:

$$\sin \zeta = \left[1 - \left(\frac{V_B}{V_T(y')} \right)^2 \right]^{1/2} \tag{4.25}$$

Finally, from Eqs. (4.18), (4.21), and (4.25) we can solve for x_0:

$$x_0 = V_B \left[t - \int_0^{y'} \frac{dy}{V_T(y)} \right] \left[1 - \left(\frac{V_B}{V_T(y')} \right)^2 \right]^{1/2} \tag{4.26}$$

and from Eqs. (4.19), (4.21), and (4.24)

$$y_0 = y' + \frac{V_B^2}{V_T(y')} \left[t - \int_0^{y'} \frac{dy}{V_T(y)} \right] \tag{4.27}$$

Before proceeding further, notice that $\zeta = 90° - \phi$ (see Fig. 4.10). Thus from Eq. (4.24) we have

$$\cos (90° - \phi) = \sin \phi = V_B/V_T(y')$$

Comparison with Eq. (4.2) indicates that $\phi = \delta$, provided that δ is now interpreted as the value of the cone angle corresponding to the track etch rate at the position y'. Thus, ϕ is the critical angle of etching for the corresponding track etch velocity. It may be shown that, for the etchant path which corresponds to the minimum time, the line $Q'P$ is normal to the tangent to the pit wall at P. Therefore if we have experimentally obtained a complete profile of the etched track (by means of, say, a microtome cut

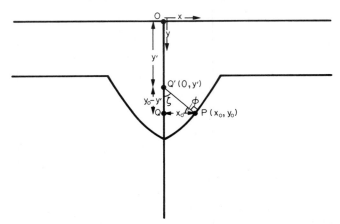

FIG. 4.10   *Track geometry parameters for varying along-the-track. velocity of etching,* V_T. *The track shown has* $\theta = 90°$ *with the top surface, and* V_T *slowing down towards the end of the track. To reach an arbitrary point P on the wall of the etch pit, the etchant could take any route, e.g. travel from O to Q' (an arbitrary point lying on the track) at a variable velocity* $V_T(y)$ *and then from Q' to P at a constant bulk-etching velocity* V_B; *or it could proceed along OQP, where* $O\hat{P}Q = 90°$; *etc. The actual route followed by the etchant is the optimum one, i.e. the one that takes the minimum time in travelling from O to P.*

through an etched track or an SEM image of a track replica (see §4.3)) we can find the track etch rate at any point Q' as follows. Construct the normal from Q' to the pit wall. The normal cuts the pit wall at P. Then construct the perpendicular PQ to the track axis (Fig. 4.10). The angle between PQ and $Q'P$, i.e. $Q\hat{P}Q$, is $\phi = \delta = \arc \sin [V_B/V_T(y')]$, where y' is the co-ordinate of Q'. Provided that V_B is known, $V_T(y')$ may thus be readily calculated by measuring ϕ. This method is useful in certain circumstances for mapping V_T variations along a track.

Returning to Eqs. (4.26) and (4.27), we may introduce some simplification by means of the relationships $V(y') = V_T(y')/V_B$ and $h = V_B t$. Also if, with Somogyi and Szalay,[14] we write

$$h(y') = V_B \int_0^{y'} \frac{dy}{V_T(y)} = \int_0^{y'} \frac{dy}{V(y)}$$

(where, physically, $h(y')$ is the thickness of the surface layer removed during the time taken by the etchant to penetrate along the track to y'), then we may re-express Eqs. (4.26) and (4.27) in the form:

$$x_0 = [h - h(y')]\left[1 - \frac{1}{V^2(y')}\right]^{1/2}$$

i.e.

$$x_0 = [h - h(y')]\left[\frac{\sqrt{(V^2(y') - 1)}}{V(y')}\right] \qquad (4.28)$$

and

$$y_0 = y' + \frac{1}{V(y')}[h - h(y')] \qquad (4.29)$$

Our final step is to use these equations to calculate track diameters d for various values of surface removal h, given some function $V_T(y')$.

Consider the case when point P (Fig. 4.10) lies on the final etched surface, i.e. it represents any point on the edge of the etched-track opening. This means that, during the time that it takes the etchant to descend a distance h below the original surface at the bulk etching rate V_B, another route taking the same total time is for the etchant to first travel along the track a certain distance y' at a variable velocity V_T and then to proceed to point P (lying on the final surface opening) at the bulk etching velocity V_B.

When point P lies on the edge of the surface opening, we have: $y_0 = h$ and $x_0 = d/2$ (see Fig. 4.10). By substituting h for y_0 in Eq. (4.29) and simplifying, we obtain:

$$h = \frac{y'V(y') - h(y')}{V(y') - 1} \qquad (4.30)$$

and by substituting this value of h in Eq. (4.28)(and with $d = 2x_0$), we easily obtain, after some simplification:

$$d = 2[y' - h(y')]\sqrt{\left(\frac{V(y') + 1}{V(y') - 1}\right)} \qquad (4.31)$$

In both Eqs. (4.30) and (4.31), $h(y')$ is, of course, the thickness of the surface layer removed (at the bulk-etch rate V_B) while the etchant travels a distance y' along the track (at a variable velocity V_T). (The above equations apply to the case of normally incident tracks on a detector surface; Somogyi and Szalay[14] give the more generalized expressions for other angles of incidence.)

For a given function $V(y')$, we take different positions y' along a normally incident track until the right hand side of Eq. (4.30) becomes equal to the desired value of h. This defines the value of y', which may then be substituted into Eq. (4.31) to obtain the track diameter d corresponding to this value of surface removal h. This procedure must, generally, be carried out numerically; but we can easily show that it generates the correct value of d for the case $V(y') = V = $ constant. Thus

$$h(y') = \int_0^{y'} dy/V(y')$$

becomes

$$h(y') = y'/V$$

Upon substituting this into Eq. (4.30), we obtain

$$h = y' \frac{V + 1}{V}$$

or

$$y' = \frac{hV}{V + 1}$$

These values of y' and $h(y')$ can be substituted into Eq. (4.31) to yield

$$d = 2h \sqrt{\left(\frac{V - 1}{V + 1}\right)}$$

which is identical to Eq. (4.13), and to Eq. (4.7a) or (4.10a) in Table 4.2 for the case $\theta = 90°$ (with $d = D$).

In this subsection our discussion has been restricted to the earliest etching phases (before the etching has totally removed the surface layers containing the original latent damage trail); and we have only considered tracks at normal incidence to the surface. Nevertheless, the general form of the calculations outlined above applies to other cases also. One difference in the more generalized case, however, is that for angles of incidence other than 90° one does not obtain explicit expressions of the form of Eqs. (4.30) and (4.31) for h and d—though the equations are still solvable by numerical methods.

As an example of etch pit evolution in the case of varying V_T, Fig. 4.11 shows calculated track profiles for 6.1 MeV α-particles in CR-39 for various values of surface removal thickness h. Figure 4.12 shows calculated track profiles for α-particles of various energies in LR 115. It is clear from this latter figure that if the manner in which V ($= V_T/V_B$) varies along a track is known (for a given incident particle and plastic type), then the form of the etched tracks can be used to determine the energy of the incident particles.

One method of extracting $V(y')$ from observational data has already been mentioned above. Less direct techniques are also available. It may, for example, be possible to calculate d versus h plots for some guessed $V(y')$ function and to adjust this until agreement with experimental data is obtained.

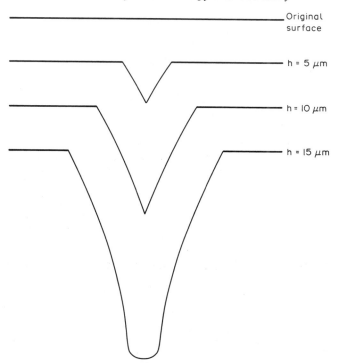

FIG. 4.11   *Calculated track profiles for 6.1 MeV alpha particles in CR-39 for surface removal thickness h of 5, 10, and 15 μm (etched in 6 M NaOH at 70°C). A variation in V along the track, given by* $V = 1 + [11.45e^{-0.339R} + 4.0e^{-0.044R}] \times (1 - e^{-0.58R})$, *has been assumed, where* $V = V_T/V_B$, *and R (in μm) is the residual range at a given point. Actually, a better fit to the data is obtained if an empirically corrected value of R, namely R′, is used in the above equation instead of R, where* $R' = R - [V/(V + 1)] \cdot (h/2)$ *(see Green et al., 1982[17b] for details).*

If the track is etched for a short time such that V varies little over the short portion of the etched-out track, then the constant-V formula can be used to determine V from diameter measurements (Eq. (4.7′)). If this is done for different residual ranges, then the function $V(R)$ can be evaluated.

If $V \gg 1$, cone length measurements are generally to be preferred for determining $V(y')$. Indeed, it is often sufficiently accurate to write

$$L_e \simeq L = V_T t \qquad (V \gg 1, \text{ and constant})$$

(where $L_e = (V_T - V_B)t$; see Fig. 4.2); and (cf. Eqs. (4.11) and (4.5), with $r_1 \simeq 0$)

$$S = L \cos \theta$$

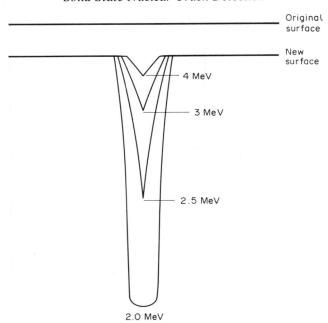

Original
surface

New
surface

4 MeV

3 MeV

2.5 MeV

2.0 MeV

FIG. 4.12 *Calculated track profiles for alpha particles in LR 115 for energies of 2, 2.5, 3, and 4 MeV and a surface removal of* h = 1 μm (etched in 2 M NaOH at 60 °C). *A variation in the etching rate ratio* V *along the track of the form* $V = 1 + [100e^{-0.446R} + 5.0e^{-0.107R}] (1 - e^{-R})$ *has been used, where* R, *the residual range, is in μm. Once again, a better fit to the experimental data on* V(R) *is obtained if an empirically corrected value of* R (viz. R') *is used in the above equation, where* $R' = R - [V/(V + 1)] \cdot (h/2)$ *(Durrani and Green, 1984.[17c])*

where S is the etched track projection on the detector surface. Provided that θ is known, $\bar{V}_T$ (or $\bar{V}$), the average value of V_T (or V) along the etched-out portion of the track, can be obtained directly from a measurement of the projected length S. In the case where V_T is varying more rapidly, but is still much larger than V_B, we can again use $L_e \simeq L$. If we consider an incremental etch time dt in which the track length increases from L to $L + dL$, we have

$$dL = V_T(L)\, dt$$

where $V_T(L)$ is the track etch rate applying to the incremental segment of track between L and $L + dL$. Thus

$$\frac{dL}{dt} = V_T(L)$$

or, since $dt = dh/V_B$,

$$\frac{dL}{dh} = V(L)$$

where $V(L) = V_T(L)/V_B$.

Hence the slope of a plot of L versus t or h can be used to obtain values for the track etching velocity at various positions along the track.

When V is not much greater than unity, it is diameter measurements which are made in order to obtain results of greater accuracy. A method proposed by Paretzke *et al.*[17] makes use of the fact that

$$\frac{d \text{ (diameter)}}{dh} = \sqrt{\left(\frac{V(y') - 1}{V(y') + 1}\right)}$$

Thus from the slope of the d versus h curve, a value of $V(y')$ at a given surface-removal h is obtained. From Eqs. (4.30) and (4.31) the position y' to which this etch rate ratio applies may be found. The determination of $V(y')$ profiles from track parameter measurements is discussed at length in reference 17.

4.2.5   *Track etching geometry in anisotropic solids*

In the preceding subsections it has been assumed that the etchant attacks the bulk material at the same rate in all directions, i.e. isotropically. This results in track-opening shapes—for constant V_T—which are smooth ellipses. In many minerals this assumption is clearly false. For example, a close examination of the openings of etched fission tracks in muscovite mica reveals that these are perfect rhomboids in shape (see Fig. 4.13), and it is clear that the etchant is strongly directional in its rate of attack (see also references 6 and 18). Such polygonally shaped etch figures have been observed in a number of minerals for many years as a result of chemical etching of naturally occurring dislocations.

Somogyi and Szalay[14] have outlined a simple model for the calculation of etch pit parameters in anisotropic media, and this model has been extended by Dorling[5] and Bull.[19] In this model[14] it is assumed that bulk etching can proceed only in certain well defined directions and that the path followed by the etchant between any two points (in the absence of a damage trail) must consist of a series of elemental paths in these directions, resulting in pit walls having a "stepped" appearance (see Fig. 4.14). It is proposed that the etchant proceeds rapidly in directions parallel to the cleavage planes (planes of low Miller index, generally) but much more slowly in directions normal to these planes.

Let us, for simplicity, consider a case where the sample surface coincides with one such plane. We also begin by considering a two-dimensional case in which a component V_p of the bulk etch rate acts parallel to the surface,

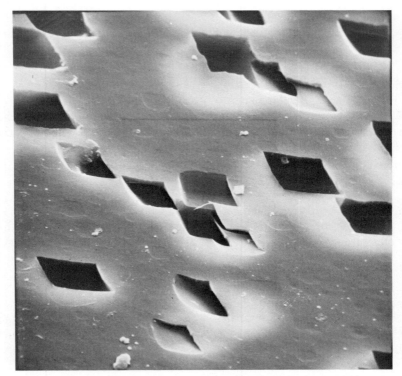

FIG. 4.13   *A scanning electron microscope (SEM) photomicrograph showing the rhomboidal openings of etched fission tracks in muscovite mica (exposed to a* ^{252}*Cf spontaneous-fission source and etched in HF). The shape results from an anisotropy in the rate of attack by the etchant, which depends on the crystal structure (see also Figs. 4.14 and 4.16).*

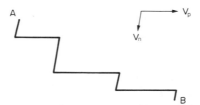

FIG. 4.14   *In anisotropic solids, the etchant may dissolve away material only—or preferentially—along certain directions. The figure shows the case where the etchant can only travel along the directions marked by the arrows, and that also at the different velocities* V_p *and* V_n *indicated by the respective lengths of the vectors. Thus the path followed by the etchant in moving from A to B must consist of elemental paths in these directions, as phenomenologically shown in the figure.*

while the chemical attack normal to the surface proceeds at a rate V_n. The thickness of the surface removed by the etchant in time t is then $h = V_n t$ (Fig. 4.15).

To reach the left-hand side of the pit at P from O, the etchant must move down the track segment OO' ($= h/\sin \theta$; see Fig. 4.15) at a rate V_T (assumed here to be constant) and then along $O'P$ at a rate V_p, taking a total time t. Hence

$$t = \frac{h}{V_n} = \frac{h}{V_T \sin \theta} + \frac{r_1}{V_p}$$

(Note that this equation is applicable except for such low angles of inclination with the cleavage plane that

$$\frac{h}{V_n} < \frac{h \operatorname{cosec} \theta}{V_T} + \frac{h \cot \theta}{V_p}$$

in which case it is quicker for the etchant to travel *across* the cleavage planes than along the track first and then along a cleavage plane.)

If we put $V = V_T/V_n$, then on solving for r_1 we obtain

$$r_1 = h \frac{V_p}{V_n} \left(1 - \frac{1}{V \sin \theta} \right)$$

It is obvious that an identical analysis applies to r_2. With the major (or longest) axis of the surface opening $D = r_1 + r_2 = 2r_1$, we therefore have:

$$D = 2h \frac{V_p}{V_n} \left(1 - \frac{1}{V \sin \theta} \right) \tag{4.32}$$

In fact this analysis is also valid for determining d, the "width" of the surface

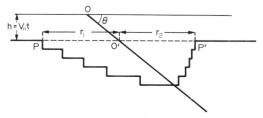

FIG. 4.15 *Etch-pit geometry in a solid with the bulk-etch-rate components $\mathbf{V}_p$ (parallel to the sample surface, assumed to be one of the cleavage planes) and $\mathbf{V}_n$ (normal to the surface, i.e. across a cleavage plane). A track making an angle θ with the sample surface is shown to yield an etch pit with extremities P and P'; the sample thickness dissolved in an etching time t is $h = V_n t$. To reach point P, the etchant must travel down the track segment OO' at a rate $\mathbf{V}_T$ (the track-etching velocity), and then along O'P at a rate $\mathbf{V}_p$, taking a total time t; similar remarks apply to point P'. Other points on the etch pit are reached in like manner (the total time taken being always t); but the figure is a phenomenological representation, and steps in the pit wall have been exaggerated.*

opening (i.e. maximum axis at right angles to D)—assuming that V_p is the same in all directions along a cleavage plane—, so that

$$d = 2h \frac{V_p}{V_n}\left(1 - \frac{1}{V \sin \theta}\right) \tag{4.33}$$

If, however, the parallel (i.e. along-the-cleavage-plane) etch rate component V_{p1} in the direction at right angles to the "vertical" plane containing the track is different to the component V_{p2} along the direction parallel to the projection of the track onto the (horizontal) detector surface (assumed to be one of the cleavage planes), we should obtain

$$D = 2h \frac{V_{p2}}{V_n}\left(1 - \frac{1}{V \sin \theta}\right) \tag{4.34}$$

and

$$d = 2h \frac{V_{p1}}{V_n}\left(1 - \frac{1}{V \sin \theta}\right) \tag{4.35}$$

A number of interesting consequences arise out of this model of track etching. Consider a crystal surface in which two quite different etch-rate components V_{p1}, V_{p2} operate (see Fig. 4.16). Then a track with projected direction (onto the detector surface) parallel to V_{p1} has V_{p2} as its transverse component, so that the width of the track, which is essentially determined by d, is governed by V_{p2}. Conversely, a track with projection parallel to V_{p2} has its width

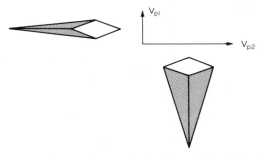

FIG. 4.16 *Schematic diagram showing the variation, with azimuthal direction, in the shape of etched tracks, as seen in the case of a solid which has two different etching-velocity components, V_{p1} and V_{p2}, both parallel to a sample surface (assumed to be one of the cleavage planes). Since V_{p2} is assumed to be much larger than V_{p1}, therefore a track with a projection (on to the sample surface) parallel to V_{p2} will have an "opening" parallel to V_{p1} which is much smaller than for a track with a projection parallel to V_{p1}. In an extreme case (e.g. for short etching time, or as a result of annealing), many of the former tracks may fail to become observable. This will show up in counting tracks as a function of the azimuthal angle (with reference to some arbitrary direction). Neglect of this phenomenon can, for instance, lead to false values in the case of fission track dating of mineral crystals that exhibit anisotropic behaviour.*

determined by V_{p1}. If $V_{p2} > V_{p1}$, then the tracks in the $\pm V_{p1}$ directions (i.e. parallel and antiparallel to V_{p1}) will appear broadened compared to those in the $\pm V_{p2}$ direction. In cases where the difference between V_{p1} and V_{p2} is large, this can lead to situations where tracks in one direction appear to be well developed and those in other directions (especially at right angles to that direction) are essentially invisible, with a consequent loss of the apparent etching efficiency, although this may recover on extended etching.

Such behaviour has been observed in the case of bronzite, $(Mg, Fe) SiO_3$, etched in 48% HF.[6] Distributions of track density with azimuthal angle (with reference to some arbitrary direction) then show a characteristically double-peaked form, with peaks corresponding to directions $\pm V_{p1}$ of Fig. 4.16. These effects can often be reduced (although not always totally eliminated) by the choice of a more appropriate etchant.

Anisotropic etching is often a problem in fission track dating (see Chapter 8). For example, sphene[19a] exhibits a degree of anisotropy in its etching behaviour which depends on the amount of (fossil) α-recoil damage stored in it, the anisotropy being absent in highly damaged crystals.

4.3 Some Special Techniques for Track Parameter Measurements

The most commonly used method of measuring track parameters is to use an optical microscope with a micrometer eyepiece attachment, employing transmitted light.* Sometimes other methods are adopted; and these will be briefly described here.

The measurement of the major and minor axes of an etch pit opening may be made using a scanning electron microscope (SEM). The minimum resolvable width w for an optical microscope is given by

$$w = \frac{\lambda}{2n \sin \alpha}$$

where λ is the wavelength of the light, α is the half-angle subtended at the object point by the objective aperture, and n is the refractive index of the medium between the object viewed and the objective ($n \sin \alpha$ is characteristic of a particular objective and is called the "numerical aperture", seldom exceeding 1.6); for a derivation of this formula for w, see any standard textbook on optics, e.g. reference 20. For an optical microscope w is typically $\sim 0.2 \ \mu m$; whereas for a scanning electron microscope the resolution is usually limited by the width of the electron beam used to scan the sample, and is $\sim 0.01 \ \mu m$ (i.e. ~ 100 Å). The disadvantage of the SEM, however, is that, in the normal mode of operation, only surface features are observable, so that

---

* With the onset of automation, it is becoming increasingly common to project the track openings on a closed-circuit TV screen and to make measurements with a video position analyser (VPA), or with the aid of hardware and software developments; see Chapter 7 for details.

measurements on that part of the etched track which lies beneath the detector surface are not possible. This may, to some extent, be overcome by the use of replication techniques. Here, some replicating fluid, for example cellulose acetate dissolved in acetone, is spread on the surface of the etched-track detector. A strip of solid cellulose acetate film is then placed on top of the fluid and pressed down, so that the fluid penetrates into the etch pits. Upon evaporation of the acetone, a solid replica of the tracks may be easily peeled off the sample. The track hole is now revealed as a cone (or some other

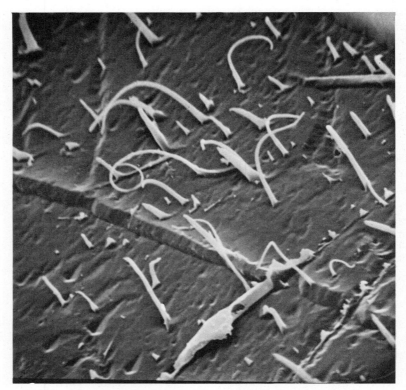

FIG. 4.17   *A scanning electron microscope (SEM) photograph of a plastic replica of fossil-fission tracks in a whitlockite (merrillite) crystal from the Bondoc meteorite at a magnification of ~ 1500 × . The replica is obtained by spreading a replicating fluid (cellulose acetate dissolved in acetone in this case; alternatively, silicone rubber could have been used) onto the surface of an etched sample and suitably pressing it into the etch pits. The apparent curvature of the track replicas is an artifact of the process and becomes severe (as here) in the case of long, thin strands extracted out of their fine, deep holes. Used in conjunction with SEM viewing, the technique facilitates measurements on the shapes, angles and depths of the etch pits.*

tapering spike) of cellulose acetate standing above the replica support, and (ideally) reproducing the angle and the length of the etched track (see Fig. 4.17). This may, for example, be examined under an SEM and the length of the track measured. Unless, however, the track is of large cone angle, the replica is likely to become twisted and distorted, so that the track dip angle, for example, may not be accurately represented.

A technique which, while not practicable for the measurement of large numbers of tracks, nevertheless offers the most direct means of evaluating $V_T(y')$ is the microtome slice method. This involves cutting very thin slices through sheets of plastic in which there are etched tracks at normal incidence to the surface. The cut is made parallel to the track direction, and the etch pit profile is thus directly exposed for microscopic observation.

Recently, Al-Najjar and Durrani[20a] have described the so-called track profile technique (TPT) to measure various track parameters in CR-39 irradiated with a variety of heavy ions, fission fragments and α-particles.

4.4 Environmental Effects on Track Etching

In this section, some of the parameters affecting the track etching process are discussed. We may subdivide these parameters into four categories:

(1) conditions operating prior to particle irradiation;
(2) conditions operating during irradiation;
(3) conditions operating after irradiation but before etching; and
(4) modifications to etching conditions.

The goal of most studies of this nature (apart from gaining an improved understanding of the track formation and etching processes themselves) is usually to improve the sensitivity of the track detectors or to control the etching behaviour in such a way as to facilitate the discrimination between tracks formed by different particles.

(1) In this category the effects of pre-irradiation of plastics with high doses of u.v. and γ-photons as well as of annealing have been examined. For example, Frank and Benton[21] find that increases in the bulk etching rate of Lexan polycarbonate occur after exposure to a few tens of megarads of ^{60}Co γ-rays. (Incidentally, it has been proposed that this property could be utilized in high-exposure γ-ray dosimetry.) The effect presumably arises from the degradation of the plastic under irradiation (see Chapter 3). Similar effects are observed in cellulose nitrate.[22] The radiation damage effects in crystals and glasses which have been subjected to high doses of charged particles, are discussed under (3) below.

Al-Najjar *et al.*[23] have given annealing treatments to CR-39 prior to α-irradiation. Both V_B and V_T are found to increase after preannealing at 100 °C for 10–15 hours; but the ratio V_T/V_B is not greatly altered.

(2) The most important effect occurring in plastics during irradiation is the increase of sensitivity caused by the presence of oxygen.[24,25] As mentioned in Chapter 3, oxygen is believed to combine with ions and radicals, preventing their recombination. O'Sullivan and Thompson[26] found that the track etch rates for various heavy ions incident on Lexan increased as the temperature during irradiation decreased (see also Chapter 6).

(3) After irradiation with track-forming particles in Lexan, exposure to u.v. is found to increase the ratio V_T/V_B. This effect has been studied by Benton and Henke,[27] Crawford et al.,[28] DeSorbo and Humphrey,[29] and DeSorbo.[30] Again, the presence of oxygen seems to be important. The energy deposited in the plastic along particle tracks by the u.v. leads to increased etchability of the track.

Durrani et al.[31] gave massive doses of 3 MeV protons to glass detectors both before and after their exposure to ^{252}Cf fission fragments and Fe ions. In both cases, but particularly when the proton irradiation was made afterwards, a significant diminution of etched-track diameters for the heavily charged particles was observed. The reduction in diameters was accompanied by a fall in the registration efficiency of the heavy particles. Both V_B and V_T were found to increase with the proton dose, but V_B increased more drastically than V_T, thus leading to an increase in the critical angle of etching. For example, a proton irradiation of 2×10^{16} cm^{-2}, made subsequently to the ^{252}Cf exposure, resulted in the critical angle for fission track registration in soda-lime glass becoming $\simeq 61° 40'$ (from $35° 30'$). This reduced the registration efficiency $f = 1 - \sin \theta_c$ to $\sim 12\%$ (from $\sim 42\%$). Subsequent annealing of the heavily irradiated glass was found by these authors[31] to restore V_B and f to their pre-irradiation values.

Radiation damage effects of high doses of solar flare and solar wind particles on lunar surface soil grains, and their consequences for the registration of very heavy cosmic rays and fission fragments in the mineral crystals, have been reported by a number of workers (e.g. Bibring et al.[32]).

(4) The influence of changes in the etching parameters on the detector sensitivity has been extensively studied. Most widely reported are the variation in the etching rate with etchant temperature and concentration, and the effect of etch products in the etching solution.

Bulk (and track) etching rates at a fixed etchant concentration are proportional to the factor $\exp(-E_a/kT)$, where E_a is an activation energy for the etching process, k is Boltzmann's constant, and T the absolute temperature of the etchant. Plots of log (etch rate) versus $1/T$ yield activation energies for the etching process: typically ~ 1 eV (see, e.g., reference 23).

The effects of etchant concentration are more complex. Frank and Benton[21] found that the bulk etch rate of Lexan in NaOH solution increased approximately as the square of the etchant molarity, whereas Enge *et al.*[33] and Hashemi-Nezhad *et al.*[34,34a] report more complex behaviour for cellulose nitrate (CA80-15).* In particular Enge *et al.*[33] find that, in Daicel cellulose nitrate**, as the etchant molarity is increased the bulk etch rate reaches a plateau value, unless the etching process is interrupted with frequent washing treatments. This effect appears to be due[33] to the formation of a colloid layer of etch products on the sample surface. Hashemi-Nezhad *et al.*[34,34a] find that V_T in cellulose nitrate etched in NaOH increases with the etchant molarity up to ~ 6 M and then begins to decline with further increase in molarity. The function $V(dE/dx)$ in all cases reaches a plateau with increasing dE/dx values, but the height of the plateau value of $V(= V_T/V_B)$ falls at M values above ~ 6.

Grabez *et al.*[35] report the effect of etchant temperatures on the etch induction time T_{ind} (see also Ruddy *et al.*[36†]) in CR-39, and the use of this parameter to differentiate between different ionizing particles. Törber *et al.*[37] present a model (the "diffusion-etch model"), proposing a two-phase development of latent tracks—in which ordinary etching of the plastic is accompanied by a diffusive chemical interaction of the etchant ions with the as-yet-unetched, but radiation-damaged, material. By enhancing the diffusion process, and depressing the etching process, through low-temperature ($-10\,°C$ to $+5\,°C$) soaking in NaOH, it is claimed[38] that the registration threshold for quasi-relativistic Fe ions in Daicel CN could be greatly improved.

Much work has recently been carried out on the dependence of the etching properties of CR-39 on the concentration and temperature of the etchant. Figures 4.18 and 4.19 (from references 39 and 40, respectively), for example, display the very strong dependence of the parameter V on the molarity of the etchant, and on the temperature of etching for a given molarity. These figures signify that it is possible to greatly alter the response of a given plastic simply by changing the molarity or temperature of the etchant. The response can thus be adjusted to optimize particle identification and energy discrimination properties of the plastic. These observations also imply that careful work is required in order to establish the optimum etching conditions for each batch of CR-39 (or other plastic) for obtaining the best discrimination between different types and energies of particles.

Finally, as mentioned at the beginning of this chapter, Peterson[4] has found that the ratio V_T/V_B in Lexan rises with increasing concentration of Lexan etch products in the etching solution. NaOH etchants saturated with these

* Made by Kodak-Pathé of France.
** Daicel (also known as DaiCell) is manufactured by Dai Nippon Co. of Japan.
† Ruddy *et al.*[36] found that particle tracks did not appear in plastic detectors until a certain finite time after the start of etching in a "continuous etch" method. This they termed the *etch induction time*, and suggested that this parameter might provide a means of distinguishing between ions of different Z (and possibly M) values.

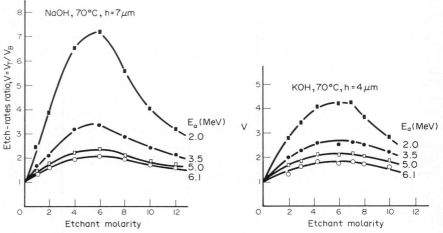

FIG. 4.18   *The etch rates ratio* $V = V_T/V_B$ *in CR-39, for α-particles of different energies* E_α, *as a function of etchant molarity. Data are plotted for NaOH and KOH etchants at 70°C and for removed-layer thicknesses* h = 7 μm *and 4 μm, respectively. The value of* V *goes up at first, but, after a broad maximum at* ∼6 M, *begins to decline again. The decline may be attributed to the slowing down of the removal of etch products from deep track holes at high etchant concentrations (see references 34 and 34a). The shapes of the* V *curves—in conjunction with others not shown here—suggest methods of optimizing the etching conditions (the type, molarity, and temperature of the etchant, as well as the thickness of the removed layer) as a means of energy discrimination of different charged particles. The curves also underscore the importance of careful calibration of each batch of the plastic detector used for particle identification and spectrometry in order to optimize the etching conditions (Figure from Green* et al.[39])

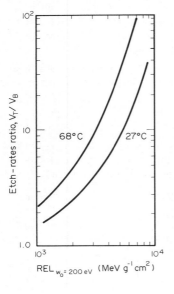

FIG. 4.19   *The etch rates ratio* $V = V_T/V_B$ *plotted as a functi⟨ of restricted energy loss (REL, see §3.3 (c) for details), with* W_o 200 eV *taken as the upper cut-off energy for the delta-ray ele trons, for tracks in CR-39 etched in NaOH at two different te peratures. Once again (see legend to Fig. 4.18), the figure impl⟨ that it is possible to greatly alter the response of a given plas⟨ by changing the etching conditions, so that it is possible to ⟨ just the detector response to optimize particle-identification a⟨ energy-discrimation properties of the plastic. The curves are bas⟨ on low-energy light ions (in the range* 6 ≤ Z ≤ 29, *with* E 9.6 MeV/nucleon) *detected with CR-39 manufactured by Persho⟨ Mouldings Ltd. (Figure from Fowler* et al.[40])

etch products are regularly used in cosmic ray work and result in a lower ionization threshold for track registration than when pure NaOH etchant is used.

Sufficient indication has been given in this section of the complex nature of the etching process. No doubt, as a greater understanding of the track etching phenomenon is gained, optimum etching conditions for any given material would be arrived at in a more systematic manner. The observations outlined above have been mainly concerned with track etching of plastics. Although etch rates for minerals and glasses certainly depend on etchant temperature and concentration, no such extensive array of environmental effects has as yet been catalogued for these materials.

References

1. H. G. Paretzke, T. A. Gruhn & E. V. Benton (1973) The etching of polycarbonate charged-particle detectors by aqueous sodium hydroxide. *Nucl. Instrum. Meth.* **107**, 597–600.
2. T. A. Gruhn, W. K. Li, E. V. Benton, R. M. Cassou & C. S. Johnson (1980) Etching mechanisms and behaviour of polycarbonates in hydroxide solution: Lexan and CR-39. In: *Proc. 10th Int. Conf. Solid State Nucl. Track Detectors*, Lyon, & Suppl. 2, *Nucl. Tracks*. Pergamon, Oxford, pp. 291–302.
3. G. Somogyi (1977) Processing of plastic track detectors. *Nucl. Track Detection* **1**, 3–18.
3a. G. Somogyi & I. Hunyadi (1980) Etching properties of the CR-39 polymeric nuclear track detector. In: *Proc. 10th Int. Conf. Solid State Nucl. Track Detectors*, Lyon, and Suppl. 2, *Nucl. Tracks*. Pergamon, Oxford, pp. 443–52.
4. D. D. Peterson (1970) Improvement in particle track etching in Lexan polycarbonate film. *Rev. Sci. Instr.* **41**, 1214–55.
5. G. W. Dorling (1974) Etching studies of glasses and minerals. M. Sc. thesis, Physics Department, University of Birmingham.
6. S. A. Durrani, H. A. Khan, R. K. Bull, G. W. Dorling & J. H. Fremlin (1974) Charged-particle and micrometeorite impacts on the lunar surface. In: *Proc. Fifth Lunar Sci. Conf., Geochim. Cosmochim. Acta*, Suppl. 5. Pergamon, New York, pp. 2543–60.
7. D. Lal, A. V. Muralli, R. S. Rajan, A. S. Tamhane, J. C. Lorin & P. Pellas (1968) Techniques for proper revelation of etch-tracks in meteoritic and terrestrial minerals. *Earth Planet. Sci. Lett.* **5**, 111–19.
8. P. H. Fowler, D. L. Henshaw, C. O.'Ceallaigh, D. O'Sullivan & A. Thompson (1978) Measurement of the cosmic ray element abundances between ≃ 300 and ≃ 750 MeV/N in the region from nickel to krypton using Lexan track detectors. In: *Proc. 9th Int. Conf. Solid State Nucl. Track Detectors*, Munich, and Suppl. 1, *Nucl. Tracks*. Pergamon, Oxford, pp. 1017–21.
9. W. Enge, K. Grabisch, L. Dallmeyer, K. P. Bartholomä & R. Beaujean (1975) Etching behaviour of solid state nuclear track detectors. *Nucl. Instrum. Meth.* **127**, 125–35.
10. H. A. Khan (1973) An important precaution in the etching of solid state nuclear track detectors. *Nucl. Instrum. Meth.* **109**, 515–19.
11. R. L. Fleischer, P. B. Price & R. M. Walker (1975) *Nuclear Tracks in Solids: Principles and Applications*. University of California Press, Berkeley.
12. R. L. Fleischer, P. B. Price & R. T. Woods (1969) Nuclear particle track identification in inorganic solids. *Phys. Rev.* **88**, 563–7.
13. R. P. Henke & E. V. Benton (1971) On geometry of tracks in dielectric nuclear track detectors. *Nucl. Instrum. Meth.* **97**, 483–9.
14. G. Somogyi & S. A. Szalay (1973) Track-diameter kinetics in dielectric track detectors. *Nucl. Instrum. Meth.* **109**, 211–32.
15. A. Ali & S. A. Durrani (1977) Etched-track kinetics in isotropic detectors. *Nucl. Track Detection* **1**, 107–21.

16. R. L Fleischer & P. B. Price (1964) Glass dating by fission fragment tracks. *J. Geophys. Res.* **69**, 331–9.

16a. S. A. Durrani & D. A. Hancock (1970) Effect of strain on fission-track ages of tektites. *Earth Planet. Sci. Lett.* **8**, 157–62.

16b. H. A. Khan, R. Brandt, N. A. Khan & K. Jamil (1983) Track-registration-and-development characteristics of CR-39 plastic track detector. *Nucl. Tracks* **7**, 129–39.

16c. P. H. Fowler (1977) Ultra heavy cosmic ray nuclei—Analysis and results. *Nucl. Instrum. Meth.* **147**, 183–94.

16d. H. A. Khan & S. A. Durrani (1972) Prolonged etching factor in solid state track detection and its applications. *Radiat. Effects* **13**, 257–66.

17. H. G. Paretzke, E. V. Benton & R. P. Henke (1973) On particle track evolution in dielectric track detectors and charge identification through track radius measurement. *Nucl. Instrum. Meth.* **108**, 73–80.

17a. G. Somogyi, R. Scherzer, K. Grabisch & W. Enge (1978) A spatial track formation model and its use for calculating etch-pit parameters of light nuclei. In: *Proc. 9th Int. Conf. Solid State Nucl. Track Detectors*, Munich, and Suppl. 1, *Nucl. Tracks*. Pergamon, Oxford, pp. 103–18.

17b. P. F. Green, A. G. Ramli, S. A. R. Al-Najjar, F. Abu-Jarad & S. A. Durrani (1982) A study of bulk-etch rates and track-etch rates in CR-39. *Nucl. Instrum. Meth.* **203**, 551–9.

17c. S. A. Durrani & P. F. Green (1984) The effect of etching conditions on the response of LR 115. *Nucl. Tracks.* **8**, 21–4.

18. G. W. Dorling, R. K. Bull, S. A. Durrani, J. H. Fremlin & H. A. Khan (1974) Anisotropic etching of charged-particle tracks in crystals. *Radiat. Effects* **23**, 141–3.

19. R. K. Bull (1976) Studies of charged particle tracks in terrestrial and extraterrestrial crystals. PhD thesis, Department of Physics, University of Birmingham.

19a. A. J. W. Gleadow (1978) Anisotropic and variable etching characteristics in natural sphenes. *Nucl. Track Detection* **2**, 105–17.

20. F. A. Jenkins & H. E. White (1957) *Fundamentals of Optics*. McGraw-Hill, New York.

20a. S. A. R. Al-Najjar & S. A. Durrani (1984) Track Profile Technique (TPT) and its applications using CR-39. I: Range and energy measurements of alpha-particles and fission fragments; II: Evaluation of *V* versus residual range. *Nucl. Tracks* **8**, 45–49 (I) and 51–56 (II).

21. A. L. Frank & E. V. Benton (1969) Dielectric plastics as high exposure gamma ray detectors. In: *Proc. Int. Top. Conf. Nucl. Track Registration in Insulating Solids & Applications.* Clermont Ferrand, **V**, 84–92.

22. E. V. Benton (1968) A study of charged particle tracks in cellulose nitrate. *US Naval Radiological Defense Laboratory, San Francisco, Report* NRDL-TR-68-14.

23. S. A. R. Al-Najjar, R. K. Bull & S. A. Durrani (1980) Some chemical and electrochemical etching properties of CR-39 plastic. In: *Proc. 10th Int. Conf. Solid State Nucl. Track Detectors*, Lyon, and Suppl. 2, *Nucl. Tracks*. Pergamon, Oxford, pp. 323–7.

24. K. Becker (1968) Tne effect of oxygen and humidity on charged particle registration in organic foils. *Rad. Res.* **36**, 107–18.

25. E. V. Benton (1970) On latent track formation in organic nuclear charged particle track detectors. *Radiat. Effects* **2**, 273–80.

26. D. O'Sullivan & A. Thompson (1980) The observation of a sensitivity dependence on temperature during registration in solid state nuclear track detectors. *Nucl. Tracks* **4**, 271–6.

27. E. V. Benton & R. P. Henke (1969) Sensitivity enhancement of Lexan nuclear track detector. *Nucl. Instrum. Meth.* **70**, 183–4.

28. W. T. Crawford, W. DeSorbo & J. S. Humphrey (1968) Enhancement of track etching rates in charged particle-irradiated plastics by a photo-oxidation effect. *Nature* **220**, 1313–14.

29. W. DeSorbo & J. S. Humphrey, Jr. (1970) Studies of environmental effects upon track etching rates in charged particle irradiated polycarbonate film. *Radiat. Effects* **3**, 281–2.

30. W. DeSorbo (1979) Ultraviolet effects and aging effects on etching characteristics of fission tracks in polycarbonate film. *Nucl. Tracks* **3**, 13–32.

31. S. A. Durrani, H. A. Khan, S. R. Malik, A. Aframian, J. H. Fremlin & J. Tarney (1973) Charged-particle tracks in Apollo 16 lunar glasses and analogous materials. In: *Proc.*

Fourth Lunar Sci. Conf., Geochim. Cosmochim. Acta, Suppl. 4. Pergamon, New York, pp. 2291–2305.

32. J. P. Bibring, J. Chaumont, G. Comstock, M. Maurette, R. Meunir & R. Hernandez (1973) Solar wind and lunar wind microscopic effects in the lunar regolith (abstract). In: *Lunar Science—IV (Abstracts),* Lunar Science Institute, Houston, pp. 72–4.

33. W. Enge, K. Grabisch, R. Beaujean & K.-P. Bartholomä (1974) Etching behaviour of a cellulose nitrate plastic detector under various etching conditions. *Nucl. Instrum. Meth.* **115,** 263–70.

34. S. R. Hashemi-Nezhad, P. F. Green & S. A. Durrani (1980) Effect of etchant normality on the response of CA80-15 cellulose nitrate to heavy ions. In: *Proc. 10th Int. Conf: Solid State Nucl. Track Detectors,* Lyon, and Suppl. 2, *Nucl. Tracks.* Pergamon, Oxford, pp. 245–50.

34a. S. R. Hashemi-Nezhad, P. F. Green, S. A. Durrani & R. K. Bull (1982) Effect of etching conditions on the bulk-etch rate and track-etching response of CA 80-15 cellulose nitrate. *Nucl. Instrum. Meth.* **200,** 525–31.

35. B. Grabez, P. Vater & R. Brandt (1981) The etch-induction time (T_{ind}) and other registration properties in CR-39 detectors for well-defined ions. *Nucl. Tracks* **5,** 291–97.

36. F. H. Ruddy, H. B. Knowles, S. C. Luckstead & G. E. Tripard (1977) Etch induction time in cellulose nitrate; a new particle identification parameter. *Nucl. Instrum. Meth.* **147,** 25–30.

37. G. Törber, W. Enge, R. Beaujean & G. Siegmon (1982) The diffusion-etch-model Part I; Proposal of a new two phase track developing model. In: *Proc. 11th Int. Conf. Solid State Nucl. Track Detectors,* Bristol, and Suppl. 3, *Nucl. Tracks.* Pergamon, Oxford, pp. 307–10.

38. I. Milanowski, W. Enge, G. Sermund, R. Beaujean & G. Siegmon (1982). The diffusion-etch-model Part II: First application for quasi-relativistic Fe-ion registration in Daicel cellulose nitrate. In: *Proc. 11th Int. Conf. Solid State Nucl. Track Detectors,* Bristol, and Suppl. 3, *Nucl. Tracks.* Pergamon, Oxford, pp. 311–14.

39. P. F. Green, A. G. Ramli, S. R. Hashemi-Nezhad, S. A. R. Al-Najjar, C. M. Ooi, F. Abu-Jarad, R. K. Bull & S. A. Durrani (1982) On the optimisation of etching conditions for CR-39 and other plastic track detectors. In: *Proc. 11th Int. Conf. Solid State Nucl. Track Detectors,* Bristol, and Suppl. 3, *Nucl. Tracks.* Pergamon, Oxford, pp. 179–82.

40. P. H. Fowler, S. Amin, V. M. Clapham & D. L. Henshaw (1980) Track recording properties of the plastic CR-39 for non-relativistic ions in the charge range $6 \le Z \le 29$. In: *Proc. 10th Int. Conf. Solid State Nucl. Track Detectors,* Lyon, and Suppl. 2, *Nucl. Tracks.* Pergamon, Oxford, pp. 239–44.

41. R. L. Fleischer & P. B. Price (1963) Tracks of charged particles in high polymers, *Science* **140,** 1221–2.

42. R. L. Fleischer, P. B. Price & R. M. Walker (1965) Tracks of charged particles in solids. *Science* **149,** 383–93.

43. B. G. Cartwright, E. K. Shirk & P. B. Price (1978) A nuclear track recording polymer of unique sensitivity and resolution. *Nucl. Instrum. Meth.* **153,** 457–60.

44. P. B. Price & R. M. Walker (1962) Chemical etching of charged particle tracks. *J. Appl. Phys.* **33,** 3407–12.

45. S. R. Hashemi-Nezhad & S. A. Durrani (1981) Registration of alpha-recoil tracks in mica: the prospects for alpha-recoil dating method. *Nucl. Tracks* **5,** 189–205.

46. R. L. Fleischer & P. B. Price (1963) Charged particle tracks in glass. *J. Appl. Phys.* **34,** 2903–4.

47. N. Bhandari, S. G. Bhat, D. Lal, G. Rajagopalan, A. S. Tamhane & V. S. Venkatavaradan (1971) Fission fragment tracks in apatite: Recordable track lengths. *Earth Planet. Sci. Lett.* **13,** 191–9.

48. A. J. W. Gleadow, A. J. Hurford & R. Quaife (1976) Fission track dating of zircons: improved etching techniques. *Earth Planet. Sci. Lett.* **33,** 273–6.

49. S. Krishnaswami, D. Lal, N. Prabhu & A. S. Tamhane (1971) Olivines: revelation of tracks of charged particles. *Science* **174,** 287–91.

50. C. W. Naeser & E. H. McKee (1970) Fission-track and K-Ar ages of Tertiary ash-flow tuffs, north-central Nevada. *Bull. Geol. Soc. Am.* **81,** 3375–84.

CHAPTER 5

Thermal Fading of Latent Damage Trails

When irradiated track detectors are held at high temperatures for a sufficient length of time, it is found that either the latent damage trails are totally removed from the detector, so that they cannot be revealed by subsequent etching, or their development by etching is impaired. This thermal annealing process is of considerable importance for a number of reasons.

In fission track dating it is this removal of tracks at elevated temperatures in nature that effectively resets the fission track "clock", so that a fission track age actually measures the length of time which has elapsed since the rock last cooled through a temperature known as the "closure temperature". This is the temperature at which quantitative fission track retention begins; its significance is discussed at greater length later in this chapter. In addition to this effect of total track removal, environmental temperatures over geological time periods can also result in modifications in the properties of the tracks, as alluded to above (e.g. their sizes and number densities undergo reductions); this partial annealing must be allowed for if an accurate fission track age is to be determined. A positive consequence of track size reduction by heat is that this effect can be used to draw inferences about the thermal history of a geological sample.

Such thermal effects are important in other applications; for example they impose limitations on the type of reactor environment in which a given track detector (e.g. plastic or glass) could be reliably used for neutron flux determination.

Finally, annealing data provide some information on the nature of the latent damage trail itself.

5.1 The Nature of the Annealing Process

The annealing of the damage trails at elevated temperatures presumably occurs via the diffusion of atomic defects through the crystal lattice or the movement of molecular fragments within a polymer. Interstitial atoms can then recombine with lattice vacancies, and broken molecular chains may rejoin and various active species recombine. The details of such processes are not fully understood. In the case of crystals, low-angle X-ray scattering experiments indicate that particle tracks apparently consist of extended defects surrounded by numerous point defects,[1,2] and that during annealing these point defects are more readily removed than the extended defects.

A typical annealing experiment would consist of the following procedure. A number of crystals containing fission tracks (resulting from either spontaneous or neutron-induced fission) are heated at various temperatures T for varying lengths of time t. After this heating or annealing step, the crystals are etched and the track densities $\rho(T, t)$ measured and compared with those

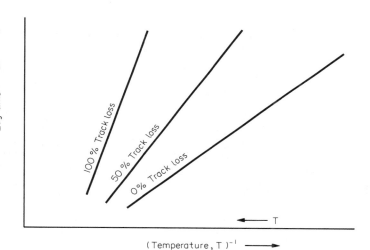

FIG. 5.1 *Schematic diagram showing an Arrhenius-type plot for annealing of fission tracks. On such a diagram, all time-temperature combinations leading to the same degree of track-loss lie along a straight line. Such a plot has to be constructed on the basis of experiments carried out on each mineral of interest. The slopes (E_a/k, where k is Boltzmann's constant) of these straight lines yield the corresponding "activation energies", E_a, for the repair process. These energies are found, typically, to be of the order of a few eV and to increase as the degree of track-damage repair, and hence of loss of etchable tracks, increases. Arrhenius plots of the above type, based on annealing experiments at several temperatures over various periods of time, are a prerequisite for the calculation of "closing temperatures", since the value of F necessary for the calculation (see §5.4) must be taken from such diagrams. For such determinations to form a reliable basis for calculation, the track etching conditions must be kept constant. The value of T corresponding to 50% track loss over a geological timespan (typically $\sim 10^6 - 10^8$ yr) of isothermal heating—which, in a given mineral, produces approximately the same degree of track-density loss as does steady cooling through the temperature domain of partial track retention by the mineral over the same timespan—gives a rough measure of the effective closing temperature T_c for the mineral (see §5.4).*

in unannealed control samples, viz $\rho(0)$. The data may be plotted on an Arrhenius-type plot such as that shown in Fig. 5.1, where the groups of (T, t) combinations for samples showing the same fractional track loss yield a fan-like array of lines when log t is displayed against $1/T$.

In general terms, track repair during annealing can, as indicated above, be viewed as being the result of diffusion processes. The diffusion coefficient has a temperature dependence of the form

$$D = D_0 \exp\left(-E_a/kT\right)$$

where D_0 is a constant (of dimensions (length)^2/time), and E_a is an "activation energy" for the repair process in question. The timescale for the diffusion of track-forming defects will be proportional to the inverse of D, so that its magnitude will be proportional to $\exp(E_a/kT)$. Thus a plot of log t versus $1/T$ produces straight lines of slopes E_a/k. In this way, activation energies for the repair process have been deduced and are typically of the order of a few electron-volts. Such activation energies are typical of the diffusion of atomic or molecular, rather than electronic, defects in solids. Annealing data thus lend support to the models of track structure outlined in Chapter 3.

An interesting feature displayed by graphs such as Fig. 5.1 is that activation energies for track repair tend to increase (with the slope of the plot lines) as more of the track damage is removed.[3] Fleischer, Price, and Walker[4] have proposed that during the early stages of annealing the high distortion energy (or stress) of the disturbed lattice aids the return of interstitial atoms to normal lattice sites, whereas during the latter part of the annealing process this distortion is much reduced.

5.2   The Effects of Pre-annealing on the Etched Tracks

In this section we examine the observable effects produced by annealing on tracks which are subsequently etched.

Although some general features of the annealing process are reasonably well understood, the problem of calculating the changes in track structure as a function of annealing time and temperature is a complex one, which has not yet been solved. It seems reasonable to suppose, however, that, as annealing proceeds, the defect concentration along a portion of the track length falls owing to a diffusion of defects away from the core and as a result of the filling of vacancies with atoms from interstitial sites. In plastics, some repair of broken molecular chains may occur. As the defect concentration falls, the track etching velocity for that portion of the track will (for a given set of etching conditions) decrease, until a level of defect concentration is reached at which no preferential etching (compared to the general or bulk etching of the medium) is possible. This process of loss of preferential etching occurs first at those portions of the track which were originally most weakly etchable. For fission fragment tracks this corresponds to the end of the fission fragment trajectory; whereas for, say, energetic Fe ions (neglecting the relatively small last portion of the track between the Bragg peak of ionization and the end of the Fe ion range), it corresponds to the portion of trajectory farthest from the point at which the ion was brought to rest. This is because the rate of ionization or energy loss is small at these early positions on the trajectory of an energetic heavy ion. This effect is clearly seen in Fig. 5.2, which shows the track-etch rates along Kr tracks in augite after various annealing treatments[5].

Thus, healing-up of the track takes place at the extremities of the etchable damage, first leading to a reduction in the length over which preferential etching takes place. For internally produced tracks, as the etchable range R falls, so as a consequence does the number of tracks crossing a given surface (i.e. the number density of etched tracks). In the case of a sample irradiated externally with fission fragments (from a ^{252}Cf source in close proximity to the detector, for example), the track density will not, of course, depend on the range, since the number of etched tracks is simply equal to the number of particles from the source which cross the sample surface at an angle greater than its critical angle of etching, θ_c (see Chapter 4). In such a case, the observed reduction in track density with increased temperature and time of annealing is not due to this range-shortening effect, but is caused, rather, by changes in θ_c.

As the track etch velocity V_T is reduced by further annealing, so is the critical angle θ_c increased, since $\theta_c = \sin^{-1} V_B/V_T$. The bulk-etching velocity V_B remains constant, provided that the heating treatment is not severe enough to produce basic changes (such as phase transformations) in the physical properties of the detector. Increases in θ_c result in reductions in track density for both internally and externally produced tracks; and for the latter (where track registration efficiency for a thin external source, for example, is $f = 1 - \sin \theta_c$), the increase in θ_c is the principal cause of track density decrease. When annealing is extremely severe, the complete healing-up of the radiation damage, resulting in the complete disappearance of tracks, is tantamount to V_T becoming equal to V_B. This means $1 - \sin \theta_c = 0$ as well as $\cos^2 \theta_c = 0$—making both the external and the internal detection efficiency nil (see §4.2.3).

As mentioned above, those portions of particle tracks which correspond to the most intense ionization by the track-forming particle are the most resistant to annealing. A further consequence of this effect is that heavy-ion tracks are more resistant to annealing than those of lighter ions.[6] This

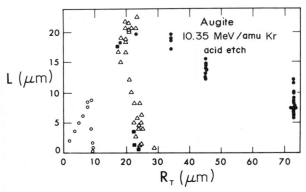

• 30 min at 525 °C
△ 30 min at 650 °C
■ 50 h at 600 °C
○ 30 min at 700 °C

FIG. 5.2   *The "L–R_T plot" for 10.35 MeV/nucleon Kr ions in the silicate mineral augite (a clinopyroxene), using an acid etchant ($2HF: 1H_2SO_4: 4H_2O$ for 30 min) and the "angled-polishing technique" (see Fig. 6.7 for details of the technique, and §6.1.1 for a general description of the L–R plot). In the figure, L is the length of the etched cone, and R_T is the distance from the point of entry of the ion into the new surface of the crystal (produced by grinding and polishing of the irradiated crystal) to the end of the etchable range of the ion. As a result of the angled polishing, all values of the residual range from 0 to the total range R_0 of the particle in the medium are available in a single sample, with all tracks etched for the same time t. The figure depicts the effect of different annealing treatments on the subsequent track etch rates (since the mean etch rate $\bar{V}_T = L/t$) as a function of R_T. The diagram shows that the regions of the track at the highest residual range, where track etch rates are low, are more readily removed by annealing than the regions at low residual range where the radiation damage density due to ionization, and hence the track etch rates, are highest. Thus, even 700 °C for 30 min has failed to render the tracks unetchable at small R_T values, whereas none have survived at high R_T values save for those subjected to the lowest degree of annealing (note that for such large values of R_T, a great degree of scattering in the etched cone length L is common—cf. the three columns of filled black dots—because of local variations in regions of low ionization rate). (Figure after Price et al.[5])*

result has been used in attempts to separate out VVH ($Z \gtrsim 30$) from VH ($20 \leq Z \lesssim 28$)* cosmic-ray tracks in lunar crystals.[2,7]

In crystals, annealing is affected by the anisotropy of certain crystal structures. Diffusion coefficients are dependent upon crystallographic direction; and as a result rates of track-annealing depend on the orientation of the track relative to the crystal axes. It seems reasonable to assume that dif-

---

* Note that the demarcation between the VH and VVH components of cosmic rays differs somewhat in the literature. We have adopted approximate values of their charge domains.

fusion of defects perpendicular, rather than parallel, to the track axis is the most important process in track annealing; consequently, tracks for which a high diffusion rate for defects acts in a direction nearly at right angles to the track length will anneal out more rapidly than those for which the diffusion rates perpendicular to the track direction are small. This effect was first observed in mica.[8] Green and Durrani[9] found that both external ^{252}Cf-fission and internal ^{238}U fission tracks lying nearly parallel to the c-axis of the hexagonal structure of apatite were more resistant to annealing than tracks in other directions; this supports the assumption that diffusion is greater in the crystal planes parallel to the c-axis of the prism than in the basal plane.

In experimental studies of annealing effects, two approaches have generally been adopted. The first is known as isochronal annealing (where the annealing time is kept fixed, but the temperature is varied—being usually increased in steps), and the second as isothermal annealing (fixed temperature, annealing time varied). It is possible, of course, to combine the two approaches, and use different time steps at gradually increasing temperatures, for instance. Usually, different samples are used for each annealing step, because cumulative effects of track reduction are difficult to interpret.

Figures 5.3a and 5.3b show schematic representations of typical curves for, respectively, isochronal and isothermal annealing for fission track densities (or lengths). At constant time (Fig. 5.3a), no reduction effects are observed until some characteristic temperature is reached, which is not strongly dependent on the annealing time. At constant temperature (Fig. 5.3b), track densities (and lengths) fall rapidly with time at first (provided that the temperature is high enough), then more slowly. This is consistent with the observation, noted earlier in this chapter, that the early stages of annealing are characterized by lower activation energies for diffusion than are the later stages. It should be noted, however, that Burchart *et al.*[10] have recently suggested that this rapid loss during the early stages of an isothermal annealing experiment always occurs regardless of the previous thermal history of the sample. If this proves to be correct, then the above simple interpretation would need to be modified. Also it would have important implications for the extrapolation of laboratory annealing data to geological time scales.

5.3 Typical Annealing Temperatures for Fission Tracks in Various Materials

The stability of fission tracks has been extensively studied, over the years, in view of their importance for fission track dating; fission fragments are, moreover, readily available, and have the ability of being recorded by all track detectors. It is found that the stability of fission tracks in different materials varies greatly. Tracks in plastics are affected by temperatures as low as

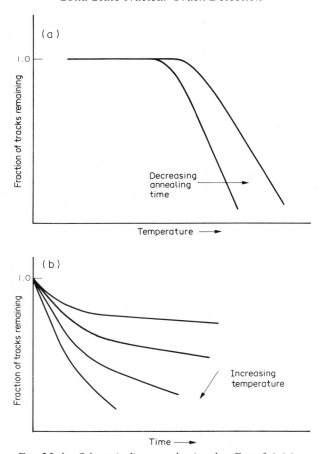

FIG. 5.3a,b *Schematic diagrams showing the effect of (a) iso-chronal annealing, and (b) isothermal annealing, on fission track density (or length). In case (a), for a given (constant) time of annealing, no reduction effects are observed, until some character-istic temperature is reached, which is not strongly dependent on the annealing time. In case (b), at a given (constant) tempera-ture, track densities (and lengths) fall at first rapidly with in-creasing time of annealing (provided that the temperature is high enough), then more slowly. See text for discussion.*

80–100 °C applied for one hour; whereas in some minerals, such as clino-pyroxene, quartz, zircon, and sphene, fission tracks are stable for long periods at temperatures of ~ 500 °C or even higher. In Table 5.1 the track retention properties of different track detectors are displayed. The parameter given is the temperature for virtually complete fading of fission tracks in 1 h.

TABLE 5.1   *Track retention*
characteristics of some common
detectors. Typical temperatures for
100% loss of fission tracks (FT) in
1 h of annealing are shown.

Material	100% FT loss in 1 h (°C)*
Plastics	
Cellulose nitrate	80–100
CR-39	∼250**
Lexan	>185
Makrofol	165
Glasses	
Soda-lime glass	350–400
Tektite glass	∼500
Mineral Crystals	
Apatite	350–400
Clinopyroxene	500–600
Epidote	625–725
Feldspar (plagioclase)	700–800
Merrillite	∼450
Mica	500–600
Olivine	400–500
Orthopyroxene	450–500
Quartz	1000
Sphene	650–800
Zircon	750–850

* These temperatures should be regarded only as rough guides. The retention temperatures of both minerals and plastics depend on their exact composition as well as on the etching conditions employed. Many of the mineral names, in particular, cover a wide range of compositions. For a more complete list see references 4, 11 and 11a.
** At this temperature, CR-39 develops extensive cracks and becomes discoloured.

5.4   Closing Temperatures

In Chapter 8, fission-track dating will be discussed at some length. At this point it is probably sufficient to remark that, over geological time scales, a U-bearing crystal will accumulate considerable numbers of spontaneous fission tracks through the fission of ^{238}U. If the uranium content of the crystal is known, then the number of these tracks can be used to deduce an "age" for this crystal. In interpreting these ages it is important to establish exactly what the ages measure. From the preceding sections of this chapter it will be clear that the tracks counted in a crystal are only those which have accumulated since the crystal last cooled below temperatures that were high enough to produce track annealing. If a rock was formed and cooled virtually instantaneously to low values of temperature at some time t years ago, then it is obvious that the fission track age measures this time t, and that the same age should be obtained for all minerals from this rock. For a rock which cooled down steadily over a period of time not insignificant compared to the total time since rock formation (or since an event which may have significantly reheated the rock), the interpretation of this age is not clear-cut. Furthermore, different ages may be obtained for different minerals which have different track retention temperatures.

If a rock cools down as shown in Fig. 5.4, we must ask as to what point on the cooling curve is it that represents the effective start of track retention? The temperature at which the fission track "clock" is "switched on" is known variously as the *closing temperature* or the *closure temperature*; it depends on the type of crystal under consideration and on the cooling rate of the rock.

Closing temperatures may be calculated using the analysis outlined below, which closely follows (but has a different notation than) an analysis presented by Haack.[12] It should be noted, however, that several other, somewhat different, approaches to the problem have also been made.[13,14,14a]

For a steadily cooling rock, the manner in which fission tracks accumulate may be simulated by replacing the continuous function $T(t)$ by a series of steps, each of constant temperature (see Fig. 5.5). Now consider a small time interval dt within one of these isothermal steps. The rate at which fission tracks are produced per unit volume of crystal is given by

$$\frac{dn}{dt} = N_8 \lambda_{f8}$$

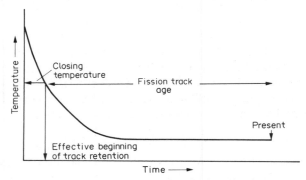

FIG. 5.4 *Idealized cooling curve for a rock. "Closing temperature" is defined as that temperature on the cooling curve at which a given type of crystal begins effectively to retain its spontaneous-fission tracks. The effective fission track age is, then, the time elapsed since that juncture to the present moment (assuming no episode of reheating in the meantime).*

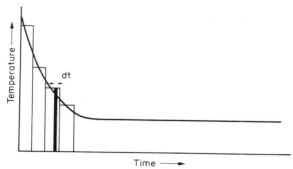

FIG 5.5 *Replacement of a continuous cooling curve (of the type shown in Fig. 5.4, representing a continuous function T(t)) by a series of isothermal annealing steps represented by the hollow rectangles. Each step may be further subdivided into a large number of tiny time-intervals dt, during which an element of fission track density proportional to dt is accumulated by the rock—out of which a fraction dependent on the instantaneous temperature (and also proportional to dt) is lost by annealing. Such analysis is used in estimating the closing temperature and thus calculating the effective fission track age of the rock.*

where λ_{f8} is the fission decay constant of ^{238}U, and N_8 is the number of ^{238}U atoms per unit volume. The incremental fission track density (in the absence of annealing) on an internal surface of the crystal is given by

$$d\rho = N_8\lambda_{f8}f(R, \theta_c)\,dt$$

where $f(R, \theta_c)$ is a geometrical factor which relates the number of tracks per unit volume to the surface density. It has the value $R\cos^2\theta_c$ for internal tracks with constant V_T along their length (see Chapter 4). Here R is the range of a *single* fission fragment and θ_c is the critical angle of etching.

In any time interval at constant temperature, provided that the degree of annealing is not so great that a different (and higher) effective activation energy for annealing becomes applicable, the fractional reduction in track density per unit time will remain constant. The loss term for the time interval dt may then be written as $\alpha(T)\rho\,dt$, where $\alpha(T)$ is a sensitive function of temperature T. The net change in track density is therefore

$$d\rho = N_8\lambda_{f8}f(R, \theta_c)\,dt - \alpha(T)\rho\,dt \tag{5.1}$$

It has to be emphasized that the second term on the right-hand side of Eq. (5.1) does not necessarily imply a reduction in the number of fission tracks per unit volume (i.e. elimination of fission tracks), but rather refers to a reduction in the observed track density at a surface due to changes in the value of $f(R, \theta_c)$ brought about by annealing. For timescales short compared to the (overall) half-life of ^{238}U, the track production rate (per unit time) recorded per unit area may be regarded as a constant, P, and we can write

$$\frac{d\rho}{dt} = P - \alpha(T)\rho \qquad (5.2)$$

This equation may then be integrated over the whole of the isothermal interval of duration t to give

$$\rho(T) = \frac{P}{\alpha(T)} - C \exp\{-\alpha(T)t\} \qquad (5.3)$$

The constant of integration C may be found by inserting the appropriate boundary conditions as follows.

There are two groups of tracks to be considered. The first are those which were present at the start of the isothermal interval under consideration, having a density ρ_0. For these we have the condition $P = 0$, if we regard them as quite distinct from those produced during the time interval in question. From Eq. (5.3), then:

$$\rho_0(t) = -C \exp\{-\alpha(T)t\}$$

Since

$$\rho_0(t) = \rho_0 \quad \text{at } t = 0, C = -\rho_0$$

hence

$$\rho_0(t) = \rho_0 \exp\{-\alpha(T)t\} \qquad (5.4)$$

Thus the tracks present at the beginning of an interval of isothermal annealing are reduced exponentially with the length of the time interval.

The fraction of tracks retained after a time t at temperature T may be found from data such as those displayed in Fig. 5.1 (the Arrhenius plot). If we denote this fraction by F, then

$$F = \rho_0(t)/\rho_0 \qquad (5.5)$$

and from Eq. (5.4)

$$F = \exp\{-\alpha(T)t\} \qquad (5.6)$$

This part of the analysis has assumed that all of the tracks are "fresh" ones at the start of the interval. In fact, of course, some of these will be the residual portion of earlier track populations, by now much reduced in number by annealing. Since these tracks have undergone annealing, therefore presumably a higher activation energy should prevail during any subsequent annealing that they undergo, making them more resistant to fading. This follows from the discussion in §5.1, where it was mentioned that activation energies for annealing increase as the degree of track reduction is increased. This point has been emphasized by Bertagnolli *et al.*;[15] and it should be borne in mind that the analysis presented here may lead to the estimation of closing temperatures which are too low.[15]

The second group of tracks mentioned above consists of tracks denoted by ρ', which are produced only during the isothermal interval under consideration; and here $\rho'(0) = 0$ at the beginning of that interval. Therefore, from Eq. (5.3) with $t = 0$:

$$C = \frac{P}{\alpha(T)}$$

and

$$\rho'(t) = \frac{P}{\alpha(T)}(1 - \exp\{-\alpha(T)t\}) \qquad (5.7)$$

The total track density at the end of an interval of length t is given by

$$\rho(t) = \rho_0(t) + \rho'(t)$$

Therefore, using Eqs. (5.5), (5.6), and (5.7), we obtain

$$\rho(t) = \rho_0 F + \frac{P}{\alpha(T)}(1 - F)$$

Upon substituting for $\alpha(T)$ from Eq. (5.6), we get

$$\rho(t) = \rho_0 F + \frac{Pt(1 - F)}{-\ln F}$$

In the light of the foregoing, the track density ρ_k at the end of the kth interval is given by

$$\rho_k = \rho_{k-1} F + \frac{Pt(1 - F)}{-\ln F} \qquad (5.8)$$

since the starting track density for this interval is the track density ρ_{k-1} at the end of the previous interval.

For any given continuous form of temperature function $T(t)$, the procedure given above can be used, along with experimental annealing data for the mineral of interest, to compute the track density as a function of time. For a given cooling rate, the length of the isothermal interval (t) is determined by selecting the size of temperature step between consecutive intervals. Starting with $\rho_{k-1} = 0$, at a temperature higher than that at which track retention begins, Eq. (5.8) is applied to each successive time interval in turn. As k increases, the temperature falls and the fraction of tracks retained (F) rises. At each step, F is taken from the appropriate temperature/time point

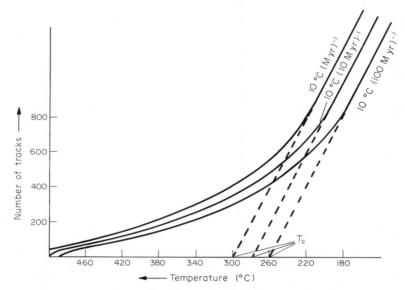

FIG. 5.6   *The growth in the number of tracks (solid lines) as the temperature falls (from left to right), and with it the time increases (also from left to right, but at different rates for the three curves), for garnet in a rock cooling steadily (in a stepwise manner) at the three different cooling rates, viz. suffering a fall of 10 °C over an isothermal time interval of 1 million, or 10 million, or 100 million years, respectively. The growth follows Eq. 5.8, and the production of fresh tracks per respective time interval is taken as 100. The fraction* $F = \rho_0(t)/\rho_0$ *(Eq. 5.5) of tracks surviving after each interval is read from a figure for garnet of the type Fig. 5.1. Towards the end of the cooling period (right-hand of Fig. 5.6), the curves become linear, i.e. no fading takes place and the track density simply grows at the rate of production,* P. *These straight lines, when extended downwards, give the effective closing temperatures* T_c *indicated on the abscissa. The closing temperature is thus a function of the cooling rate of the crystals in the rock. (After Haack.[12])*

of an Arrhenius diagram, such as Fig. 5.1 for the mineral concerned, and ρ_k is calculated, and plotted against the total elapsed time. This iterative process yields a track growth curve similar to that shown in Fig. 5.6. Figure 5.6 (after Haack[12]) shows track density plotted against decreasing temperature for a continuous linear cooling of garnet at three different cooling rates (10, 1, and 0.1 °C Myr^{-1}, respectively).*

* Note that the track production rates *P*, for the three curves in Fig. 5.6, have been so chosen as to compensate for the variation in the time scale with the cooling rate implicit in the figure (along the abscissa), thus yielding parallel straight lines in the non-fading regime. In reality, for a given rock, and hence a given U content, the *P* value should be constant (cf. Eq. (5.1)), which would give parallel straight lines in the non-fading region of a ρ versus *t* plot.

In Fig. 5.6, at low temperatures, i.e. towards the end of the cooling period, the curves become linear: annealing no longer plays a part and the track density simply increases at the rate of production P. If the linear regions of these curves are extrapolated back to zero track density, they intersect the temperature axis at points T_c, which correspond to the effective closing temperatures for garnet at these cooling rates. Thus the final track density is the same as if track production had suddenly begun to operate (at a constant rate P) at these temperatures T_c, with no attendant annealing. For garnet, a change in cooling rate from $10°C$ Myr^{-1} to $0.1°C$ Myr^{-1} only changes this "switch on" temperature from $300°C$ to $260°C$ according to Haack.[7]* This weak dependence on cooling rate is fortunate, since without supporting evidence this rate may not be well known for a sample on which fission track dating is to be carried out.

The calculation of closure temperatures depends upon extrapolation of annealing data obtained over laboratory timescales (typically hours or days) to geological timescales (many millions of years).[15a] Recently, several workers[16-18] have measured track densities in apatites from various depths within deep drill holes. The temperatures of such holes, as a function of depth, may be readily measured; and a knowledge of the local geology allows estimates of the duration of the high-temperature regime for apatites to be made. Annealing data extending to geological timescales have thus been obtained. These results tend to confirm the previous estimates of effective closing temperatures, although Gleadow and Duddy[18] suggest that the range of temperatures over which track retention in drill-hole apatites varies from 0% to 100% is less than would be expected from laboratory annealing data.

Closing temperatures obtained for apatite are, typically, within the range 80–120 °C (depending, of course, upon the cooling rate). Sphene begins to retain tracks already at ~ 300 °C.

Before leaving this subject, it may be worthwhile to make a brief reference to the case of "α-recoil tracks" and their thermal stability. Huang and Walker[19] pointed out the possibility of using, for dating purposes, tracks produced in minerals such as mica by the heavy residual nuclei recoiling during the α-decay of U and Th isotopes contained in the rock. These tracks are called α-recoil tracks (α-RTs), and can be enlarged by etching. (The α-particle tracks themselves, being below the etchability threshold of the silicate minerals are, of course, not registered.) Huang and Walker[19] gave equations, analogous to those for fission-track dating equations, which could, in principle, give the age of the geological (or archaeological) object, with the advantage that the ratio of α-decay to fission events is very high (e.g. $\sim 10^6$ for ^{238}U).

---

* In the case of slower cooling, tracks have faded for longer times. This means that low temperatures can produce a higher degree of fading; which is tantamount to saying that tracks are effectively retained at a lower temperature (T_c).

Hashemi-Nezhad and Durrani[20] have, however, recently re-examined the subject and pointed out that, because of the relatively small amount of energy possessed by the α-recoiling nuclei (typically ~ 0.1 MeV total energy), the α-RTs are expected to anneal out more easily than fission tracks, when subjected to elevated temperatures. The unpublished data of Hashemi-Nezhad[21] indicate that, for a given annealing temperature, the annealing time required to completely erase fossil α-RTs in biotite mica is at least an order of magnitude smaller than that for fossil fission tracks. The lower thermal stability of α-RTs, however, also implies[21,21a] that their closing temperatures are much lower than those for fission tracks. This property of α-RTs can be put to good use as a complementary, and a more sensitive, technique to the fission track method in studying the thermal history of geological samples. The cooling-down history of minerals capable of recording α-RTs could, thus, be extended to much more recent times (because of their greater abundance than fission tracks), and down to much lower temperatures (because of their greater thermal instability), than are possible by studying the annealing behaviour of fission tracks in those samples.

5.5   Annealing Correction Methods

So far we have assumed that once the temperature of a rock falls below the closing temperature for a given mineral, the tracks in that mineral will remain perfectly stable thereafter. However, the situation is frequently not so simple. For example, the rock may undergo a transient heating episode after the primary cooling of the system. If the fission tracks are only partially removed during this transient temperature-excursion, then the fission track age will reflect neither the time elapsed since the primary cooling nor that since the secondary event. Fortunately, however, in such circumstances some of the tracks will be shortened (or have reduced diameters in the case of etched tracks in glasses), and this factor can be used to diagnose and to correct for such annealing effects. This point will be discussed in greater detail in Chapter 8. Wagner[22] has recently reviewed some of the implications and the geological significance of the different cooling-down behaviours of rocks (fast, slow, or mixed). It has also been shown by Green[22a] that significant track shortening can occur when tracks are stored over geological timescales at temperatures well below the closing temperatures.

Perhaps it should be mentioned here in passing that a controversy exists between various fission track dating groups as to whether or not an initial 10–15% reduction in the lengths of fission tracks in minerals produces a commensurate reduction in their number densities (and hence in the apparent age of the rock). Thus Green[23] reports experimental data that indicate that there is a linear one-to-one correspondence in the reductions of mean etchable range (TINT[24] length) and of the fission track density on internal surfaces of

Durango apatite crystals. On the other hand, Gleadow and Duddy[18] report their observation on apatites from Otway Group in S. Australia, which show that fully etched confined tracks (TINTs and TINCLEs[24]) exhibit no track density reduction for the initial $\sim 15\%$ naturally-produced reduction in the fossil-fission track lengths. These latter observations are very similar to the results of laboratory annealing experiments on apatite carried out by Nagpaul *et al.*[25] in order to obtain a relationship between confined-track lengths and track densities in these crystals as well as in other minerals. Whether these differences can be attributed to crystallographic anisotropies of etching[26,27] and of annealing[27a], or to the effect of prolonged etching[28,29] (see also §4.2.3.1), or to other causes as yet not understood, is a question that will, hopefully, be resolved by future work. Further discussion of age-correction methods (e.g. the isochronal and isothermal plateau-age correction) will be found in Chapter 8 (§§8.6.1; 8.6.2).

5.6 Track Seasoning

When the annealing properties of fossil tracks in lunar samples are compared with those of artificially produced tracks in the same materials, it is found that the fossil tracks are significantly more difficult to anneal than the fresh tracks.[5,7,30] Since fossil tracks have usually been partially annealed by low or moderate environmental heating over many millions of years, it would be expected that the activation energies applicable to the surviving portions of these tracks would be higher than those relating to the early stages of fresh-track annealing; these surviving parts would, thus, be expected to require higher temperatures for their elimination.[31] Nevertheless, the differences are large ($\sim 150\,°\mathrm{C}$ between the total-removal temperatures for fresh Fe tracks in feldspar and fossil, mainly cosmic ray, tracks in the same mineral[7]) and cannot be attributed to this effect alone. It seems that some "hardening" (or "seasoning") of the tracks occurs during long storage at slightly elevated temperatures—an effect which may be enhanced by the presence of radiation fields such as occur at the lunar surface.[7,32,33] This effect has not yet been fully reproduced in the laboratory. Perhaps over long periods of time extended defects grow and stabilize at the expense of point defects.[2]

In closing this chapter it should be pointed out that environmental effects other than temperature and radiation field can be responsible for track removal and modification. For example, the passage of a shock wave through a crystal during impact between rocky bodies (such as a meteorite colliding with the lunar surface) can lead to the segmentation of tracks, as sections of the crystal move with respect to each other along slip planes.[34] However, it seems that temperature is by far the most important parameter affecting the storage of tracks in nature.

References

1. E. Dartyge, J. P. Duraud & Y. Langevin (1977) Thermal annealing of iron tracks in muscovite, labradorite and olivine. *Radiat. Effects* **34**, 77–9.
2. E. Dartyge, J. P. Duraud, Y. Langevin, & M. Maurette (1978) A new method for investigating the past activity of ancient solar flare cosmic rays over a time scale of a few billion years. In: *Proc. Lunar Planet. Sci. Conf. 9th*. Pergamon, New York, pp. 2375–98.
3. C. W. Naeser & H. Faul (1969) Fission-track annealing in apatite and sphene. *J. Geophys. Res.* **74**, 705–10.
4. R. L. Fleischer, P. B. Price & R. M. Walker (1975) *Nuclear Tracks in Solids: Principles and Applications*. University of California Press, Berkeley.
5. P. B. Price, D. Lal, A. S. Tamhane & V. P. Perelygin (1973) Characteristics of tracks of ions $14 \leq Z \leq 36$ in common rock silicates. *Earth Planet. Sci. Lett.* **19**, 377–95.
6. M. Maurette (1970) Some annealing characteristics of heavy-ion tracks in silicate minerals. *Radiat. Effects* **5**, 15–19.
7. R. K. Bull & S. A. Durrani (1975) Annealing and etching studies of fossil and fresh tracks in lunar and analogous crystals. In: *Proc. Lunar Sci. Conf. 6th*. Pergamon, New York, pp. 3619–37.
8. Ya. E. Geguzin, I. V. Vorob'eva & I. G. Berzina (1968) Thermal stability of uranium fission fragment tracks in muscovite single crystals (effect of anisotropy) *Soviet Phys. Solid State* **10**(6), 1431–4.
9. P. F. Green & S. A. Durrani (1977) Annealing studies of tracks in crystals. *Nucl. Track Detection* **1**, 33–9.
10. J. Burchart, J. Gałązka-Friedman & J. Král (1981) Experimental artifacts in fission-track annealing curves. *Nucl. Tracks* **5**, 113–20.
11. R. L. Fleischer, P. B. Price & R. M. Walker (1965) Solid state track detectors: Applications to nuclear science and geophysics. *Ann. Rev. Nucl. Sci.* **15**, 1–28.
11a. G. A. Wagner (1978) Archaeological applications of fission-track dating. *Nucl. Track Detection* **2**, 51–64.
12. U. Haack (1977) The closing temperature for fission track retention in minerals. *Am. J. Sci.* **277**, 459–64.
13. G. A. Wagner & G. H. Reimer (1972) Fission track tectonics: The tectonic interpretation of fission track apatite ages. *Earth Planet. Sci. Lett.* **14**, 263–8.
14. E. Märk, M. Pahl, E. Purtscheller & T. D. Märk (1973) Thermische Ausheilung von Uranspaltspuren in Apatiten, Alterskorrekturen und Beiträge zu Geothermochronologie. *Tschermaks Min. Petr. Mitt.* **20**, 131–54.
14a. K. James & S. A. Durrani (1985) Fission track closure temperatures. *Paper presented at the 13th Int. Conf. Solid State Nucl. Track Detectors*, Rome (and to be published in *Nucl. Tracks* **12**, 1986).
15. E. Bertagnolli, E. Märk, E. Bertel, M. Pahl & T. D. Märk (1981) Determination of palaeotemperature of apatite with the fission track method. *Nucl. Tracks* **5**, 175–80.
15a. P. Mold, R. K. Bull & S. A. Durrani (1984) Fission-track annealing characteristics of meteoritic phosphates. *Nucl. Tracks* **9**, 119–28.
16. C. W. Naeser & R. B. Forbes (1976) Variation of fission track ages with depth in two deep drill holes (Abstract) *Trans. Am. Geophys. Union* **57**, 353.
17. C. W. Naeser (1979) Fission track dating and geologic annealing of fission tracks. In: *Lectures in Isotope Geology* (eds. E. Jäger and J. C. Hunziker). Springer-Verlag, Heidelberg, pp. 154–69.
18. A. J. W. Gleadow & I. R. Duddy (1981) A natural long-term track annealing experiment for apatite, *Nucl. Tracks* **5**, 169–74.
19. W. H. Huang & R. M. Walker (1967) Fossil alpha-particle recoil tracks: a new method of age determination. *Science* **155**, 1103–6.
20. S. R. Hashemi-Nezhad & S. A. Durrani (1981) Registration of alpha-recoil tracks in mica: The prospects for alpha-recoil dating method. *Nucl. Tracks* **5**, 189–205.
21. S. R. Hashemi-Nezhad (1981) Experimental and theoretical studies of nuclear track detectors: Application to geochronology, radioactive haloes and related radiation phenomena. PhD thesis, University of Birmingham, England.

21a. S. R. Hashemi-Nezhad & S. A. Durrani (1983) Annealing behaviour of alpha-recoil tracks in biotite mica: Implications for alpha-recoil dating method. *Nucl. Tracks* **7**, 141–6.

22. G. A. Wagner (1981) Fission-track ages and their geological interpretation. *Nucl. Tracks* **5**, 15–25.

22a. P. F. Green (1980) On the cause of shortening of spontaneous fission tracks in certain minerals. *Nucl. Tracks* **4**, 91–100.

23. P. F. Green (1981) "Track-in-track" length measurements in annealed apatites. *Nucl. Tracks* **5**, 121–8.

24. D. Lal, R. S. Rajan & A. S. Tamhane (1969) Chemical composition of $Z \geq 22$ in cosmic rays using meteoritic minerals as detectors. *Nature* **221**, 33–7.

25. K. K. Nagpaul, P. P. Mehta & M. L. Gupta (1974) Annealing studies on radiation damage in biotite, apatite and sphene, and corrections to fission track ages. *Pure Appl. Geophys.* **112**, 131–9.

26. G. W. Dorling, R. K. Bull, S. A. Durrani, J. H. Fremlin & H. A. Khan (1974) Anisotropic etching of charged-particle tracks in crystals. *Radiat. Effects* **23**, 141–3.

27. S. A. Durrani, H. A. Khan, R. K. Bull, G. W. Dorling & J. H. Fremlin (1974) Charged-particle and micrometeorite impacts on the lunar surface. In: *Proc. Lunar Sci. Conf. 5th.* Pergamon, New York, pp. 2543–60.

27a. S. Watt, P. F. Green & S. A. Durrani (1984) Studies of annealing anisotropy of fission tracks in mineral apatite using Track-IN-Track (TINT) length measurements. *Nucl. Tracks* **8**, 371–5.

28. H. A. Khan & S. A. Durrani (1972) Prolonged etching factor in solid state track detection and its applications. *Radiat. Effects* **13**, 257–66.

29. S. R. Hashemi-Nezhad & S. A. Durrani (1981) Correction of thermally-lowered fission-track ages of minerals and glasses: The importance of the prolonged etching factor. *Nucl. Tracks* **5**, 101–11.

30. I. D. Hutcheon, P. P. Phakey & P. B. Price (1972) Studies bearing on the history of lunar breccias. In: *Proc. Lunar Sci. Conf. 3rd.* Pergamon, New York, pp. 2845–65.

31. H. A. Khan & I. Ahmad (1975) "Seasoning" of latent damage trails in lunar samples. *Nature* **254**, 126–7.

32. J. P. Bibring, J. Chaumont, G. Comstock, M. Maurette, R. Meunir & R. Hernandez (1973) Solar wind and lunar wind microscopic effect in the lunar regolith (Abstract). In: *Lunar Sci.* **IV**. Lunar Science Inst., Houston, pp. 72–4.

33. S. A. Durrani, H. A. Khan, S. R. Malik, A. Aframian, J. H. Fremlin & J. Tarney (1973) Charged-particle tracks in Apollo 16 lunar glasses and analogous materials. In: *Proc. Lunar Sci. Conf. 4th.* Pergamon, New York, pp. 2291–305.

34. R. L. Fleischer & H. R. Hart, Jr. (1973) Mechanical erasure of particle tracks, a tool for lunar microstratigraphic chronology. *J. Geophys. Res.* **78**, 4841–51.

CHAPTER 6

The Use of Dielectric Track Recorders in Particle Identification

In a large number of applications of nuclear detectors it is not sufficient merely to count the numbers, or to measure the energy, of particles impinging on the detector. It is necessary also, as far as possible, to determine the nature of these nuclear particles—usually characterized by their charge Z and mass M.

Generally speaking, a detector system will record signals which depend on particle properties such as the energy E, velocity v (or $\beta = v/c$, where c is the velocity of light), rate of energy loss dE/dx in the medium, or some combination of these properties. A successful system must collect a sufficient subset of these parameters to allow Z and M to be determined. For example, if we are able to measure the rate of energy loss, total energy, and velocity of a particle with sufficient accuracy then, since dE/dx in the given medium is a function only of the particle charge and velocity, and the energy is a function of its mass and velocity, the nature of the particle may be precisely determined.

In the case of an etched track, essentially two parameters are measured: the track etching rate V_T and the residual range R at which this etch-rate value holds.[1,2] Now the range of a particle is, for a given stopping medium, a function of the charge, mass, and velocity of that particle. As we saw in Chapter 3, variations in track etchability are not quite adequately described in terms of changes in the rate of energy loss dE/dx. A better fit to the data is obtained by using parameters such as primary ionization J or the restricted energy loss $(dE/dx)_{W < W_0}$. The values of both of these quantities in a given stopping medium, however, like dE/dx, depend on particle charge and velocity. Thus by measuring $R(Z, M, \beta)$ and $V_T(Z, \beta)$, we have only two equations for three unknowns, and at first sight a complete identification of the particle would seem to be impossible. However, the dependence of R upon M is weaker than its dependence on Z; and in addition we can usually place restrictions on the possible values of M for a given Z. For example, in the case of studies of cosmic ray particles in the region of Fe we have[3]

$$2Z - 3 \le M \le 2Z + 3 \tag{6.1a}$$

and this restriction, along with the two relationships

$$V_T = f(Z, \beta) \tag{6.1b}$$

114

and

$$R = g(Z, M, \beta) \tag{6.1c}$$

is sufficient to allow a particle to be identified.

In practice it is often possible to measure V_T and R at a number of points along the particle trajectory, and so a series of equations $V_T(R_i) = f(Z, \beta_i)$ and $R_i = g(Z, M, \beta_i)$ are obtained. Ideally, this series of equations would be solved to obtain Z and M. Unfortunately, the functions $f(Z, \beta)$ which are found to best describe the track etching process contain constants (such as K in Eq. (6.3), below, for primary ionization) that are not calculable directly, but rather serve as fitting parameters. It is therefore necessary to use $V_T(R)$ data for ions of known charge and mass in order to empirically establish the relationship

$$V_T = f(Z, \beta)$$

6.1   Calibration

The actual method of calibration and data reduction used varies considerably among different research groups. However, a typical scheme might be described as follows.

6.1.1   The L–R plot

The detector material is irradiated with beams of several types of ions of known charge and mass—preferably of charge, mass, and energy comparable to those which one may expect to encounter in the course of the actual experiment in which the detector concerned is to be used. Track etch rates are then measured as a function of residual range for the tracks produced by these ions.

A number of possible schemes exist to make such measurements. The most straightforward is illustrated in Fig. 6.1. A particle passes through a number of layers of plastic, each of which is etched under the same conditions. A series of cone lengths may be measured, and pairs L_i, R_i obtained (where L is the etched-cone length and R is the residual range). The time taken to etch out a cone of length L is given by

$$t = \int_{R_{0i}}^{R_{0i} + L_i} \frac{dR}{V_T(R)} \tag{6.2}$$

where R_{0i} is the distance of the low-energy end of the ith etch cone from the end of the particle trajectory (see Fig. 6.1), and L_i is the length of that etch

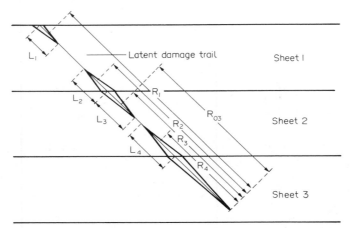

FIG. 6.1 *Track-etch rate* V_T *vs residual range* R. *The figure illustrates the relation between the etched-track length* L *and the residual range* R *for an ion which has passed through plastic sheets 1 and 2 and come to rest in sheet number 3. On etching all three sheets under the same conditions (including the etching time), one obtains a series of etch cones (at either surface when the particle has gone right through), which are measured to yield pairs* L_i, R_i. *Provided that the etch rate does not vary rapidly over the cone length, a mean track-etch rate* $\bar{V}_{Ti} = L_i/t$ *(where* t *is the common etching time for all cones) may be taken to be the value of* V_T *corresponding to a residual range* R_i *reckoned from the midpoint of the ith etch cone. (For clarity, the layer removed by etching from each surface is not shown).* R_{03} *is one example of the distances* R_{0i} *from the low-energy end of an etch cone to the stopping point of the particle. The data can then be plotted as in Fig. 6.2. By virtue of known range–energy relations one can thus, from the measured* V_T *versus* R *data, obtain* V_T *versus* J *(primary ionization function) curves–since each* R *value implies a unique* E, *and hence a* J, *value. See text for discussion.*

cone. The mean etch rate obtaining over the etched-out portion of track is of course $\bar{V}_T = L/t$. Provided that the etch rate does not vary rapidly over the cone length, an acceptable form of data reduction consists in plotting a mean track etch rate $\bar{V}_{Ti}$ at a residual range given by $R_i = R_{0i} + (L_i/2)$.

It must be borne in mind, however, that when V_T is varying rapidly over lengths $\simeq L$, then it will be necessary to fit a $V_T(R)$ function which will generate the correct L, t values via Eq. (6.2).

Using this method, a series of sets of data consisting of points on a V_T–R plot for each ion is constructed (Fig. 6.2). We can then construct a function $f(Z, \beta)$ to give the best fit to these data. For the purposes of this discussion

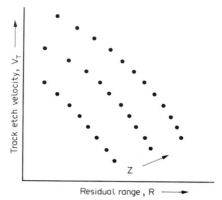

FIG. 6.2 *Schematic diagram showing the type of data obtained from measurements such as those shown in Fig. 6.1. The track etch velocity V_T falls with increasing residual range R (except at very small values of R, close to the end of the ion's path), as dE/dx falls with increasing R (cf. Fig. 6.1). For a given residual range, ions of higher charge-number Z, of course, produce a higher ionization damage density, and hence lead to a higher track etch velocity. The above are essentially L–R plots (cf. Fig. 6.5).*

it will be assumed that the data are fitted using the primary ionization function J given essentially by:

$$J = a \frac{Z_{\text{eff}}^2}{\beta^2} \left[\ln \left(\frac{\beta^2}{1 - \beta^2} \right) + K + \cdots \right] \qquad (6.3)$$

where Z_{eff}, the effective charge, is given[4] by:

$$Z_{\text{eff}} = Z[1 - \exp(-130\beta/Z^{2/3})] \qquad (6.4)$$

and where a and K are constants; β is, of course, the ratio of the particle velocity to that of light. The fitting procedure is similar to the above if, for example, the REL function (see §3.3(c)) is used instead of J.

By using a suitable range–energy relationship (see references 5, 6, for example), one can obtain the energy and thus the ion velocity corresponding to each data point on the curve V_T–R. Employing Eqs. (6.3) and (6.4), we could then calculate J exactly—except that the constants a and K are not known. The lack of knowledge of a is not important for the practical purpose of fitting the experimental data, since it acts only as a scaling factor. K, however, influences the form of the function $f(Z, \beta)$ and needs to be determined from the experimental data.

Having chosen a certain value of K, the function $J(K, Z, \beta)$ may be calculated for each value of R, and the data then plotted as V_T–J. For an appropriate value of K, V_T is a smoothly varying function of J which is single-valued over the whole range of J (i.e. a value of J should correspond to only one value of V_T regardless of the combination of Z and β which gives rise to it). This is illustrated in Fig. 6.3.

The actual form of the relationship $V_T = f(Z, \beta)$ varies considerably from one detector material to another, and even for a single detector material when different etching conditions are used. Generally, functions of the type

$$V_T \propto J^\alpha$$

where $\alpha \simeq 2$, are obtained. K shows an equally wide range of values.[7] For some detectors, $K \to \infty$ is used. Inspection of Eq. (6.3) shows that in the limit of very high K values the term in brackets becomes a constant, independent of β, and therefore we have

$$V_T = f\left(\frac{Z_{eff}^2}{\beta^2}\right)$$

Figure 6.4 shows typical V_T–J curves for some types of plastic detectors. These response curves can be used, along with the range–energy relationship, to generate V_T–R profiles for any ion, since each J corresponds to a certain β (and therefore to an E) and hence to an R value. Such V_T–R (or the equivalent L–R) profiles are shown in Fig. 6.5. The point of interest arising from Figs. 6.4a and 6.5a is that, since the response curve V_T–J for cellulose nitrate

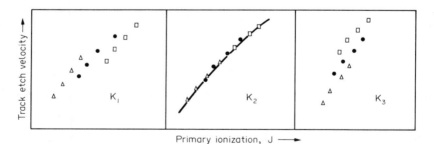

FIG. 6.3 *Idealized plots of the track etch velocity* V_T *versus the primary ionization* J, *where* J *has been calculated from measurements such as those shown in Figs. 6.1 and 6.2 (viz. from* V_T *versus* R *data; each* R *value giving a unique* E—*i.e. a unique* β—, *and hence a* J, *value), using three different values for the fitting constant* K *(Eq. 6.3). The symbols represent the experimental points (and the corresponding calculated* J *values) for three different ions used for calibration purposes. It is seen from these plots that when* J *is calculated using* $K = K_2$, *all of the data for the various ions fall on a single smooth curve.* K_2 *is therefore the appropriate value of* K *for this case (e.g. a given type or types of mineral crystal or plastic).*

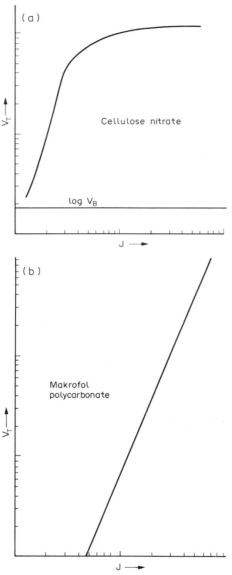

FIG. 6.4a,b   *The figures show the general form of the "response function"* $V_T(J)$ *for two types of plastic, viz. (a) cellulose nitrate, and (b) Makrofol polycarbonate. The function generally has the form* $V_T \propto J^{\alpha}$, *where* α *is a constant, though in the case of cellulose nitrate* V_T *assumes a plateau value at high values of J. These response curves can be used, along with the range–energy relations, to generate* V_T *versus R profiles for any ion (see Fig. 6.5). The figure shown here have been adapted from the work of Tripier and Debeauvais,[8] and although in that work* V_T *had been plotted against* dE/dx, *our substitution of J for the latter does not significantly affect the form of the curves. Note that the data have been plotted on the log–log scale.*

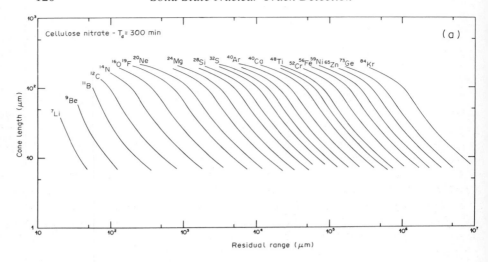

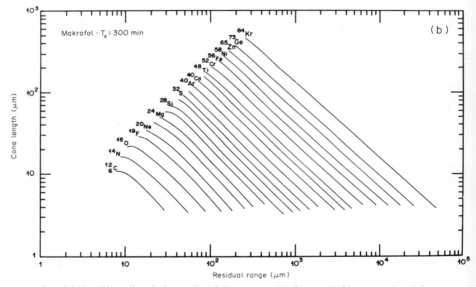

FIG. 6.5a,b   *Plots of etched cone length* L *versus residual range* R *for various ions (from* $^{7}_{3}$ Li *to* $^{84}_{36}$ Kr *in (*a*) cellulose nitrate, and (*b*) Makrofol polycarbonate. These plots have been constructed using the response curves of Fig. 6.4, each* J *in that figure corresponding to a particular value of* R *in the present figure; and similarly each* V_T *corresponds, for a given etching time* t*, to a particular value of* L *(where* $L = V_T \cdot t$*). In the figures,* T_d *stands for the etching time (300 min.). Note that the saturation effect in the case of cellulose nitrate (cf. Fig. 6.4(*a*)) leads to a crowding together of the* L–R *profiles for heavy ions at high etch rates* V_T *(and hence high* L *values). (From Tripier and Debeauvais.*[8] *)*

reaches a plateau at high J, this leads to a crowding together of $L-R$ profiles for this detector at high Z with low range values (which condition corresponds to high primary ionization). This will clearly result in a much reduced charge resolution in the plateau region.

Once the functional relationship between V_T and R is known, it is possible to compute cone length L as a function of R using Eq. (6.2). This is the form of the calibration which is of most direct relevance to experimental data, viz. an $L-R$ plot.

At this point we must consider the measurement of cone lengths L in greater detail. We have seen in Chapter 4 that the most easily measured quantity, viz. the projected length S, is related to the etched-out length L of the track by the formula:

$$S = L \cos \theta - \frac{h}{\tan \theta} + \frac{\tan \delta [L - (h/\sin \theta)]}{\sin \theta - \cos \theta \tan \delta} \tag{6.5}$$

where θ and δ are the dip and half-cone angles, respectively, and h is the thickness removed from the detector surface (see Eqs. (4.5) and (4.11) and Fig. 6.6). This equation for L is valid, provided that V_T is approximately constant along the etched-out portion of the track. In many cases, particularly for heavy ions in plastic (and crystal) track detectors, δ is very small (a few degrees of arc), and so is h; so that, to a fair degree of accuracy,

$$S \simeq L \cos \theta \tag{6.6}$$

For calibration ions, θ will generally be known; but in an actual experiment in cosmic ray or nuclear physics this may not be the case. Two possibilities then exist for measuring θ. If several sheets of plastic have been penetrated, then the dip angle may be readily obtained from measurements of the projected distance S_{12} between the intersections of the track with the top and bottom surfaces of the plastic sheet, and the sheet thickness d, where $\tan \theta = d/S_{12}$. If the track crosses only one surface, then it may be necessary to focus the microscope successively on the top and the bottom of the cone, taking readings from the calibrated fine-focus drum of the microscope on each occasion and making allowance for the refractive index of the plastic. A $\tan \theta$ value may thus be obtained in a way similar to that just described.

The values of L obtained by the above procedures are then used to construct an $L-R$ plot, which may then be compared with the calibration data.

Another approach to constructing an $L-R$ plot, devised by Price and co-workers[8a] for use with crystal detectors, has been alluded to in Chapter 5 (see caption to Fig. 5.2). This is the so-called "angled-polishing technique". Here, a crystal, irradiated at an angle (say 45°) with a collimated beam of accelerated heavy ions, is embedded in epoxy resin at a small angle with the surface. A single operation of grinding can then yield a succession of residual

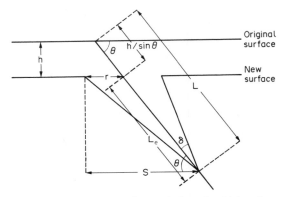

FIG. 6.6   *Some parameters of an etched track, which makes an angle θ with the detector surface (see also Fig. 4.4). Here L is the total length of the etched cone, S its projected length on the final (or new) surface of the detector, δ the semi-cone-angle, h the thickness of the removed layer, and r the distance of the etch pit opening from the crossing point of the latent track with the new surface (denoted by O' in Fig. 4.4, where r_1 stood for r). The length of the latent track removed during the etching is seen to be h/sin θ. It has been shown in Chapter 4 (see Eqs. (4.5) and (4.11)) that, in the "conical phase",*

$$S = L_e \cos \theta + r_1$$

$$= L_e \cos \theta + \frac{\tan \delta [V_T - (V_B/\sin \theta)]t}{\sin \theta - \cos \theta \tan \delta}$$

On substituting $V_T t = L$ and $V_B t = h$, and remembering that $L_e = L - h/\sin \theta$ (where L_e is the cone length below the new surface and L that below the old surface), one obtains the relation:

$$S = L \cos \theta - \frac{h}{\tan \theta} + \frac{\tan \delta [L - (h/\sin \theta)]}{\sin \theta - \cos \theta \tan \delta}$$

If δ and h are small (as is often the case for crystals and plastics), the above equation reduces to:

$$S \simeq L \cos \theta$$

ranges from 0 to R_0 in the detector, where R_0 is the total range of the heavy ion in the crystal (see Fig. 6.7a). When the polished crystal is subjected to etching for a certain length of time t, the following situation emerges. At positions on the crystal surface where the residual range is small, the etchant will reach the end of the latent track within the etching time t. As one proceeds from one end of the crystal to the other (i.e. from right to left in the figure shown), the length of the etched cone, L, will gradually increase, until it reaches a maximum value L_{max}, at which position the etchant is just able to reach the end of the latent damage trail in the time available. Beyond that

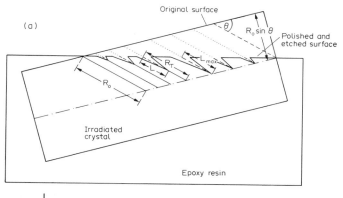

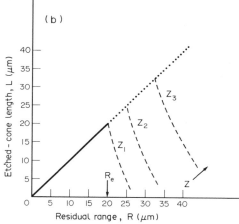

FIG. 6.7a,b   *The L–R_T plot by the "angled-polishing technique".
The method consists in mounting the irradiated crystal at a shallow
angle in epoxy resin and then carefully grinding it to remove a
layer of depth up to $\sim R_0 \sin \theta$, where R_0 is the total range of
the heavy ion in the medium and θ is the angle of irradiation (see
figure a). The crystal is then polished and etched for time t. This
results in a series of etched cones of varying length L, depending
on the total residual range R_T at a given point on the polished
surface. At low residual ranges the etchant is able, in times $<$t,
to reach the end of the etchable track, so that $L = R_T$. Eventually
(proceeding leftwards in the figure), a residual range is reached
at which the etchant is just able to reach the end of the track
in the total available time, t; this yields a maximum etched cone
length L_{max}. At greater residual ranges (further to the left in the
figure), progressively shorter etched cones are obtained as the etch
rate at, and just below, the polished surface decreases (because of
lower dE/dx values). The cone length L is now equal to the local
etch rate multiplied by the etching time; the measured cone length
thus yields a value for the track etch rate $\bar{V}_T$ at an effective residual
range $R = R_T - L/2$. Measurements of L versus R_T yield plots
such as those shown schematically in figure b. The effect of increasing ionization rate with increasing ionic charge Z is depicted
at the far end of the (45°) L–R_T curve. Tracks of more highly
ionizing particles possess greater etchability, and the position of
the maximum etched cone length is shifted to higher residual
ranges. This allows particles of different species to be distinguished
from each other. (Figure after Price et al.[8a])*

position (i.e. further to the left in the figure), the ion path crossing the polished surface represents a region of lower etch rate (corresponding to a higher energy of the heavy ion and hence lower dE/dx); the etchant can, therefore, no longer reach the end of the latent track in time t. The etched track length is now equal to $\bar{V}_T t$, where $\bar{V}_T$ is the average track-etch velocity in the region. The etched track length will therefore decrease as the residual range increases (and the average etch rate $\bar{V}_T$ falls).

The etched cone length L therefore yields the local track etch rate, and this etch rate applies to an effective residual range $R = R_T - L/2$. The position of the maximum etched-cone length L_{max} shifts to longer residual ranges (and attains a larger value) as the ion charge Z—and along with it the etch rate V_T—increases (see Fig. 6.7b). This method can be used to determine the values of etch rate as a function of residual range in various crystals. It may be noted that variations in mineral composition,[8a] crystallographic effects,[8b] and statistical fluctuations in both the ionization and the etching processes[8a] introduce considerable dispersion in the L–R plots, particularly at high residual ranges (cf. Fig. 5.2).

The above procedures and considerations can be used in establishing criteria for distinguishing between particles of different species on the basis of carefully constructed calibration curves.[2,8a,c] It must, however, be remembered that such correlations are better for some minerals than others; e.g. according to Price *et al.*,[8a] the correlation between the etch rate and ionization rate is smoothest for feldspars and clinopyroxenes, and roughest for olivine.

6.2 Charge Assignment

When the detector system has been calibrated, L–R plots are constructed for the unknown ions. These experimental data must then be compared with the calibration L–R curves in order to assign a charge Z to the track-forming particle. There are various methods of doing this.

In one scheme,[9] a computer program has been set up to compare the set of L–R data points for each track with the curves obtained from the calibration. For each calibration L–R curve, the function χ^2 given by

$$\chi^2 = \sum_{i=1}^{n} \frac{(L_{im} - L_{ic})^2}{\Delta L_{im}} \tag{6.7}$$

is evaluated, and a Z assignment is chosen such that χ^2 is minimized. Here L_{im} is the measured length of the ith cone; L_{ic} is the cone length obtained from the calibration curve at the same value of R; and ΔL_{im} is the error in the cone length measurement.

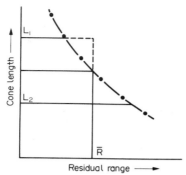

FIG. 6.8   *One scheme for reducing the L–R data to a single parameter* $\bar{R}$. *Here the black dots represent the experimental points for the etched-cone length* L *versus the residual range* R *for an unknown ion. These L–R data are fitted with a curve, which is integrated between fixed limits of* L_1 *and* L_2, *as shown here), yielding a value for the mean range* $\bar{R}$, *which corresponds to a mean cone length. This mean range* $\bar{R}$, *representing the L–R data set between the given limits, is then compared with values generated from the calibration results to determine the charge of the unknown ion. (After Siegmon et al.[11])*

In the method due to Enge and co-workers,[10,11] the $L–R$ data for each unknown ion are fitted with a curve which is then integrated between fixed limits of L so as to replace each set of data by a single number, viz. the mean range $\bar{R}$ (see Fig. 6.8). These mean ranges are then compared with values generated from the calibration results.

6.3   Low-energy Particles

In some instances (most frequently, close to the end of a particle trajectory: when its rate of ionization increases towards the Bragg-peak region), the etch rate will vary rapidly over short sections of the track. This introduces a number of practical and more fundamental difficulties into the procedures described above.

On the practical side, particles of low energy may not traverse more than one sheet of plastic. Commercially available plastics have thicknesses typically ~ 20–$100\ \mu m$; and, for example, Fe ions of 1 MeV/nucleon will be stopped in $\sim 15\ \mu m$ of plastic. In order to obtain information on the variation of V_T along such a track, it may be necessary to measure cone lengths as a function of etch time. If only very short tracks (less than $\sim 10\ \mu m$ long) are available, or if low V_T ($\lesssim 2 - 3 \times V_B$) values are present, then more accurate information on V_T may be obtained by utilizing track-diameter, or major and minor axis, measurements.

More fundamentally, the constant-V_T cone-length equations used in data reduction in §6.1.1 are no longer valid. Also, the method of plotting a mean $\bar{V}_T = L/t$ at a mean residual range ceases to be accurate when V_T is varying rapidly over the etched track length. Clearly, a more complex procedure is necessary. A more suitable scheme may be as follows. Plots of diameter (or major axis) D or projected length S as a function of etch time t (or of surface-removal thickness h) are utilized to estimate V_T as a function of R using the constant-V_T track geometry formulae. This $V_T(R)$ is then used as a first guess in calculating expected cone lengths or track diameters, utilizing the varying-V_T formulae[12] (see §4.2.4). The $V_T(R)$ relationship is then adjusted until a fit to experimental $S(h)$ or $D(h)$ data is obtained.

An alternative method is to determine $V_T(R)$ from direct measurements on the etched-track profile,[2,7] as discussed in Chapter 4. The examination of track replicas with a scanning electron microscope (and the track profile technique (TPT)[40b]) are particularly useful methods for such measurements.

Other modifications to the basic $L-R$ method are discussed below in sections describing various applications of the particle identification properties of track detectors.

6.4 Charge and Mass Resolution

A number of factors limit the charge and mass resolution attainable with track detectors.

Since the detection of particles essentially comes down to the measurement of etch cones, it is the errors in these latter measurements that limit the attainable accuracy. Cone-length measurements will have errors associated with them which will depend to some extent on the individual scanner, and on the quality of the optics employed.

As mentioned in §4.3, the limit of microscope resolution is $\sim \lambda/2n \sin \alpha$, where λ is the wavelength of the light used, n is the refractive index of the medium between the sample and the objective, and α is half of the angle subtended by the objective lens at the object. The term $n \sin \alpha$ is called the numerical aperture of the lens. For white light, the effective wavelength is about 560 nm. The largest achievable numerical aperture is about 1.6, and so the smallest resolvable separation of objects is around 0.18 μm. Under ideal conditions this represents the minimum uncertainty in track length measurements; and for cones of $\sim 30-40$ μm this constitutes a fractional error of 0.005. The subject of resolving power of microscopes is discussed in standard textbooks on optics (see, for example, reference 13). Measurement errors will also affect the determination of the residual range R. Overall errors are reduced by choosing such etch times as will maximize the cone lengths, and also by measuring many cones along the particle trajectory.

Other experimental factors which can impair charge and mass resolution include the presence of non-uniformities in the composition of the detector material and also in the etching process. The former factor can only be minimized by careful selection and testing of the detector material. Some cellulose nitrates show a tendency to possess such non-uniformities. The newly discovered track-etch polymer, CR-39, is, because of its great sensitivity, potentially a most valuable detector in particle identification. Early studies[14] indicated that it was extremely uniform and isotropic, although other workers have since found evidence for troublesome non-uniformities in response.[15,16] Work is now being carried out to improve the curing process for this plastic in order to resolve these difficulties.[16a] Non-uniformities in the etching process can arise if temperature or concentration gradients in the etchant solution are allowed to develop. These effects are usually overcome by mechanical stirring, or even by ultrasonic agitation of the solution during etching.[10]

A more fundamental limitation on the performance of the detector is set by the variations in the degree of solid-state damage (and hence in the track etch rate) caused by statistical variations in the ionization rate of the track-forming particle. Ahlen has suggested[17] that, since plastic track detectors seem to respond essentially to "distant" collisions (leading to low δ-ray energies) between the track-forming ion and the target electrons—i.e. collisions some nanometres away—hence they should have an intrinsic resolution which is superior to that of other particle detectors (in the absence of effects such as non-uniformity of the plastic, discussed above). Henke and Benton[18] have discussed the optimization of track etching parameters for the resolution of isotopes when using plastic detectors, and they conclude that plastics are potentially superior to electronic detectors in the charge region $Z > 20$.

A number of values for both the charge and the mass resolution of plastic track detectors have been reported. Price et al.[19] obtained a charge resolution ΔZ of ~ 0.3 charge units, and a mass resolution of ~ 2 amu in Lexan for cosmic rays in the region of $12 \leq Z \leq 30$. Siegmon et al.[10] reported a mass resolution of ~ 0.7 amu for Fe isotopes in the cosmic rays using u.v.-sensitized Lexan. Cartwright et al.[14] have suggested that charge and mass resolutions of < 0.1 charge units and ~ 0.3 amu are possible in CR-39. If such results for CR-39 are confirmed for large-area stacks, then this detector will prove to be of the utmost value in cosmic ray studies.

6.5   Some Applications of Particle Identification Techniques

We now discuss some of the more important areas in which the particle identification properties of track detectors have been utilized.

6.5.1   Cosmic ray physics

One of the earliest fields of application for plastic detectors was the determination of the charge spectrum of the heavy nuclear content of the primary cosmic rays.[1,20-22]

Passive detector assemblies are particularly useful in cosmic ray work, because the necessity to lift the detector above the bulk of the atmosphere which so effectively shields the earth from these particles places severe constraints on the complexity of the detector systems that can be used. Some of the earliest studies of heavy primary cosmic rays employed stacks of nuclear emulsions (see Fowler *et al.*[23,24]). These have now been largely superseded by plastic track detectors; and although ever-increasing ingenuity in the construction of electronic detectors has led to their increased use in cosmic ray work, plastic track detectors still have a rôle to play. This has been particularly demonstrated in the study of ultra-heavy cosmic rays, where very large detector areas must be used. Integrating systems, such as plastic track detectors or nuclear emulsions, have obvious advantages over real-time electronic devices in low-flux applications exemplified by cosmic ray work. In addition to exposures during balloon flights, plastic detectors have been flown aboard the Skylab;[25] and long-term exposures have been made using the Long Duration Exposure Facility (LDEF) launched by the NASA Space Shuttle[26,27] in the spring of 1984.

The charge and mass spectra of the cosmic rays reflect both the nuclear processes which produced the cosmic rays and the interactions which they have undergone prior to interception by the detector. The study of these spectra yields information on the nucleosynthesis of the elements and on the path length of the cosmic rays in the interstellar medium. It is beyond the scope of this book to review cosmic ray results obtained by employing the track technique; however the use of plastic track detectors in cosmic ray physics presents some specific problems which are of interest.

While calibration of the plastics, using accelerator irradiations in the usual way (cf. §§6.1 and 6.2 above), is often employed, frequently a different approach is adopted. In any region of the cosmic ray spectrum, there will be one or two charges which are known to predominate. For example, beyond $Z \gtrsim 10$, iron ($Z = 26$) is by far the most abundant species. Any data obtained from an experiment designed to examine this region would be dominated by Fe tracks. Assuming that the constant K in Eq. (6.3) is known, the function $V_T(J)$ is chosen such that the $L-R$ profile calculated for Fe passes through the region of the experimental $L-R$ plot which is most densely populated with data points. This "internal" calibration is valuable for the reasons outlined below.

Plastic track detectors are prone to changes in response as a result of various environmental factors (see for example DeSorbo,[28] and O'Sullivan and Thompson,[29] and also the discussions in Chapters 3 and 4). Thus, O'Sullivan and Thompson[29] found that the response of Lexan and CR-39 was affected by the temperature during irradiation* (which would be the case with a balloon flight). Therefore the etch rates of tracks of a given ion obtained in the laboratory may be different from those for the same ion incident on a detector exposed in space. The use of the above-described "internal" calibration technique avoids the systematic errors which would result from a laboratory calibration procedure. It should be noted, in passing, that changes in detector response brought about by fluctuations in temperature during irradiation represent a further potential source of impairment of charge and mass resolution in cosmic ray experiments.**

The scanning of large areas (many square metres) of plastics for cosmic ray studies presents a considerable practical problem. If ~ 10 m^2 of plastic is to be analysed for, say, tracks from ultra-heavy nuclei, then it would clearly be very time-consuming if all the tracks present in a plastic stack (possibly many millions of events) were to be measured. The procedure usually adopted[30] is to take one or more sheets of plastic from the stack (consisting, possibly, of several hundred sheets in all), and to etch them for such a time that etch cones formed by the most heavily ionizing energetic particles (which offer an etch channel at each side of the sheet) join together within the body of the sheets to form holes passing right through it. Once such holes have been obtained, they may be easily located without the necessity of scanning the sheet under a high-power microscope. In one method[31] every fourth sheet of heavily etched plastic (e.g. etched for ~ 160 h) is placed on a sheet of sensitized paper and ammonia vapour is passed over the plastic. The vapour passing down the etched hole comes into contact with the paper, and produces a large, bright blue spot which can be observed at very low magnifications.

These spots are then used to locate the positions of ultra-heavy particle tracks in the stack, and the etch rate data for the identification of these tracks may be collected by etching the remaining plastics in the stack in the usual manner.

---

* This effect has some similarity to the dependence of thermoluminescence (TL) response on the temperature of irradiation,[29a] which has a bearing on the TL produced in, say, meteorites while they were at low temperatures (~ 100–200 K) in space. Further work on the track registration efficiency of meteoritic crystals at low temperatures (to simulate their in-orbit state) should also be of interest, and is in progress in the authors' laboratory.

** Similarly, the chemical reactivity of a latent track in Lexan, stored in air, increases roughly logarithmically with the time delay between irradiation and etching.[27a,b] This would cause "old" tracks to appear to have been produced by more highly ionizing particles than "young" tracks. There is some evidence (K. G. Harrison and A. R. Weeks, private communication, 1986) that, in the case of CR of CR-39 stored in air, the track registration sensitivity falls by a few % per month during the time elapsed prior to its exposure to radiation (i.e. it exhibits "aging"); however, it does not seem to show "fading" of latent tracks in the time period between irradiation and etching.

The production of a through-hole will, of course, depend on the velocity and angle of entry of the particle as well as its charge. The efficiency with which events are detected in this manner will depend on the charge Z of the incident particle: the efficiency rising from zero at some Z_{min}, such that a particle with this charge will produce a through-hole, under the etching conditions used, only when it is near to its maximum ionization rate and enters at right angles to the surface of the plastic. This efficiency will rise rapidly as a function of increasing Z, and must be calculated accurately if a reliable charge spectrum is to be derived from the track data.

A more fundamental problem, often encountered in cosmic ray applications, centres on the determination of particle velocity given by β ($= v/c$). When a particle comes to rest in a stack of plastics, one can measure the total range, and hence obtain β, in effect, from range–energy relationships. At high energies, however, the particle may pass right through the detector assembly. For example, a heavy particle with an energy of 500 MeV/nucleon (quite low by cosmic ray standards) will have a range equivalent to a thickness of several g cm^{-2} of plastic. At first sight it might appear that this problem could be easily overcome by making ever thicker plastic-stack assemblies. However, the interaction lengths for heavy nuclei are also in the g cm^{-2} range; and so adding on further layers of plastic is of limited use, since the particle will interact before coming to rest, and generate secondary products.

This problem can be overcome by using a hybrid detector system incorporating, for example, a Cherenkov scintillator in order to obtain velocity information.[31,32] The Cherenkov light is recorded by a fast photographic film. Another possibility is to utilize the velocity selection afforded by the earth's magnetic field.[7]

It is possible, however, in certain circumstances, to identify the particles solely from the track data. As the ionizing particle slows down in passing through the plastic stack, the etch rate increases; and the rate of increase depends on the initial particle-velocity. The knowledge of fractional increase of etch rate over a given thickness of plastic traversed, along with the values of the etch rates themselves, is sufficient to allow particle identification.

Fowler[33] has defined an etch rate gradient G given by

$$G = -\frac{1}{V_T}\frac{dV_T}{dx} \tag{6.8}$$

The usefulness of this parameter can be seen from the following considerations.

For many plastic detectors (cf. §6.1):

$$V_T = c_1 J^{n/2}$$

and

$$J \simeq c_2 (Z_{eff}/\beta)^2$$

therefore

$$V_T = a(Z_{\text{eff}}/\beta)^n$$

where c_1, c_2, n and a are constants.

Upon differentiation, we see that

$$G = \frac{n}{\beta} \frac{d\beta}{dx} \tag{6.9}$$

By differentiating the non-relativistic expression for the kinetic energy of a particle of mass M, viz. $E = \frac{1}{2}Mv^2 = \frac{1}{2}M\beta^2 c^2$, we have

$$\frac{dE}{dx} = Mc^2\beta \frac{d\beta}{dx}$$

Hence, upon substitution into Eq. (6.9), we obtain

$$G = \frac{n}{\beta^2} \frac{1}{Mc^2} \frac{dE}{dx}$$

or, since $dE/dx \simeq kZ_{\text{eff}}^2/\beta^2$:

$$G = \frac{n}{\beta^4} \frac{kZ_{\text{eff}}^2}{Mc^2} \tag{6.10}$$

where k is a constant.

Therefore the gradient G also depends, like the etch rate, on Z_{eff} and β, but (for $n \sim 4$) has a weaker Z_{eff}-dependence. If two etch rates, V_{T1} and V_{T2}, are measured at two points separated by Δx, then G is given by

$$G = \frac{2(V_{T2} - V_{T1})}{(V_{T1} + V_{T2})\Delta x}$$

This gradient is associated with an effective etch rate $V_{T\,\text{eff}}$ given by[34]

$$V_{T\,\text{eff}} = \frac{V_{T1} \times V_{T2}}{(V_{T1} + V_{T2})/2}$$

Figure 6.9 shows curves of $V_{T\,\text{eff}}$ plotted against G. It is clear that calculated values for different ions plot along quite separate curves; and thus this method allows particles of different charge to be identified. A more detailed discussion of this approach is given in reference 34.

An extensive description of the application of track detectors to cosmic ray physics is given in the book by Fleischer, Price, and Walker.[7]

6.5.2 *Nuclear physics*

Once again, we will not attempt to present a review of all the areas of nuclear physics in which the track technique has been applied; but will, rather,

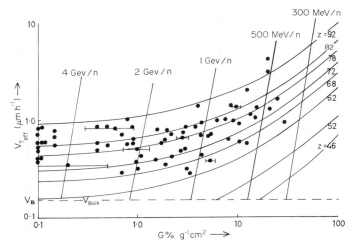

FIG. 6.9   *Effective track-etch rate*

$$V_{T_{eff}} = \frac{V_{T1} \cdot V_{T2}}{(V_{T1} + V_{T2})/2}$$

plotted against the percentage gradient

$$G = \frac{V_{T2} - V_{T1}}{\frac{1}{2}(V_{T1} + V_{T2}) \cdot \Delta x} \times 100\%$$

for several ions in Lexan polycarbonate. Here V_{T1} and V_{T2} are the measured etch rates at two points separated by Δx. See text (and reference 34) for discussion. Solid curves represent the theoretical curves for different ions, of charge Z, so that by plotting the experimental points (solid dots) it may be possible to distinguish between ions of different charge. The slanting lines indicate the value of the energy per nucleon of the ions which corresponds to the values of $V_{T_{eff}}$ and G at the point of intersection of such a line with any given ionic curve. (After Fowler.[33])

concentrate on a few specific problems which arise in the application of track detectors in nuclear physics, and will describe some of the special techniques which have been adopted to solve them.

Particle energies encountered in laboratory nuclear physics experiments are usually much lower than those typical of cosmic rays. Frequently, therefore, the most straightforward application of the *L–R* method, viz. that of measuring cone lengths at the surface of several successive plastic foils, is not possible. Tracks are often contained in only one foil; so that the methods described in §6.3 and in Chapter 4 must be used. Most of the techniques for low-energy particle analysis involve many measurements on the same track at different etch times. Quite apart from the problem of relocating tracks, this approach can also be slow and tedious for routine application.

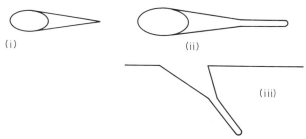

FIG. 6.10 *The development of heavy-ion tracks in plastic using the multistage procedure of Stern and Price.*[35] *Following a normal etch (i), the plastic is irradiated with u.v. and re-etched. The ultraviolet treatment greatly enhances the track etch velocity* V_T *relative to the bulk-etch velocity* V_B. *The cone angle (twice arcsin* V_B/V_T) *for the post-u.v. etching step is, therefore, much smaller (see ii, and its cross-sectional view iii); and so it is easy to distinguish between the two stages of etching. Hence, measurements on the initial etch cone and the residual range can be made simultaneously on the fully revealed track, leading to the identification of the particle responsible for the track, based on calibration data. The method is less accurate than those involving many steps of etch cone development.*

One neat method of avoiding having to make repeated measurements involves the use of ultraviolet light.[35] As was mentioned in §4.4, ultraviolet light has the property of increasing track etching rates in some polycarbonate plastics. The basic method consists of three stages. First, the plastic is etched, using standard conditions, such that cones are developed which do not reach the end of the damage trail. The plastic is then irradiated with u.v., thus enhancing the etch rates of the unetched portions of the tracks. Finally, these residual portions of the tracks are also etched out (see Fig. 6.10). The initial cone length (say, L_1) yields the etch rate; and this initial cone is easily distinguishable on the final etched track—the total length of which is used to determine the residual range at which this initial etch rate applies. Thus both the etch rate and the residual-range measurements are made simultaneously on the same track.

Of course, since only one etch rate value is obtained, the charge resolution achieved will be considerably worse than with methods involving many repeated measurements on the same track or the measurement of several etch cones at different plastic interfaces.

In nuclear physics, the plastic foil is most often used as a detector of products of reactions taking place in a target external to the plastic itself (see Fig. 6.11a). However, in some cases (Fig. 6.11b) the plastic fulfils both rôles, i.e. that of target as well as of detector.

One disadvantage of these detectors is that tracks may generally be

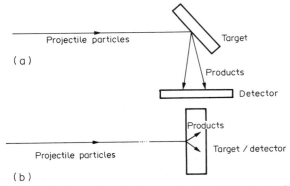

(a)

(b)

FIG. 6.11a,b *In nuclear physics experiments the plastic track detector may be used in two modes: (a) to detect reaction products emitted from an external target, or (b) as a combined target/ detector to detect "intrinsic" reaction products generated by an incident particle (say, a neutron) interacting with the constituents of the plastic itself.*

revealed only when they cross the top or bottom (i.e. the front or the rear) surface. This problem has been overcome to some extent by Benton[36] who used foils perforated by pre-etched heavy-ion tracks. These allow the etchant to penetrate below the surface and to etch out tracks lying in the body of the detector. The method is analogous to the Track-IN-Track (TINT) technique

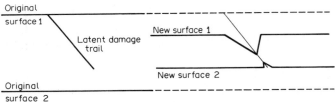

FIG. 6.12 *A technique for studying tracks due to relatively low-energy charged particles. The plastic foil is etched from both sides. In the case shown, the damage trail starts at the top surface (1), but does not reach the bottom surface (2). The track parameters (length and surface opening) are related to the track etch velocity V_T in the usual way (see the etch cone with its opening at the new surface 1). The bottom etch cone will not start to emerge until a sufficiently thick layer has been removed for the new surface 2 to cross the lower end of the latent damage trail. Hence, for a given dip angle, the size of the surface opening (and the length) of the bottom etch-cone is a sensitive function of the range (and hence the energy) of the incident particle. This technique is useful for energy spectrometry, and for distinguishing between different isotopes of an element, which have the same energy but different ranges (see, e.g., reference 39).*

devised by Lal *et al.*[37] to examine ancient cosmic ray tracks in meteoritic minerals (see the following subsection).

Another technique for the examination of tracks due to rather low energy particles is illustrated in Fig. 6.12. Consider tracks that just fail to get through the detector. They cross the first (top) surface of the detector but not the second (bottom) surface. The cone length at surface 1 is determined by the track etch velocity in the usual way. At surface 2, no track will appear until the "new" surface 2 starts to cross the end of the damage trail. The length or diameter of this track will be strongly dependent on the amount of time for which etching has been effective in enlarging the track; and hence, for a given dip angle, the track size is a sensitive function of the range of the incident particle. Thus, by measuring, say, track diameters on each surface, it is possible to obtain $V_T(R)$ data for individual tracks from a single set of measurements. Lück[38] and Balcázar-García and Durrani[39] used bottom-surface diameter measurements alone as a means of energy spectrometry for, respectively, α-particles, and ^4He and ^3He particles. The latter authors covered the irradiated side of the plastic with epoxy resin to ensure that only the reverse side was etched. Another variant of this method has been proposed by Somogyi.[40]

Yet another method of analysis for low-energy particles is the track-profile method, which has already been described in §§6.3 and 4.2.

Al-Najjar and Durrani[40a] have used the "track profile technique" (TPT) in CR-39 to perform range and energy measurements on α-particles and fission fragments. Here $\sim 600–800$ μm thick CR-39 sheets were bombarded edgewise with (i) accelerated ^4He particles of energies from ~ 15 to ~ 40 MeV, (ii) α-particles from an ^{241}Am source (and degraded to various energies by air), and (iii) fission fragments and α-particles from a ^{252}Cf source. The etching of the tracks can be facilitated in this technique by making a surface groove on the plastic sheet close to the edge, which acts like a cleavage in the TINCLE method of revelation (see §6.6). The etchant can thus proceed in both directions ("forward" and "backward") along the particle track (in addition to the etchant entering edgewise). The profiles of the particle tracks can be clearly seen by this method, and useful range-energy relations and $V_T(R)$ values established for a given plastic. The same technique was used by these authors[40b] on heavy-ion tracks (e.g. ^{56}Fe); and, by measuring various parameters connected with the track profile geometry (e.g. the rate of change of its curvature), they were able to calculate instantaneous values of $V(= V_T/V_B)$ as a function of the residual range of the heavy ion. In this method the track profile retains useful information about the variable track etch rate even after the track has been fully etched. The approach is especially useful for low-energy particles.

Hashemi-Nezhad *et al.*[40c] have constructed response curves of track etch rate V_T versus primary ionization in biotite mica, using heavy ions ($10 \leq$

$Z \leq 54$, of energy $E \leq 9.6$ MeV/nucleon) from the LINAC and the UNILAC heavy-ion accelerators at the University of Manchester, England, and at GSI, Darmstadt, Fed. Rep. Germany, respectively. These curves are useful in particle identification. They report, however, that the response curve of biotite reaches a plateau at the primary ionization (J) values of about 30. From this they conclude, *inter alia*, that biotite mica, which is one of the media often used[40d] for the search of vestiges of superheavy elements (SHE), is unlikely to prove helpful in this regard so far as the track etching approach is concerned. The major mode of decay of the SHE is believed to be through spontaneous fission. From their studies,[40c] these authors surmise that the detection of superheavy elements through charge assignment to the fission fragment tracks would not be possible, since such tracks would have etch rates in the saturation region of the V_T–J curve.

The above is by no means an exhaustive account of the various track etching techniques that can be applied to particular problems in particle identification; but it is hoped that the above discussions will have given some indication of the versatility of the method.

In many nuclear physics experiments it is, of course, not necessary to identify the particles incident on a detector, but merely to count them. These applications are considered further in Chapter 7.

6.6   The Ancient Cosmic Rays

We conclude this chapter with a brief account of some attempts that have been made to apply particle identification techniques to the tracks stored in meteoritic and lunar crystals as a result of their exposure to the ancient cosmic rays. Such studies could provide information on the composition of cosmic rays in the distant past.

Mineral crystals are extremely insensitive particle detectors; thus, typical meteoritic minerals such as olivine will not record ions with a lower ionization rate than that of a stopping Ca $(Z = 20)$ nucleus. Even Fe nuclei (the most abundant cosmic ray species in the charge region above $Z = 20$) will produce tracks in silicate crystals only over the last ~ 10–20 μm of their trajectory (although these will be somewhat longer in plagioclase feldspar—the most sensitive of the common meteoritic or lunar minerals). Thus only short segments of etchable track are available for investigation, so that many of the methods described in the earlier part of this chapter are not applicable.

The earliest attempts at particle identification in these mineral crystals did not rely on track etch rate measurements at all, but rather on the concept of the "total etchable range".[41,42,37]

This concept arose from the idea that any given detector possessed an ionization threshold J_c, below which no track could be formed. On this model a particle such as that giving rise to curve 1 in Fig. 6.13 never forms

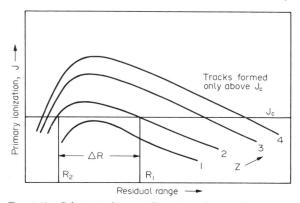

FIG. 6.13 *Schematic diagram illustrating the so-called total or maximum etchable range ΔR (sometimes termed R_{max}, cf. ref. 8a) for ions of different charge Z. It is assumed (though it is not strictly true: for etching conditions, etc., may affect the criterion) that tracks become etchable when the primary ionization J exceeds a critical value J_c for a given medium. On this criterion, ion 1 is never able to form an etchable track in the medium in question: for even at Bragg peak, $J < J_c$. Ion 2, however, makes an etchable track over the range interval $\Delta R = R_1 - R_2$, where the latter two values of the residual range correspond, respectively, to the high-energy and the low-energy end of the particle path where J falls below the critical value J_c. The total etchable range ΔR is an increasing function of the ion charge Z. These considerations have a bearing on the next figure (Fig. 6.14).*

a track, since even at the ionization peak it has $J < J_c$. The particle corresponding to curve 2 behaves differently. Although it will not form a track at high residual ranges, yet, as it slows down towards its peak ionization region, the rate of ionization J will first exceed J_c at a residual range R_1, dropping below it again at a residual range R_2. The difference $R = R_1 - R_2$ is known as the "total etchable range", and a cursory examination of Fig. 6.13 shows that it increases with increasing Z. If a sample containing tracks evenly distributed throughout its volume is ground and polished to reveal an internal surface, then, as may be easily proved, this surface crosses any point on the particle trajectories with equal probability (Fig. 6.14a). If this sample is now etched until all tracks reach the point where $J = J_c$ (i.e. when no further preferential etching is possible), then the distribution of track lengths L should show a step-function form, provided that only one charge is present (cf. Fig. 6.14b). This distribution will have an upper limit $L = \Delta R$, where ΔR is the total etchable range for the charge state forming the tracks.

If more than one species is present, each will exhibit its own rectangular distribution terminating at ΔR_i (corresponding to the species i); and the observed spectrum will consist of a superposition of all these distributions (Fig. 6.14c). Provided that a calibration has been carried out such that the

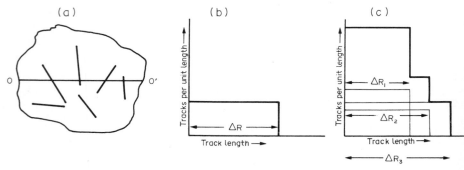

Fig. 6.14a,b,c *If latent tracks are oriented randomly in a sample, e.g. a track-storing mineral, then any section taken through it will intersect tracks along any part of their length with equal probability (see (a)). Suppose first that only one type of ion were present (and that any crystallographic anisotropy of etching could be ignored). Then if the etched track lengths were measured and the number of tracks plotted per unit length interval, the length distribution would show a square function terminating at ΔR (see (b)), where ΔR is the total etchable range of Fig. 6.13. If, however, several species of ions were present, then (see case (c)) the etched track length distribution would be a superposition of all the distributions (i.e. of square functions of lengths ΔR_1, ΔR_2, ΔR_3 ...) due to the various individual ions. Note that the etching conditions, as well as crystallographic anisotropy (see Chapter 4, Figs. 4.14, 4.16), can affect the distributions shown above.*

function $\Delta R(Z)$ is known, the charge spectrum may be unfolded from such a distribution. This method was used by Price et al.[43].

The need to unfold the charge spectrum from the length distribution, however, reduces the accuracy of this method. Lal et al.[37] devised a number of elegant schemes for overcoming this difficulty (see Fig. 6.15). Essentially, these consist in allowing the etchant to penetrate the detector via long surface-tracks due to heavy ions (the Tracks-IN-Track or "TINT" method) or via cleavages (the Tracks-IN-CLEavage or "TINCLE" method) so that tracks wholly contained within the body of the detector but crossing a surface-track or a cleavage, may be developed. In these "confined length" measurement methods, the whole of the total etchable range ΔR for these tracks may thus be measured. This technique, along with a suitable calibration, has been used in attempts to study the abundances of VH group (viz., charges in the range $20 \leq Z \lesssim 28$) in the ancient cosmic rays.[44-46] Perron[47] has recently described a calibration technique, using ~ 190 MeV/nucleon ^{238}U ions from the Bevalac accelerator of the Lawrence Berkeley Laboratory of the University of California, in which he used large olivine crystals (several millimetres in size), which were sawn at ~ 2 mm from the "front" surface, polished, and reassembled prior to the irradiation. After irradiation the two parts of the crystal were again separated so that the etching could be carried out more conveniently from the boundary, thus avoiding the blocking of the etchant along a very long sparsely ionized track. He reports a total etchable range of relativistic ^{238}U ions which is very much longer than those predicted by

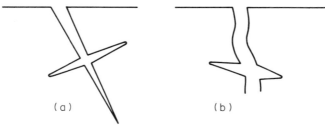

(a)   (b)

FIG. 6.15   TINT *and* TINCLE *methods of track revelation. Tracks contained within the body of a crystal may be developed by the etchant reaching them either (*a*) by passing along tracks crossing the surface (whether by chance or by design, e.g. when etchant channels are deliberately provided by irradiating the crystal surface with highly-ionizing particles, creating "host tracks" penetrating the crystal interior), or (*b*) by travelling along natural cracks or cleavages in the crystal surface (or, sometimes, by deliberately making a cut below the top surface, which is a variant of the track profile technique.[40a,b;46a]) In both cases (*a*) and (*b*), one is able to make sure of revealing the full length of all tracks crossing the host tracks or the cleavage and contained entirely within the body of the crystal, unlike the case of Fig. 6.14, where the revealed length varies with equal probability from 0 to full etchable range. Figs. 6.15 (*a*, *b*), represent, respectively, the Tracks-IN-Track (*TINT*) and Tracks-IN CLEavage (*TINCLE*) methods of revelation of Lal et al.[27] The fully etched track in each case encompasses both segments extending on either side of the etch channel. These are also referred to as "confined length measurements".*

models based on the criterion of primary-ionization threshold for track registration (i.e. ~ 800 μm in meteoritic minerals; reference 35). Perron's measurements by the above method[47] give a total etchable range of ~ 190 MeV/nucleon ^{238}U ions in olivine of at least ~ 1350 μm (or possibly over 2,000 μm, if prolonged etching is carried out). This technique has obvious advantages for etching tracks of ultra-heavy cosmic rays in meteoritic crystals, and is a variant of the TINCLE method.

It has become evident, over the years (Price *et al.*[8a]), that the concept of a sharp ionization threshold for track registration in minerals is not strictly valid. Track etch rates tend to approach the bulk etch velocity asymptotically near to the apparent ionization threshold J_c. One consequence of this is that ΔR will be an increasing function of the etching time. Also, it is to be expected that the track length spectrum will not consist of a series of sharp peaks at each value of ΔR but will, rather, take on a more "smeared-out" appearance.*

* *Note added in proof.* Perron & Maury[47a] report an "etch induction time" for heavy-ion tracks in olivine crystals similar to that known for plastic detectors (see reference 36 in Chapter 4).

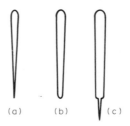

(a)   (b)   (c)

FIG. 6.16   *Krätschmer and Gentner*[48] *apply a multistage etch to track-in-track* (TINT) *events. A first treatment with the normal (or standard) etchant develops the track as shown in (a). In the second stage, an etchant is applied which does not preferentially etch the damage trail: it therefore does not significantly increase the etched length, but simply makes the etched track fatter (b). In the last stage, the track is again subjected to the normal etchant: this results in the continuation of the along-the-track etching, yielding an etch cone at the lower end of the track as shown in (c). If the direction of etching is towards the high-energy end of the damage trail, then the additional cone yields a track etch rate* V_T *at that juncture, and the distance from the low-energy end (the upper end in the figures shown) of the track yields the corresponding residual range* R. *Hence these two parameters (*V_T *and* R *) may be determined from a single series of measurements, without the need for relocation of the tracks.*

These difficulties have led some workers[48,49,8c] to consider methods of charge assignment based on measurements of track etch rate as a function of residual range.

Two approaches have been tried. Krätschmer and Gentner[48] have developed a two-stage etching method. Here, in the first stage the tracks are etched in a standard etchant, which develops etch cones (Fig. 6.16a); and in the second stage, an etchant that does not preferentially etch out the damage trail (or does so at a rate not much faster than the bulk etch rate) is used. This stage acts as a "marker" on the track (Fig. 6.16b). Finally, the track is again etched in the standard solution (Fig. 6.16c). The length of track developed during each of the standard etch stages is apparent at the end of the procedure owing to the "marking" effect. This is rather similar to the u.v. irradiation method adopted for low-energy particles in Lexan, described in §6.5.2. After all the etching stages have been completed, the track etch rate and residual range may be determined from a single series of measurements, and no relocation of the tracks is required.

In the second approach,[49,8c] tracks are repeatedly relocated and the cone lengths measured as a function of etch time, until the track end is reached.

In both methods the response function of the mineral is determined by means of heavy-ion accelerator irradiations.

These methods have now been applied to VH $(20 \leq Z \lesssim 28)$ and VVH $(Z \gtrsim 30)^*$ tracks stored in meteorites[8c] and lunar rocks.[48] As yet no well-established differences between the ancient and contemporary cosmic ray compositions have been observed.

Partial annealing of tracks in lunar and meteoritic crystals has also been used[50,51] as a means of particle identification. Here, the tracks formed by lighter ions are preferentially removed, whilst leaving behind tracks formed by heavier ions.

References

1. P. B. Price, R. L. Fleischer, D. D. Peterson, C. O'Ceallaigh, D. O'Sullivan & A. Thompson (1967) Identification of isotopes of energetic particles with dielectric track detectors. *Phys. Rev.* **164**, 1618–20.
2. P. B. Price & R. L. Fleischer (1971) Identification of energetic heavy nuclei with solid dielectric track detectors: Applications to astrophysical and planetary studies. *Ann. Rev. Nucl. Sci.* **21**, 295–334.
3. W. Enge (1977) Isotopic composition of cosmic ray nuclei. *Nucl. Instrum. Meth.* **147**, 211–20.
4. H. H. Heckman, B. L. Perkins, W. G. Simon, F. M. Smith & W. H. Barkas (1960) Ranges and energy-loss processes of heavy ions in emulsion. *Phys. Rev.* **117**, 544–56.
5. R. P. Henke & E. V. Benton (1968) A computer code for the computation of heavy-ion range–energy relationships in any stopping material. *US Naval Radiological Defense Laboratory, San Francisco, Report* NRDL TR-67-122.
6. L. C. Northcliffe & R. F. Schilling (1970) Range and stopping power tables for heavy ions. *Nucl. Data Tables* **A7**, 233–63.
7. R. L. Fleischer, P. B. Price & R. M. Walker (1975) *Nuclear Tracks in Solids: Principles and Applications.* University of California Press, Berkeley.
8. J. Tripier & M. Debeauvais (1977) Calibration of two plastic detectors and application on study of heavy cosmic rays. *Nucl. Instrum. Meth.* **147**, 221–6.
8a. P. B. Price, D. Lal, A. S. Tamhane & V. P. Perelygin (1973) Characteristics of tracks of ions of $14 \leq Z \leq 36$ in common rock silicates. *Earth Planet. Sci. Lett.* **19**, 377–95.
8b. P. F. Green & S. A. Durrani (1977) Annealing studies of tracks in crystals. *Nucl. Track Detection* **1**, 33–9.
8c. P. F. Green, R. K. Bull & S. A. Durrani (1978) Particle identification from track-etch rates in minerals. *Nucl. Instrum. Meth.* **157**, 185–93.
9. J. Sequeiros, J. Medina, A. Durá, M. Ortega, A. Vidal-Quadras, F. Fernandez & R. T. Thorne (1976) Low energy heavy cosmic ions charge discrimination with plastic detectors. *Nucl. Instrum. Meth.* **135**, 133–8.
10. G. Siegmon, D.-P. Bartholomä & W. Enge (1976) Composition of Fe-isotopes in cosmic-rays. In: *Proc. 9th Int. Conf. Solid State Nucl. Track Detectors*, Munich, and Suppl. 1, *Nucl. Tracks.* Pergamon, Oxford, 1059–68.
11. G. Siegmon, H. J. Köhnen, D.-P. Bartholomä & W. Enge (1976) The dependence of the mass-identification scale on different track formation models. In: *Proc. 9th Int. Conf. Solid State Nucl. Track Detectors*, Munich, and Suppl. 1, *Nucl. Tracks.* Pergamon, Oxford, pp. 137–43.
12. G. Somogyi (1980) Development of etched nuclear tracks. *Nucl. Instrum. Meth.* **173**, 21–42.
13. F. A. Jenkins & H. E. White (1957) *Fundamentals of Optics.* McGraw-Hill, New York.
14. B. G. Cartwright, E. K. Shirk & P. B. Price (1978) A nuclear-track-recording polymer of unique sensitivity and resolution. *Nucl. Instrum. Meth.* **153**, 457–60.

* Different authors use the notations VH and VVH to cover slightly different charge domains. We have therefore adopted slightly approximate Z values to indicate the boundary of the two groups.

15. P. H. Fowler, V. M. Clapham, D. L. Henshaw, D. O'Sullivan & A. Thompson (1980) The effect of temperature-time cycles in the polymerisation of CR-39 on the uniformity of track response. In: *Proc. 10th Int. Conf. Solid State Nucl. Track Detectors*, Lyon, and Suppl. 2, *Nucl. Tracks*. Pergamon, Oxford, pp. 437–41.

16. A. Thompson, D. O'Sullivan & C. O'Ceallaigh (1980) Development studies of CR-39 for cosmic-ray work. In: *Proc. 10th Int. Conf. Solid State Nucl. Track Detectors*, Lyon, and Suppl. 2, *Nucl. Tracks*. Pergamon, Oxford, pp. 453–7.

16a. D. L. Henshaw, N. Griffiths, O. A. L. Landen, S. P. Austin & A. A. Hopgood (1982) Track response of CR-39 manufactured in specifically controlled temperature-time curing profiles. In: *Proc. 11th Int. Conf. Solid State Nuclear Track Detectors*, Bristol, and Suppl. 3, *Nucl. Tracks*. Pergamon, Oxford, pp. 137–40.

17. S. P. Ahlen (1980) Theoretical and experimental aspects of the energy loss of relativistic heavily ionizing particles. *Revs. Mod. Phys.* **52**, 121–73.

18. R. P. Henke & E. V. Benton (1977) Isotope resolution using plastic detectors. *Nucl. Track Detection* **1**, 93–7.

19. P. B. Price, D. D. Peterson, R. L. Fleischer, C. O'Ceallaigh, D. O'Sullivan & A. Thompson (1970) Composition of cosmic rays of atomic number 12 to 30. In: *Proc. 11th Int. Conf. on Cosmic Rays*, Budapest, pp. 417–22.

20. P. B. Price, D. D. Peterson, R. L. Fleischer, C. O'Ceallaigh, D. O'Sullivan & A. Thompson (1968) Plastic track detectors for identifying cosmic rays. *Can. J. Phys.* **46**, S1149–53.

21. P. B. Price, R. L. Fleischer, D. D. Peterson, C. O'Ceallaigh, D. O'Sullivan & A. Thompson (1968) High resolution study of low energy cosmic rays with Lexan track detectors. *Phys. Rev. Lett.* **21**, 630–3.

22. P. B. Price & R. L. Fleischer (1970) Particle identification by dielectric track detectors. *Radiat. Effects* **2**, 291–8.

23. P. H. Fowler, R. V. Adams, V. G. Cowen & J. M. Kidd (1967) The charge spectrum of very heavy cosmic ray nuclei. *Proc. Roy. Soc. Lond.* **A301**, 39–45.

24. P. H. Fowler, V. M. Clapham, V. G. Cowen, J. M. Kidd & R. T. Moses (1970) The charge spectrum of very heavy cosmic ray nuclei. *Proc. Roy. Soc. Lond.* **A318**, 1–43.

25. J. H. Chan & P. B. Price (1975) Composition and energy spectra of heavy nuclei of unknown origin detected on Skylab. *Phys. Rev. Lett.* **35**, 539–42.

26. D. O'Sullivan, A. Thompson, J. Daly, C. O'Ceallaigh, V. Domingo, A. Smit & K.-P. Wenzel (1980) A solid state track detector array for the study of ultraheavy cosmic ray nuclei in earth orbit. In: *Proc. 10th Int. Conf. Solid State Nucl. Track Detectors*, Lyon, and Suppl. 2, *Nucl. Tracks*. Pergamon, Oxford, pp. 1011–19.

27. J. H. Adams Jr., M. M. Shapiro, R. Silverberg & C. H. Tsao (1980) The "Heavy Ions in Space" experiment. In: *Proc. 10th Int. Conf. Solid State Nucl. Track Detectors*, Lyon, and Suppl. 2, *Nucl. Tracks*. Pergamon, Oxford, pp. 1011–19.

27a. D. D. Peterson (1969) A study of low energy, heavy cosmic rays using a plastic charged particle detector. Ph.D. thesis, Rensselaer Polytechnic Institute.

27b. R. P. Henke, E. V. Benton & H. H. Heckman (1970) Sensitivity enhancement of plastic nuclear track detectors through photo oxidation. *Radiat. Effects* **3**, 43–9.

28. W. DeSorbo (1979) Ultraviolet effects and aging effects on etching characteristics of fission tracks in polycarbonate film. *Nucl. Tracks* **3**, 13–32.

29. D. O'Sullivan & A. Thompson (1980) The observation of a sensitivity dependence on temperature during registration in solid state nuclear track detectors. *Nucl. Tracks* **4**, 271–6.

29a. S. A. Durrani, P. J. Groom, K. A. R. Khazal & S. W. S. McKeever (1977) The dependence of thermoluminescence sensitivity upon the temperature of irradiation in quartz. *J. Phys. D: Appl. Phys.* **10**, 1351–61.

30. P. B. Price, P. H. Fowler, J. M. Kidd, E. J. Kobetich, R. L. Fleischer & G. E. Nichols (1971) Study of the charge spectrum of extremely heavy cosmic rays using combined plastic detectors and nuclear emulsions. *Phys. Rev.* **D3**, 815–23.

31. L. S. Pinsky (1972) A study of heavy trans-iron primary cosmic rays ($Z \geq 55$) with a fast film Čerenkov detector, *NASA Tech Memo* X-58102: Manned Spacecraft Center, Houston.

32. E. K. Shirk, P. B. Price, E. J. Kobetich, W. Z. Osborne, L. S. Pinsky, R. D. Eandi & R. B. Rushing (1973) Charge and energy spectra of trans-iron cosmic rays *Phys. Rev.* **D7**, 3220–33.

33. P. H. Fowler (1977) Ultra heavy cosmic ray nuclei—Analysis and results. *Nucl. Instrum. Meth.* **147**, 183–94.

34. P. H. Fowler, C. Alexandre, V. M. Clapham, D. L. Henshaw, C. O'Ceallaigh, D. O'Sullivan & A. Thompson (1977) High resolution study of nucleonic cosmic rays with $Z \geq 34$. *Nucl. Instrum. Meth.* **147**, 195–9.

35. R. A. Stern & P. B. Price (1972) Charge and energy information from heavy ion tracks in "Lexan". *Nature Phys. Sci.* **240**, 82–3.

36. E. V. Benton (1971) A method for development of volume tracks in dielectric nuclear track detectors. *Nucl. Instrum. Meth.* **92**, 97–9.

37. D. Lal, R. S. Rajan & A. S. Tamhane (1969) Chemical composition of nuclei of $Z > 22$ in cosmic rays using meteoritic minerals as detectors. *Nature* **221**, 33–7.

38. H. B. Lück (1975) Investigations on energy resolution in detecting alpha particles using a cellulose nitrate detector. *Nucl. Instrum. Meth.* **124**, 359–63.

39. M. Balcázar-García & S. A. Durrani (1977) ^{3}He and ^{4}He spectroscopy using plastic solid-state nuclear track detectors. *Nucl. Instrum. Meth.* **147**, 31–4.

40. G. Somogyi (1977) Processing of plastic track detectors. *Nucl. Track Detection* **1**, 3–18.

40a. S. A. R. Al-Najjar & S. A. Durrani (1984) Track profile technique (TPT) and its applications using CR-39. I: Range and energy measurements of alpha-particles and fission fragments. *Nucl. Tracks* **8**, 45–50.

40b. S. A. R. Al-Najjar & S. A. Durrani (1984) Track profile technique (TPT) and its applications using CR-39. II: Evaluation of V versus residual range. *Nucl. Tracks* **8**, 51–6.

40c. S. R. Hashemi-Nezhad, S. A. Durrani, R. K. Bull & P. F. Green (1984) Particle identification and measurements of track-etch rate as a function of residual range of heavy ions in biotite mica. *Nucl. Tracks* **8**, 91–4.

40d. R. V. Genetry (1970) Giant halos: indication of unknown radioactivity? *Science* **169**, 670–3.

41. R. L. Fleischer, P. B. Price & R. M. Walker (1965) Solid-state track detectors: applications to nuclear science and geophysics. *Ann. Rev. Nucl. Sci.* **15**, 1–28.

42. M. Maurette (1966) Etude des traces d'ions lourds dans les minéraux naturels d'origine terrestre et extra-terrestre. *Bull Soc. Franc. Min. Crist.* **89**, 41–79.

43. P. B. Price, R. S. Rajan & A. S. Tamhane (1968) The abundance of nuclei heavier than iron in the cosmic radiation in the geological past. *Astrophys. J.* **151**, L109–L116.

44. T. Plieninger, W. Krätschmer & W. Gentner (1972) Charge assignment to cosmic ray heavy ion tracks in lunar pyroxenes. In: *Proc. Lunar Sci. Conf. 3rd*. Pergamon, New York, pp. 2933–9.

45. T. Plieninger, W. Krätschmer & W. Gentner (1973) Indications for time variations in the galactic cosmic ray composition derived from track studies on lunar samples. In: *Proc. Lunar Sci. Conf. 4th*. Pergamon, New York, pp. 2337–46.

46. R. K. Bull & S. A. Durrani (1976) Cosmic-ray tracks in the Shalka meteorite. *Earth Planet. Sci. Lett.* **32**, 35–9.

47. C. Perron (1984) Relativistic ^{238}U ion tracks in olivine. Implications for cosmic ray track studies. *Nature* **310**, 397–9.

47a. C. Perron & M. Maury (1986) Very heavy ion track etching in olivine. *Nucl. Tracks* **11**, 73–80.

48. W. Krätschmer & W. Gentner (1976) The long-term average of the galactic cosmic-ray iron group composition studied by the track method. In: *Proc. Lunar Sci. Conf. 7th*. Pergamon, New York, pp. 501–11.

49. R. K. Bull, P. F. Green & S. A. Durrani (1978) Studies of the charge and energy spectra of the ancient VVH cosmic rays. In: *Proc. Lunar Planet. Sci. Conf. 9th*. Pergamon, New York, pp. 2415–31.

50. E. Dartyge, J. P. Duraud, Y. Langevin & M. Maurette (1978) A new method for investigating the past activity of ancient solar flare cosmic rays over a time scale of a few billion years. In: *Proc. Lunar Planet. Sci. Conf. 9th*. Pergamon, New York, pp. 2375–98.

51. D. Lhagvasuren, O. Otgonsuren, V. P. Perelygin, S. G. Stetsenko, B. Jakupi, P. Pellas & C. Perron (1980) A technique for particle annealing of tracks in olivine to determine the relative abundance of galactic cosmic ray nuclei with $Z \geq 50$. In: *Proc. 10th Int. Conf. Solid State Nucl. Track Detectors*, Lyon, and Suppl. 2, *Nucl. Tracks*. Pergamon, Oxford, pp. 997–1002.

CHAPTER 7

Radiation Dosimetry and SSNTD Instrumentation

In Chapter 6 we showed how etched track detectors can be used to obtain detailed information on the charge and velocity of incident particles. In the present chapter we consider the aplications of these detectors to the measurement of the amount of radiation incident on a system (with particular reference to the human body). The quantity to be determined is usually the dose of ionizing radiation absorbed by a body, or some parameter which is proportional to the absorbed dose. In some instances, however, where the nature of the radiation field is well known, a measure of the number of incident particles may suffice.

An increasing number of people are now being exposed to ionizing radiations in the course of their work. The International Commission on Radiological Protection (ICRP) has, from time to time, made a number of recommendations regarding dose equivalent limits for radiation workers and for the general public.[1] For example, the recommended upper limit on whole-body dose equivalent to be received by radiation workers is 5 rem (50 mSv)* per year. The dose received by such workers must therefore be the subject of strict controls. The concept of maximum permissible or tolerance levels is, in fact, no longer considered desirable; the emphasis is on seeking to avoid all radiation as far as practical. In establishments where work involving radiations is performed it is necessary to carry out radiation-survey monitoring to define radiation levels at various points within the area of work. Also, it is necessary that workers carry a personal dosimetry system in order that their accumulated dose may be accurately assessed on a periodic basis. Requirements for a good personnel dosimetry system include robustness, simplicity, cheapness and ease of evaluation. It is in this area of radiation dosimetry that solid state nuclear track detectors are finding increasing application. Other areas of radiation dosimetry in which nuclear track de-

---

* The unit of dose equivalent is sievert (Sv; in J kg^{-1}). It is related to the absorbed dose of a given radiation by the relation: 1 Sv = (Absorbed dose in Gy) × Quality factor, where the SI unit of absorbed dose, gray, is defined as 1 Gy = 1 J kg^{-1}, and quality factor is a dimensionless quantity. (The quality factor is related to, but is not the same as, the relative biological effectiveness (RBE) of a given radiation. Its value is laid down, from time to time, by edicts from ICRP and ICRU (International Commission on Radiation Units and Measurements)).

Note that 1 Gy = 100 rad; and 1 Sv = 100 rem.

tectors have proved useful include the mapping, with the help of "phantoms", of dose distributions produced by radiation beams to be used in radiotherapy.

As we have noted earlier (see Chapters 1 and 3), track detectors respond mainly to highly ionizing radiations. In the language of radiation dosimetry such radiations have high linear energy transfer (LET). LET is simply the energy lost by the particle per unit path length, and is usually quoted in keV μm^{-1}. Among the types of radiation encountered in dosimetry, α-particles and heavy ions are of sufficiently high LET to be detected directly by plastic track detectors. Neutrons and energetic pions can, however, also produce a response by generating high-LET secondary radiations in passing through matter. Some plastics, e.g. CR-39, respond to protons with relatively low LET values as well. We begin our detailed discussion by considering neutron dosimetry first, in view of the importance attached to the monitoring of neutron exposure in atomic energy establishments, fission and fusion reactor environments, hospitals, etc.

7.1 Neutron Dosimetry

Neutrons, being electrically uncharged, cannot produce ionization directly, and are therefore usually detected via nuclear reaction products. The application of track detectors to neutron dosimetry was first described by Walker et al.[2] The methods used to detect neutrons depend, to a large extent, on the neutron energy; and it is convenient to consider separately thermal neutrons, and fast and intermediate-energy neutrons.

7.1.1 Thermal neutrons

Detection systems for thermal neutrons usually utilize a nuclide such as ^{235}U, which undergoes induced fission by neutrons at thermal energies. Consider a foil or section of material containing N_5 atoms of ^{235}U per unit volume. The number n_{if} of induced fissions per unit volume produced by a thermal-neutron fluence (i.e. time-integrated flux) F (which is expressed as particles per unit area) is given by:

$$n_{if} = N_5 \sigma_{th} F \qquad (7.1)$$

where σ_{th} is the thermal-fission cross-section for ^{235}U. As we saw in Chapter 4 (Eq. (4.15)), the number of fission tracks per unit area, ρ, appearing on an external surface of a thick sample is related to the number n of fission events per unit volume through:

$$\rho = \tfrac{1}{2}nR \qquad (7.2)$$

where R is the mean range of a *single* fission fragment in the sample, and where we have ignored the effect of the critical angle of etching θ_c (since $\cos^2 \theta_c \simeq 1$ for most minerals as well as for plastics used as detectors). Note that the factor $\frac{1}{2}$ replaces the factor $\frac{1}{4}$ applicable to α-emission from a thick source because two fission fragments result from each fission event, as explained in Chapter 4. Upon substituting Eq. (7.1) into Eq. (7.2), we obtain for ρ_i, the induced-fission track density on the surface of the foil (provided that the foil is thicker than a mean fission fragment range):

$$\rho_i = \tfrac{1}{2}N_5\sigma_{th}FR \tag{7.3}$$

The last equation will also represent the track density on a detector placed in close contact with the external surface of the U-bearing foil or sample. This approach is also used for thermal-neutron dosimetry in connection with fission track dating, as we shall see in Chapter 8.

For the range of neutron fluences generally encountered in personnel dosimetry, (viz. $\sim 10^7$–10^{10} neutrons cm^{-2}), a suitable converter foil might consist of natural uranium or of uranium enriched in ^{235}U.

For a "thick" foil of natural uranium adjacent to a suitable detector, such as Lexan polycarbonate, the number of tracks obtained per neutron per cm^2 can be readily calculated from Eq. (7.3). By substituting the relevant constants (viz., $\sigma_{th} \simeq 580 \times 10^{-24}$ cm^2; $R \simeq 5 \times 10^{-4}$ cm in uranium metal), and since $N_5 = I \times N_8$, where ^{235}U: ^{238}U ratio $I \simeq 7 \times 10^{-3}$ for uranium of normal isotopic composition, and the number of ^{238}U atoms per unit volume is given approximately by $(19/238) \times 6 \times 10^{23}$ (the density of uranium metal being $\simeq 19$ g cm^{-3} and 6×10^{23} being $\approx$ Avogadro's number), Eq. (7.3) yields:

$$\rho_i \simeq \tfrac{1}{2} \times (\tfrac{19}{238} \times 6 \times 10^{23} \times 7 \times 10^{-3}) \times (580 \times 10^{-24}) \times (5 \times 10^{-4})F$$

$$\text{i.e., } \rho_i \simeq 5 \times 10^{-5}\,F.$$

Thus for a just-countable (i.e. minimum practical) track density of $\simeq 50$ tracks cm^{-2}, we would require a thermal-neutron fluence of $\simeq 10^6$ cm^{-2} incident on a thick foil of natural uranium placed next to a plastic detector.

For thermal neutrons, 1 rem (10 mSv) is equivalent to $\sim 10^9$ neutrons cm^{-2}; and so a rough estimate of the lower limit to the dose equivalent which can be readily measured with a natural U + plastic assembly is ~ 1 mrem (10 μSv). With a thick foil of almost pure ^{235}U (instead of the natural abundance of $\sim 1/140$ for that isotope), this limit could be lowered by a factor of $\sim 10^2$; i.e. a thermal neutron fluence of $\simeq 10^4$ cm^{-2} and a thermal-neutron dose of $\simeq 10^{-2}$ mrem (0.1 μSv) would be the minimum practically measurable quantity.

Equation (7.3) can be written in the form

$$\rho_i = BF \tag{7.4}$$

where B is equal to $\tfrac{1}{2}N_5\sigma_{th}R$ and is clearly a constant for any converter foil plus detector combination. (As pointed out above, $\theta_c \simeq 0°$ for materials such as plastics or sheets of mica generally used as external detectors of fission fragments, so that the etching efficiency $\cos^2 \theta_c$ implicit in Eqs. (7.3) and (7.4) (cf. §4.2.3) has been taken as 1.)

In practice, B is not usually calculated from the product of its components but is normally obtained directly from calibration experiments.

For natural U foils, a lower limit to the measurable thermal-neutron dose rate will be set by the registration, in the detector, of spontaneous fission tracks from ^{238}U. In Chapter 8 we will show that the spontaneous fission track density accumulating (from both "above" and "below") on an internal surface of a thick U-bearing sample in time t is given by

$$\rho_s = \frac{\lambda_{f8}}{\lambda_{D8}} N_8 R[\exp(\lambda_{D8}t) - 1]$$

where λ_{f8} and λ_{D8} are the fission-, and the total-decay constants of ^{238}U, respectively. N_8 is the number of ^{238}U atoms per unit volume. For times t short compared with the half-life of ^{238}U ($\simeq 4.5 \times 10^9$ year), $e^{\lambda_{D8}t}$ can be replaced by $1 + \lambda_{D8}t$, and we obtain

$$\rho_s \simeq \lambda_{f8} N_8 Rt \tag{7.5}$$

For spontaneous fission fragments emitted across an external surface of a thick sample (a consideration not usually necessary in the fission track dating case), the track density in an external dosimetric detector will be reduced by a factor of 2, so that

$$\rho_s = \tfrac{1}{2}\lambda_{f8} N_8 Rt \tag{7.6}$$

Now the time t referred to here is the time from dosimeter assembly prior to thermal-neutron irradiation up to dosimeter disassembly perhaps immediately preceding the etching and readout. Suppose, however, that this time interval is very close in value to the time period during which irradiation takes place; then, from Eqs. (7.3) and (7.6),

$$\frac{\rho_i}{\rho_s} = \frac{\tfrac{1}{2}N_5 \sigma_{th} RF}{\tfrac{1}{2}\lambda_{f8} N_8 Rt} = \frac{I\sigma_{th}\phi}{\lambda_{f8}} \tag{7.7}$$

where I is the isotopic atomic ratio of ^{235}U/^{238}U for the converter foil and ϕ ($= F/t$) is the thermal-neutron flux (n cm^{-2} s^{-1}). If we require that our induced track density be at least equal to the spontaneous track density contribution (i.e. $\rho_i/\rho_s = 1$), then our lower limit of flux is given by

$$\phi_{min} = \frac{\lambda_{f8}}{I\sigma_{th}} \tag{7.8}$$

The last expression, on substituting appropriate values for the constants contained in it (viz., $\lambda_{f8} \simeq 7 \times 10^{-17}$ yr^{-1}; $I \simeq 7 \times 10^{-3}$; $\sigma_{th} \simeq 6 \times 10^{-22}$ cm^2), leads to $\phi_{min} \sim 2 \times 10^3$ neutrons cm^{-2} h^{-1} for our natural uranium foil. This corresponds (since 1 rem $\sim 10^9$ thermal n cm^{-2}) to a minimum dose rate of $\sim 2 \times 10^{-3}$ mrem h^{-1}.

This dose rate limit could be improved (i.e. further reduced) by the use of a foil enriched in ^{235}U (thus enhancing the induced-to-spontaneous fission ratio; or in other words, I in Eq. (7.8) could be greatly increased over its value of 7×10^{-3} used in the above calculation).

This problem of background track density due to spontaneous fission can be overcome entirely by replacing uranium with an isotope such as ^{10}B, which has a cross-section of $\simeq 3840$ barns (10^{-28} m^2) for the reaction ^{10}B(n, α) ^{7}Li at thermal energies of neutrons, and no spontaneous decay. Even natural boron (which has $\simeq 20\%$ ^{10}B) is good enough. Alpha-sensitive plastics, ready-painted with a layer of a boron-containing compound, are available commercially for neutron dosimetry purposes. Examples of these are: LR 115 coated with $Li_2B_4O_7$; and CA80-15 (or its recent replacement, CN 85) coated with lithium borate on one or both sides. All the above cellulose nitrate detectors are manufactured by Kodak-Pathé of France (and designated as "Type B" films when a converter screen is incorporated.*) Boron-containing compounds coated on appropriate backing materials (e.g. the Kodak BN 1 screens containing ^{10}B) can also be utilized as independent converter screens in juxtaposition with α-sensitive plastic detectors. Such plastic foil plus alphagenic converter-screen assemblies can be readily used for pocket dosimetry or environmental monitoring devices in reprocessing plants, hospitals, etc.

7.1.2 *Fast and intermediate-energy neutrons*

In the field of personnel dosimetry, moderate fluences of thermal neutrons are not so important, since they carry relatively little dose. As neutron energy rises, however, the dose equivalent in rem carried per unit neutron fluence (or the "conversion factor") increases substantially (see Fig. 7.1).** At lower energies, deposition of radiation dose arises mainly from (n, γ) reactions. At higher energies, neutrons can transfer significant amounts of energy to charged particles via (n, p) and (n, α) reactions and elastic scattering of target nuclei. These interactions lead to energy deposition with high LET; and so the relative biological effectiveness (RBE), and thus the "quality factor", of fast neutrons is high. A quality factor of 10 is currently used for fast neutrons (as recommended by the ICRP).

In fast and intermediate-energy neutron dosimetry†, it is generally necessary to detect neutrons over a wide range of energy. For example, outside

---

* A disadvantage of such ready-coated converter compounds is that the detector can accumulate α-track densities from stray thermal neutrons before its actual use for neutron monitoring. Separate converter screens are thus preferable.

** Sims and Killough[2a] give a useful summary of the current sets of neutron fluence-to-dose equivalent conversion factors as a function of neutron energy.

† Intermediate energies of neutrons are usually taken to range from 1 keV to 500 keV, and fast neutrons to have energies over 0.5 MeV (usually up to 20 MeV).

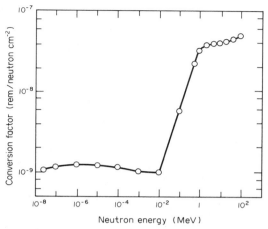

FIG. 7.1 *Dose-equivalent conversion factor for neutrons (unidirectional broad beam, normal incidence), as a function of neutron energy. After a reasonably flat value at low neutron energies, the conversion factor (rem—röntgen equivalent man—per unit neutron fluence) rises steeply by a factor of ~40 (to attain a new quasi-plateau value) as the neutron energy goes up from ~ 10 keV to ~ 1 MeV, reflecting the greater "relative biological effectiveness" (RBE, i.e. the degree of biological damage caused), and hence the higher "quality factor", of the fast neutrons. (Figure from Griffith et al.[12])*

a reactor shield, neutrons will have a broad spectrum extending from fission-neutron energies (up to ~ 10 MeV) down to thermal energies (~ 10^{-2} eV). Also, there may be energetic neutrons resulting from photo-neutron reactions of high-energy γ-rays on deuterium or beryllium (both of which are used as moderators in certain reactor types[3]).

From the form of Fig. 7.1, it is clear that a knowledge merely of the total number of neutrons present will be of relatively little use in estimating the dose equivalent, unless the shape of the neutron energy spectrum in a given environment is very well known. It will be necessary either to determine, in some way, the form of the spectrum from the detector readings or to have a detector for which the response to neutrons of a given energy is proportional to (or related, in a known way, to) the dose equivalent at that energy. In this latter case, if the response of the detector to neutrons in the energy range E to $E + dE$ is $R(E)$ per neutron, then the total response in terms of, say, tracks per cm^2 will be R_T given by

$$R_T = \int_0^\infty F(E)R(E)\, dE$$

where $F(E)\, dE$ is the fluence of neutrons with energy between E and $E + dE$.

For a dose equivalent response such that $R(E) = $ const. $D(E)$, where $D(E)$ is the dose equivalent for neutrons in the energy range E to $E + dE$, then:

$$R_T = \int_0^\infty F(E) \text{ const. } D(E) \, dE$$

$$= \text{const.} \int_0^\infty F(E)D(E) \, dE$$

Since $\int_0^\infty F(E)D(E) \, dE$ is the total dose equivalent D_T of all the neutrons in the spectrum, therefore

$$R_T \propto D_T$$

In this way, a measure of the total dose equivalent delivered is obtained from the track density produced in such a detector (for which the response is proportional to the dose equivalent in every energy interval), even though the nature of the neutron energy spectrum is not directly determined.

Spectrometric information on the neutron flux may be obtained by exposing track detectors adjacent to foils containing nuclides that have different forms of energy dependence in their fission cross-sections $\sigma_f(E)$.

A number of fissionable nuclides exist which can be used in this way. Some cross-sections of interest are shown as a function of neutron energy in Fig. 7.2a. A plastic detector placed next to a ^{235}U foil will, thus, record fission tracks from interactions with thermal, intermediate-energy, and fast neutrons. A similar assembly wrapped in Cd will only respond to neutrons with energy from the Cd cutoff at ~ 0.5 eV upwards. ^{238}U undergoes fissions only from neutrons of energy $E_n \gtrsim 1$ MeV. Thus an assembly containing a series of plastic detectors placed adjacent to a number of different nuclides can yield information on the neutron-energy spectrum. Neutron dosimetry devices involving fission foils and track detectors have been described by a number of authors (see, for example, references 4–10).

The nuclide that comes nearest to giving a direct dose equivalent response is ^{237}Np. In Fig. 7.2b, the fission cross-section for ^{237}Np is shown in juxtaposition to a plot of the ICRP dose equivalent curve for neutrons. This figure displays these curves plotted on an arbitrary ordinate scale. They clearly show a reasonable similarity of form. A ^{237}Np fission track detector system has formed the basis of several neutron dosimeters.[9,11] Griffith et al.[12] have shown that a ^{237}Np track detector system will approximate dose equivalent response to a variety of neutron spectra.

Other attempts to produce a monitor with a dose equivalent response have included (1) encapsulation of an assembly consisting of a plastic and a converter foil such as ^{235}U or ^{239}Pu inside a material which exhibits a $1/v$ neutron-absorption cross-section, e.g. ^{10}B (reference 10), and (2) utilization of the albedo principle[13,14] for detector response. In the first approach (1 above), the casing with $1/v$ response counteracts the higher fission cross-section of the ^{235}U and ^{239}Pu at lower neutron energies, thus smoothing out the detector response.

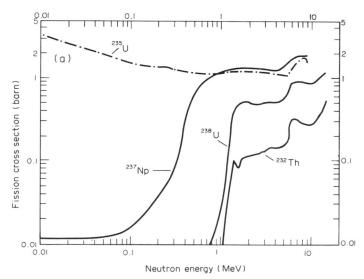

FIG. 7.2a *Some fission cross-sections (in barns (b), i.e. units of 10^{-28} m^2) as a function of neutron energy. Whereas ^{238}U and ^{232}Th have a fission threshold at ~ 1 MeV, and ^{237}Np effectively at ~ 100 keV, in neutron energy, ^{235}U displays a quasi-1/v cross-section at lower energies, and then settles down at a value of between 1 and 2 b at neutron energies above ~ 100 keV. The fission cross-section for ^{237}Np is shown in greater detail in Fig. 7.2b.*

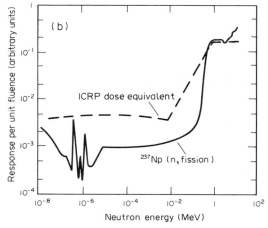

FIG. 7.2b *Comparison of the variation of the ^{237}Np fission cross-section versus neutron energy with ICRP (International Commission on Radiological Protection) dose equivalent curve for neutrons (reference 1; cf. also Fig. 7.1). If one ignores the resonances in the $\sim eV$ region, it is seen that the ^{237}Np fission cross-section roughly approximates to the ICRP curve. This property has been exploited to construct plastic track detectors incorporating a thin converter layer of ^{237}Np, whose fission products yield track densities which roughly correspond to the dose equivalent values of the incident neutron fluence. The risk to the wearer from γ rays accompanying the fission and α-decay of ^{237}Np must, however, be taken into account (e.g. a 10 mg foil of ^{237}Np can produce a wearer-dose of ~ 580 mrem ($=5.8$ mSv) per year if adequate shielding is not provided). (Figure after reference 12.)*

The albedo effect (method 2 above) depends on the fact that, when fast and intermediate-energy neutrons are incident on the human body, they are moderated and scattered back by it. This process can enable the fast incident neutrons to be detected by a monitor which responds only or mainly to thermal neutrons. Thus Tatsuta and Bingo[13] used two detectors incorporating UO_2: one with natural uranium and the other enriched in ^{235}U; each had a plastic detector on the side adjacent to the wearer's body. A Cd shield was placed on the *outside* of the two assemblies, but not between the plastic detectors and the body. The assembly with natural uranium recorded fissions produced mainly by the incident fast neutrons; whereas the enriched-uranium packet registered tracks produced predominantly by the thermalized neutrons backscattered by the body. The system gave a good approximation to dose equivalent response, except in the region of neutron energies between 10 keV and 1 MeV. The albedo process also enhances the response of ^{237}Np detectors to neutrons in the energy range 1 keV to 100 keV, where the free-air response is low.[11,14a]

Over the last few years a number of attempts have been made to develop new types of albedo dosimeters. Notable among these are those incorporating a material that emits α-particles through its interaction with thermalized albedo neutrons—the α-particles then yielding the tracks registered in the polymer detector. Thus ^6LiF embedded in Teflon with a polycarbonate detector has been used by Sohrabi.[15] Gomaa and co-workers[16-19] and Griffith *et al.*[20] have used a cellulose nitrate–$Li_2B_4O_7$ combination as well as CR-39 plastic with $Li_2B_4O_7$ or ^6LiF (n, α) radiators (with or without a CN plastic such as LR 115). The CR-39 plastic records both fast-neutron recoils (either from within itself or those from a polyethylene "proton radiator") and the albedo-neutron-induced α-tracks, thus covering, in principle, the full range of neutron energies.[20] Thermoluminescence produced in the LiF phosphor by the (n, α) reaction may be used as a complementary measurement. The tracks recorded by CR-39 may be etched either normally or electrochemically (see §7.4.4). The reported sensitivity values for CR-39 range from ~ 3–20 tracks cm^{-2} $mrem^{-1}$ for neutron energies 0.1 MeV–18 MeV impinging directly on CR-39 or from a radiator,[18,21] to $\sim 7 \times 10^3$ tracks cm^{-2} $mrem^{-1}$ for the low-energy albedo neutrons* incident on the LiF radiator.[18] The high track densities in the latter case allow measurements to be made by optical-density methods.[18] A typical albedo dosimetry set-up is shown in Fig. 7.3. (For an alternative set-up see reference 21a.)

Some advantages of track dosimeters include the following:

(1) They exhibit good retention of tracks at moderately high temperatures and show insensitivity to fogging, etc., compared to nuclear emulsions.

---

* The much larger track density per mrem in the latter case reflects the lower quality factor for thermal, compared to fast, neutrons as well as the large contribution from the thermal (n, α) reactions on LiF.

(2) The need for darkroom conditions for the development and fixing of the latent image in the case of nuclear emulsions is obviated when using track detectors.

(3) ^{237}Np has a better dose equivalent response than emulsions or thermoluminescent (TL) albedo dosimeters.

(4) A permanent record of the dosimeter data is obtained (unlike the TL dosimeters used in their normal mode).

(5) Fission tracks lend themselves to relatively simple methods of automatic counting (see §7.4).

(6) Track detectors are generally insensitive to γ-rays, which is often a great advantage. (Only if large fluences of high-energy γ-rays impinge on a detector assembly containing fissile material, and thus produce photofission, is this advantage lost.)

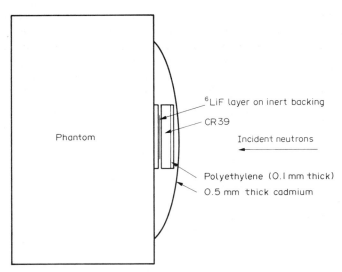

FIG. 7.3 *Albedo/threshold dosimeter assembly incorporating a CR-39 detector and a ^{6}LiF radiator with a Cd cover only on the outside; (after Gomaa et al.[18]). Neutrons are incident from the right in the figure. Thermalized neutrons return from the phantom (or human body) to produce α-tracks from the ^{6}Li(n, α) reaction, which are recorded in the CR-39 detector. Neutrons which initially possess low energies fail to penetrate the outer Cd cover. The fast neutrons, thus, produce not only recoil tracks in the CR-39 sheet (whether from interactions within the CR-39 itself or from the adjacent polyethylene "radiator") but also α-particle tracks after the thermalization in the body. The whole neutron-energy range of interest is thus covered. Thermoluminescence produced in the LiF phosphor by the (n, γ) reaction yields a possible complementary measurement of the neutron dose.*

The major disadvantage of devices utilizing fission foils is their inherent radioactivity, which can deliver a significant dose to the dosimeter wearer. A 10 mg foil of ^{237}Np can produce a wearer-dose of 580 mrem ($= 5.8$ mSv) per year,[12] although provision of some shielding between the dosimeter and the body can reduce this. This disadvantage is not important in the case of an environmental dosimeter.

It may be possible to obtain neutron spectrum information from a series of foils of nuclides yielding (n, α) reactions placed adjacent to α-sensitive plastics.[22] However, in the fast-neutron region, cross-sections, and therefore sensitivities, are too low to enable them to be used in personnel dosimetry when only moderate fluences of neutrons are expected. Such a scheme may, nevertheless, find application in high-flux environmental dosimetry. The advantage of these (n, α) foils is that they are not radioactive and do not give a wearer-dose.

Another approach that has received attention is the use of tracks produced by "intrinsic" recoils and (n, α) reactions taking place within the plastics themselves, or by similar interactions occurring in light-element "radiator foils" placed in contact with the surface of a plastic detector. Charged particles resulting from these reactions which cross the detector surface can be etched out and counted in the usual way.

The constituent elements of most plastics include C, O, and H. In the case of cellulose nitrate, nitrogen is also present. Neutrons can undergo elastic collisions with these constituent nuclei and the resulting recoils may have sufficient energy to form etchable tracks in the plastic. The *maximum* recoil energy resulting from an elastic interaction by a neutron of energy E_n is given by

$$E_{R,max} = \frac{4M_R m_n}{(M_R + m_n)^2} E_n$$

where M_R and m_n are the masses of the recoiling nucleus and the neutron, respectively. (The above formula is equivalent to saying that the *minimum* final energy of an elastically scattered particle of unit atomic mass (1 amu) impinging on a nucleus of atomic mass A is a fraction $[(A - 1)/(A + 1)]^2$ of its initial, i.e. incident, energy.) Thus, an ^{16}O nucleus can carry off a maximum fraction 0.22 of the incident neutron energy, whereas a hydrogen nucleus can carry off the entire neutron energy. Figure 7.4 shows calculated maximum ranges for recoiling C, N, O and H nuclei as a function of neutron energy. Most plastics will not record the H recoils. However, CR-39 does record tracks due to protons of energies up to several MeV.[32] The number of tracks arising from neutron interactions with any given type of atom in the plastic will depend not only on the elastic-scattering cross-sections but also on the etchable range of the recoils produced. Elastic scattering cross-sections for neutrons impinging on C, N, O nuclei are of the order of a barn

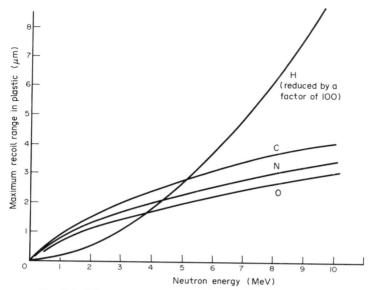

Fig. 7.4 *Maximum ranges of H, C, N and O recoils in plastic as a function of neutron energy. The curves were calculated for a plastic of density 1.4 g cm⁻³, using the range-energy tables of Northcliffe and Schilling[22a] for these ions in Mylar plastic. The elements represent the constituents of the generally used plastics for track detection, which can thus act as "intrinsic" detectors for tracks produced through the elastic collisions of these constituent nuclei with incident neutrons. The maximum fractional energy of a recoiling ion of atomic mass A in elastic collision with a neutron (1 amu) is given by $1 - [(A - 1)/(A + 1)]^2$. A recoiling hydrogen nucleus can, of course, carry away the entire energy of the incident neutron—and thus have a substantial range in plastic. This is of special significance in the case of a proton-sensitive plastic such as CR-39, where ~95% of all recorded tracks are usually due to proton recoils from the incident fast neutrons.[36a]*

$(10^{-28}$ m²), and are somewhat higher for hydrogen. The ranges of recoiling H nuclei are very large in polymers (cf. Fig. 7.4); and in plastics such as CR-39, where a large fraction of this range is etchable, recoiling protons can make a large contribution to the intrinsic track density.

Alpha-particle tracks are also produced from various "intrinsic" (n, α) re-actions, and these can contribute to the total track density observed in an irradiated plastic. Table 7.1 gives some details of the (n, α) reactions on C, N, and O nuclei; and Figs. 7.5a,b,c show the corresponding cross-sections as a function of neutron energy. The contribution from α-particles to the total track density will depend on the composition of the plastic and the range of α-particle energies over which tracks can be recorded in it.

TABLE 7.1   *Some (n, α) reactions of importance for*
neutron detection using "intrinsic" tracks in plastics

Reaction	Q-value (MeV)	Threshold energy (MeV)
$^{12}_{6}C(n, \alpha)^{9}_{4}Be$	-5.71	6.19
$^{14}_{7}N(n, \alpha)^{11}_{5}B$	-0.15	0.16
$^{16}_{8}O(n, \alpha)^{13}_{6}C$	-2.20	2.34
$^{12}_{6}C(n, n')3\alpha$	-7.27	$\sim$9.6 (effective)[22d]

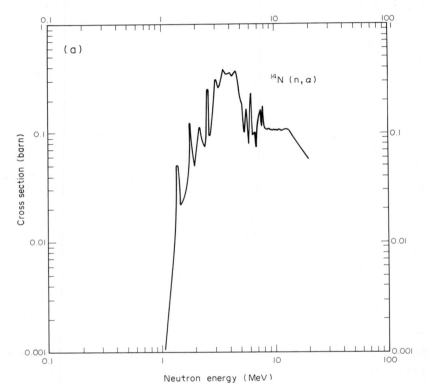

FIG. 7.5a,b,c   *(n, α) cross-sections (in barns, 10^{-28} m^2) as a function of neutron energy for neutrons incident on (a) ^{14}N, (b) ^{16}O, and (c) ^{12}C. The data for figures, (a) and (c) have been taken from ENDF/B-4,[22b], and for figure (b) from the "Barn Book"[22c]. The corresponding Q-values are shown in Table 7.1. The "intrinsic" α-particles are partially recorded in cellulose nitrate (e.g. LR 115 and CN 85, manufactured by Kodak-Pathé of France) and cellulose acetate (e.g. Daicel, manufactured by Dai Nippon of Japan) detectors, as well as polycarbonates such as Markrofol E (manufactured by Bayer A.G. of W. Germany), in addition to being fully recorded in CR-39 polycarbonate (manufactured by Pershore Mouldings Ltd. and American Acrylic Inc.). All of them also record, of course, the heavier reaction products in each case.*

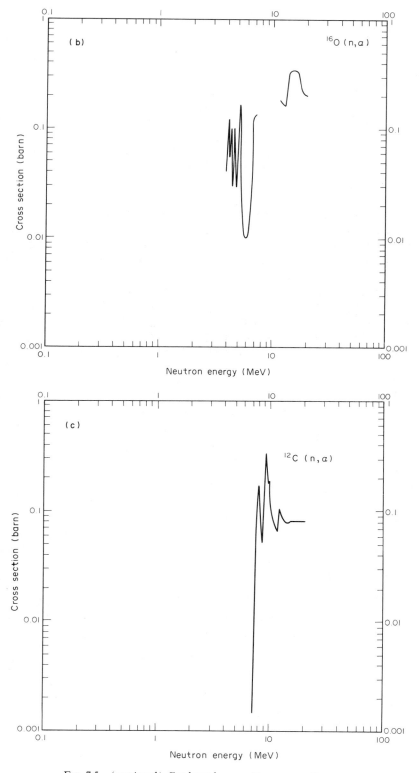

Fig. 7.5 *(continued). For legend see caption on preceding page.*

For most plastics, $\sim 10^{-4}$ to 10^{-5} intrinsic tracks are produced per incident fast neutron. At a neutron energy of 5 MeV, this corresponds to around 2–0.2 tracks cm^{-2}/mrem. The minimum neutron energy for the production of readily observable recoil tracks is about 0.5–1 MeV. The sensitivity can be increased by placing a sheet of material consisting of some light element in front of the detector. Frank and Benton[23] used a number of such "radiator foils", and found that beryllium gave good results. Becker[24] achieved a ~ 12-fold increase in sensitivity by using a gaseous He radiator. A number of workers have discussed the use of elastic recoils and (n, α) reactions in neutron dosimetry.[12,23–28] The recoil-track density in some plastics seems to show a dependence on neutron energy above ~ 0.5 MeV which is similar to that of the first-collision dose in tissue.[24] This is obviously helpful in computing the dose imparted to the human body by fast neutrons.

Recently, the thermoset plastic known commercially as CR-39 (polymerized diethylene glycol bis (allyl carbonate)) has been extensively studied.[29–31] As noted above, CR-39 will readily detect recoiling protons from neutron interactions (see Fig. 7.6). Benton *et al.*[21] have examined the neutron detection properties of this plastic, and find that neutrons of energy as low as 200 keV can be detected. They also report that, with a polyethylene radiator, about 23 tracks cm^{-2}/mrem are produced by irradiation with neutrons from an Am-Be source.

Ramli (reference 31a; see also 31b,c,d) has attempted to develop a "flat" dose equivalent dosimeter for fast neutrons using CR-39 in conjunction with hydrogenous front radiators. The aim was to construct a CR-39 detector assembly, incorporating radiators of different hydrogen content (e.g. cellulose nitrate (H, 32%), polyethylene terephthalate (H, 50%), polyethylene (H, 66.7%)) of different thicknesses placed in sequence in front of the CR-39 detector, such that the response (in terms of electrochemical etch spot density per unit dose equivalent) would be independent of the energy (and impact angle) of the incident neutron. Some degree of success in meeting these requirements was achieved, the most promising combination being an assembly consisting of a 600 μm CR detector with two front radiators: 150 μm polyimide and 900 μm polystyrene (with various procedures of pre-etching and electrochemical etching (§7.4.4); see references 31c–e for further details).

A measure of the fluence of very high-energy ($E_n \gtrsim 10$ MeV) neutrons (e.g. those used in radiotherapy, and in fusion studies) incident on a plastic can be obtained by counting the number of characteristic "3-pronged" tracks (see Fig. 7.7) resulting from the reaction $^{12}C(n, n')3\alpha$, which has a threshold at ~ 8–10 MeV.[23,33,34,34a,b] Another method for measuring the energy spectrum of neutrons with energy between 5 and 20 MeV has been explored by Frank and Benton,[35] who placed various thicknesses of Au absorber between a beryllium radiator and sheets of cellulose nitrate plastic. In such a system the absorbers prevent recoils from interactions of low-energy neutrons with

Neutrons

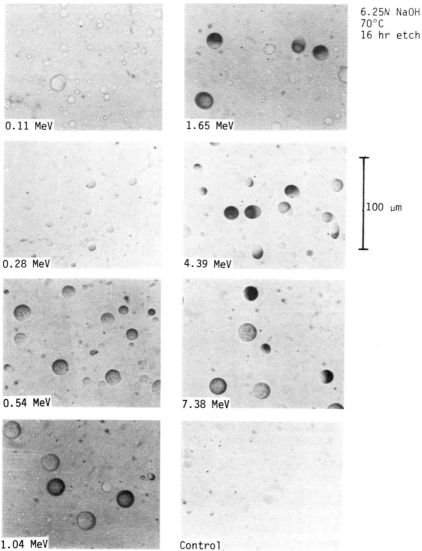

FIG. 7.6 *Etched tracks of recoil protons in CR-39 as a result of bombardment with monoenergetic neutrons of the various energies indicated on the photograph. The etching was in 6.25 M NaOH at 70 °C for 16 h. Tracks from 0.11 MeV neutrons are just observable, but are difficult to distinguish from the background. An unirradiated but etched foil, used as a control, is also shown. (Figure from Benton et al.[21])*

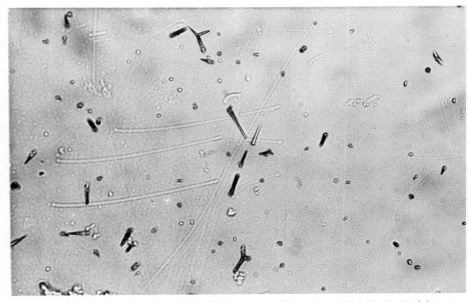

FIG. 7.7 *Photomicrograph of triple-α tracks in cellulose nitrate CA 80-15 (Kodak-Pathé), resulting from the reaction $^{12}C(n, n')$ 3α by 15.1 MeV neutrons. Since the reaction has an effective threshold of ~ 9.6 MeV, only those neutrons which have energies substantially above this value can yield triple-α tracks that can be revealed by etching. Such tracks can, thus, serve as a measure of the fluence of neutrons of energy $E_n \gtrsim 10$ MeV—e.g. Balcázar et al.[34] give values of $(8–11) \times 10^{-7}$ triple-α's per incident neutron for tracks produced on the back-surface of CA 80-15 by neutrons of energies ~ 14.7 to 19.1 MeV (cf. efficiencies of $\sim (2–3) \times 10^{-5}$ ordinary recoil tracks/neutron over the same energy range). It should be borne in mind that for a triple-α event to be revealed by etching, at least one of the α-tracks must cross the initial top or bottom etched surface of the plastic. The efficiency also appears to depend on etching conditions such as the length of etching time (since high-energy α's begin to etch out slowly at first.[34] The triple-α tracks have been used for computing the energy spectra of the high-energy incident neutrons by considering the kinematics of the reaction (Balcázar and Durrani.[34a]).*

the radiator from reaching the detector. Tuyn and Broerse[36] report the variation of recoil etch pit diameters with neutron energy. They find that the pit-diameter distribution is also dependent on the angle of incidence of the neutrons.

We have so far been primarily concerned with personnel neutron dosimetry. Track detectors have, however, also found application in other areas of neutron dosimetry. Tuyn and Broerse[26] and Becker[25] have used track detectors of depth dose studies in "phantoms" exposed to neutron beams.

Petoussi et al.[36a,b] have studied the spatial distribution of charged particles produced by high-energy collimated neutron beams in a tissue-equivalent water phantom with a view to determining the dose delivered to the human body in radiotherapy using high-energy neutrons (e.g. those generated by a beam of ~ 60 MeV protons from a cyclotron recently built near Liverpool,

England, for this purpose). The recoiling heavy ions (C, N, O) and α-particles from (n, α) reactions are of special importance in this context because of their high LET values, which may lead to intense, localized damage to human body cells. These authors have used a variety of plastic detectors (CR-39, LR 115, Lexan) in an attempt to distinguish between different types of recoiling particles (p; α; C, N and O heavy ions) on the basis of differences in the charge registration threshold of the various plastics. By deploying these plastic detectors at different coordinates within the phantom, both on and off the collimated neutron beam axis, the spatial distribution of the charged particle tracks, and hence of dose due to the different types of recoils, could be measured, and compared with theoretical predictions based on Monte Carlo calculations.

Gold et al.[37] have reviewed the application of track detector techniques to the study of neutron fluences in and around nuclear reactors. In-core measurements may require exposure of the detectors to very high temperatures. In such cases mica and quartz, rather than polymers, are used to record fission fragments emitted from various fission foils bombarded by neutrons inside the reactor core.

Green et al.[37a] describe the design of a passive environmental neutron spectrometer ("PENS") for fast-neutron measurements, which could, for instance, be deployed in reactor environment. In this design, fast neutrons produce (n, α) reactions in a ^{10}B converter, and the α-particles (which, besides sharing the positive Q value of the reaction, viz. 2.79 MeV, carry a part of the fast neutron's kinetic energy) are detected by an LR 115 foil. By interposing absorbers of different thicknesses (up to ~ 20 μm of polycarbonate plastic), the detector foils are effectively shielded from α's generated by neutrons below various threshold energies. The requirement of producing throughholes in the sensitive layer (~ 12 μm thick) of the LR 115, under standardized etching conditions, leads to the establishment of an energy-dependent response in terms of tracks per neutron. Computer modelling is used to assess the device: the α-energy (as gauged by the absorber thickness traversed) leading to the estimation of the energy of the neutron that produced the (n, α) reaction. The unfolding of the neutron spectrum in such an approach (especially when neutrons are not monodirectional) presents problems which have not been fully solved yet. For the general subject of neutron dosimetry and its current status, the reader is referred to a recent compilation of useful papers.[37b]

7.2 Alpha Particle Dosimetry and Radon Measurements

Alpha particles (unlike the neutrons) can, of course, form etchable tracks directly in a suitable plastic. The range of alpha energies over which this will occur is, however, limited. To a first approximation, plastics show a threshold,

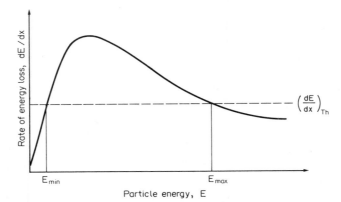

FIG. 7.8   *To a first approximation, track detectors may be characterized by a threshold $(dE/dx)_{Th}$, in the linear rate of energy loss, above which etchable tracks are formed in a given material. This behaviour leads to the existence of minimum and maximum energies, E_{min} and E_{max}, for detectable α-particles. The exact values depend on etching conditions, including the thickness of the removed layer (this is especially noticeable in the case of LR 115, where it is usually necessary to make through-holes in the top ∼6 or 12 μm-deep sensitive layer.[38a]) The highly sensitive CR-39 polycarbonate presents, practically, no upper and lower energy limits for α-particle detection.*

$(dE/dx)_{Th}$, in the rate of energy loss above which tracks are formed. Hence upper and lower limits on the energy of the α-particle are set by points at which the dE/dx versus E curve crosses the line $dE/dx = (dE/dx)_{Th}$ (see Fig. 7.8). In practice, E_{min} and E_{max} corresponding to the formation of identifiable tracks may differ somewhat from this definition. E_{min} will usually be that energy at which the range of the α-particle becomes too short for the resulting track to be readily observed. E_{max} will be the energy at which V_T, the track etching rate, becomes so low that tracks of length (or diameter) insufficiently large to be recognizable are formed in a given etch time. This upper limit will therefore be dependent on the etching conditions.[38,38a]

Table 7.2 sets out approximate E_{min} and E_{max} values for a number of α-sensitive plastics in common use. CR-39, the most sensitive plastic, will record α-particles of several tens of MeV in energy;[29] so that, for almost all practical dosimetric purposes, this upper energy limit presents no constraint. For the other polymers, E_{max} ranges from $\lesssim 1$ MeV up to ~ 6 MeV. In certain circumstances, where high-energy α-particles are to be detected, a layer of some stopping material should be included between the α-source and the plastic in order to degrade the α-energies into the domain to which the plastic is sensitive;[39] alternatively, prolonged etching may also reveal tracks of high-energy α's.

TABLE 7.2   *Limits on detectable α-particle energies for some plastics**

Type of plastic	E_{min} (MeV)	E_{max} (MeV)
Lexan and Makrofol E polycarbonates	~ 0.2	~ 3
Cellulose nitrate (CA80-15; CN 85; LR 115†; Daicel)	~ 0.1	4–6
CR-39 (homopolymer of allyl diglycol carbonate)	~ 0.1	> 20

* These upper and lower α-energy limits (E_{max} and E_{min}, respectively) are intended as rough guides only. They will depend on the etching conditions and on the exact form of the plastic used. In the case of the polycarbonates, post-irradiation exposure to u.v. light facilitates the revelation of α-particle tracks.

† In the case of LR 115, which has a sensitive layer (~ 12 μm thick in "type II" film and ~ 6 μm in "type I" film) followed by an inert backing, through-holes in the front layer can only be made if a certain thickness of the top layer is removed by etching; this places upper and lower limits on the energy of the detectable α-particles.[38a] LR 115 "type I" film is more suitable for registering low-energy α-particles from (n, α) reactions with ^{10}B and ^{6}Li.

In the natural environment, the vast majority of α-emitting nuclei are derived from the decay series of ^{238}U, ^{232}Th and ^{235}U. From the radiological point of view, the most important of these nuclides are ^{222}Rn and ^{220}Rn (from the ^{238}U and ^{232}Th series, respectively) and their α-emitting descendants.[40] Radon, being an inert gas, can readily diffuse through solid matter and enter the atmosphere; it may then be inhaled and thus present a health hazard. The non-gaseous daughter products of radon (e.g. ^{218}Po and ^{214}Po in the case of ^{222}Rn) may be deposited within the lungs. Typical doses to the population, as a result of radon emission inside dwellings, etc., range from ~ 1 mSv to ~ 10 mSv per year.[40,52b] The decay products of ^{222}Rn and ^{220}Rn (sometimes called thoron, and given the symbol Tn) are listed in Tables 7.3a,b.

Note that in the case of a long-lived radioactive parent and short-lived daughter(s), "secular equilibrium" is reached after many half-lives of the latter. If the decay constant of the parent (say ^{238}U) and the daughter (^{222}Rn) are

TABLE 7.3a   *The decay products of* ^{222}Rn, *a gaseous member of the naturally occurring radioactive series* ^{238}U → ^{206}Pb

Traditional name	Isotope	Half-life	Radiations	α-particle decay energy (MeV)
Radon	^{222}Rn	3.82 days	α	5.49
Radium A	^{218}Po	3.05 min	α	6.00
Radium B	^{214}Pb	26.8 min	β, γ	–
Radium C	^{214}Bi	19.8 min	β, γ	–
Radium C'	^{214}Po	164 μs	α	7.69
Radium D	^{210}Pb	22.3 years	β, γ	–
Radium E	^{210}Bi	5.01 days	β	–
Radium F	^{210}Po	138.4 days	α	5.30
Radium G	^{206}Pb	stable		

Solid State Nuclear Track Detection

TABLE 7.3b   *The decay products of* ^{220}Rn, *a gaseous member of the
naturally occurring radioactive series* ^{232}Th → ^{208}Pb

Traditional name	Isotope	Half-life	Radiations	α-particle decay energy (MeV)
Thoron	^{220}Rn	55.6 s	α	6.29
Thorium A	^{216}Po	0.15 s	α	6.78
Thorium B	^{212}Pb	10.64 h	β	–
Thorium C	^{212}Bi	60.55 min	$\begin{cases} β \ (66.3\%) \ (\text{to } {}^{212}\text{Po}) \\ α \ (33.7\%) \ (\text{to } {}^{208}\text{Tl}) \end{cases}$	– 6.1
Thorium C'	^{212}Po	0.3 μs	α (to ^{208}Pb)	8.78
Thorium C''	^{208}Tl	3.05 min	β (to ^{208}Pb)	–
Thorium D	^{208}Pb	stable		

designated by λ_1 and λ_2, respectively, then in secular equilibrium $\lambda_1 N_1 = \lambda_2 N_2$, where N_1 and N_2 are the respective numbers of nuclei; similarly, in the case of ^{232}Th (with parameters λ_3, N_3) in secular equilibrium with its decay products including ^{220}Rn (with λ_4, N_4), $\lambda_3 N_3 = \lambda_4 N_4$. Now we know that the value of λ_1 is roughly $3 \times \lambda_3$; so if, for simplicity (as is justified by the natural abundances of ^{232}Th and ^{238}U), we take $N_3 \simeq 3N_1$, then: $\lambda_1 N_1 = \lambda_2 N_2 \simeq \lambda_4 N_4$. This implies that, assuming secular equilibrium in both cases, the activities of radon (^{222}Rn) and thoron (^{220}Rn or "Tn") are roughly the same in nature (even though the ratio N_4(thoron)/N_2(radon) = $\lambda_2/\lambda_4 = \tau_{1/2}(\text{Tn})/\tau_{1/2}(\text{Rn}) = 55.6\text{s}/3.82\text{d} \simeq 1/6000$). Provided that the earlier members of the two natural decay chains (all of which are solid products) are in internal secular equilibrium, if the chains are broken by the loss of the gaseous members, the latter, i.e. radon and thoron, will again attain their secular equilibrium values within a few (respective) half-lives (i.e. a couple of weeks and a few minutes, respectively). The rates of production of α-particle tracks on a plastic detector exposed to them will, thus, be comparable from the decay of radon and thoron under the above conditions.

High concentrations of radon are encountered by uranium miners; and there exists a need to monitor their exposure to radiation from this source. In view of the difficulties involved in lung dosimetry, it is usual to measure the exposure to radon and its daughter products in terms of the radiological content of the air rather than absorbed dose.[41]

Essentially, two types of dosimeter are in use: active and passive. In the active form, air is pumped through a filter and a detector is used to measure the α-particles emitted by collected radon daughter nuclei.[41–46] The passive dosimeter[47–49] generally records only those decays that occur in the air. The basic concepts of radon dosimetry have been reviewed by Frank and Benton.[41]

In radon dosimetry it is useful to define two terms: the "potential energy" contained in the air, and the "working level". The potential energy (PE) is the amount of α-energy which will be eventually deposited per litre of ambient air by radon *daughters* in decaying down to the comparatively long-lived ^{210}Pb. The working level (WL) is defined as that concentration of short-lived *daughter* nuclei (in whatever proportions) per litre of air which contains a potential (α-particle) energy of 1.3×10^5 MeV.[50]*

Frank and Benton[41,46] describe an active track etch dosimeter consisting of a pump, a filter, collimators, and two degrader foils adjacent to two cellulose nitrate detectors. The first degrader foil is of such a thickness as will bring the residual energy of the 6.0 MeV α of ^{218}Po down into the range to which the cellulose nitrate is sensitive. The other (thicker) foil will bring both this α-particle and the 7.69 MeV α from ^{214}Po into the sensitive region. Frank and Benton[41] then show that the track densities ρ_1, ρ_2 per cm^2 measured in the detectors behind these foils are related to the exposure in "working level hours" WLH (i.e. exposure time in hours $\times$ concentration of daughter nuclei expressed in working levels) through the equation

$$\text{WLH} = (4.62\rho_1/\varepsilon_1 + 5.92\rho_2/\varepsilon_2)/(10^5 \times V)$$

where ε_1, ε_2 are efficiencies of track registration in the plastic, and V is the air sampling rate in litres per hour. (The numbers 4.62×10^{-5} and 5.92×10^{-5} simply come from dividing the energies of the two α-particles, viz. 6.0 MeV and 7.69 MeV, respectively, by 1.3×10^5 MeV to convert them into WLs.) This active dosimeter showed a good linear relationship between track density and exposure, in the laboratory, and was due to be tested while worn by working miners. Active dosimeters suffer from problems owing to dust clogging up the filter and reducing the air flow after extended use.

Passive radon dosimeters have also been constructed using track detectors.[47-49] Lovett[48] used a cellulose nitrate sheet mounted on a miner's helmet. Here, too, care must be taken to avoid changes in the response of the plastic due to a build-up of dust on the detector surface.

The measurement of α-particles from radon and its daughter products is of interest in areas other than uranium mines. Frank and Benton[51] have used a passive track-etch type of dosimeter to measure radon levels in houses built on uranium-ore tailings in Colorado, USA. Fremlin and co-workers[40,52] have used both active and passive devices employing track detectors to measure radon concentrations within normal domestic dwellings in the United Kingdom.

In view of the potential health hazard presented by radon emanation inside houses, mines, etc., the measurement of radon and its α-emitting descendants

---

* The numerical factor (1.3×10^5 MeV) comes from defining WL as the total α-energy ultimately delivered by short-lived radon daughters (^{218}Po, ^{214}Pb, ^{214}Bi and ^{214}Po; see Table 7.3a) all decaying in radioactive equilibrium with 100 pCi of ^{222}Rn (their progenitor) in a litre of ambient air.

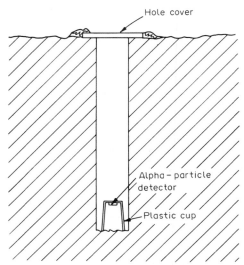

Hole cover

Alpha – particle
detector

Plastic cup

Fig. 7.9   *Radon detector arrangement used in uranium explora-*
tion. Cups containing α-particle detectors (usually cellulose nit-
rate or CR-39) are buried a little below the surface in a promising
geological area. The α-particles emitted in the decay of ^{222}Rn
and ^{220}Rn *(issuing from the uranium and thorium ores, respec-*
tively) and of their short-lived daughter products are recorded in
the detectors usually over several weeks of burial, during which
time steady-state is reached between the production and decay of
randon $(^{222}Rn\ \tau_{1/2},\ 3.82\ d;\ ^{220}Rn\ \tau_{1/2},\ 55.6\ s)$. *Detectors*
retrieved from cups, which are usually located on a grid pattern,
are etched, and the α-particle track densities found across the grid
help pinpoint the positions of the ore deposits. Various methods
exist for discriminating between "radon" (^{222}Rn) *and "thoron"*
(^{220}Rn), *e.g. on the basis of the energy of their decay-α's by*
using multi-detector devices,[57a,b,c] *or the half-lives of the two*
isotopes.[57d,e] *Similar arrangements have been employed for detec-*
ting the increased emanation of Rn caused by volcanic activity,
which, it is hoped, may be useful in predicting earthquakes. (Fig-
ure after Fleischer and Mogro-Campero.[56]*)*

(including the "plate-out" effects,[51a] i.e. deposition of the solid radioactive
daughter products) with the help of plastic track detectors has, in recent
years, grown into a vast and expanding new field.* The reader is referred to
references 52a,b for recent developments in this field.

The measurement of radon emanations has also been used as a method
of pinpointing uranium-ore deposits.[53–57a] The usual method is to place
plastic detectors inside cups and to bury these at a small distance below the

---

* Radon is estimated to constitute fully ∼40% of the average annual dose received by the
general population from *all* sources:[52b] it represents by far the strongest single source of radiation exposure for humans.

surface of the soil in a grid pattern covering the area to be explored, which may have been designated by geologists as a promising uranium-bearing tract. Radon gas, issuing from the uranium and thorium ores, and its daughter products are detected via α-particle tracks recorded in the plastic (usually cellulose nitrate or CR-39). The distribution of track-counts across the grid helps in pinpointing the positions of the richer ore deposits. There is evidence that the radon transport is faster than can be accounted for simply on the basis of diffusion of the emanation, and some bulk flow of radon-bearing air must be occurring.[56] A typical radon-detector arrangement used in such work is shown in Fig. 7.9. Incidentally, Fleischer and co-workers[56,58] have also used this technique in attempts to predict earthquakes, on the premise that unusual subterranean activity and possibly attendant fissures and strains result in the release of larger quantities of radon gas from the U and Th content of the earth's crust than at normal times. Radon has also been used as a tracer for studying exchanges in the troposphere.[58a]

7.3 Charged Particles other than Alphas

A variety of applications of track detectors to the dosimetry of heavily ionizing particles have been explored. Many of these are described in the book by Fleischer, Price, and Walker.[59] Two of the more interesting applications are described below.

7.3.1 Dosimetry of high-LET radiations in space

During space flights (e.g. Apollo missions), and over prolonged excursions in space, such as those undertaken by the crews of the Skylab and Cosmos spacecraft, high-LET primary cosmic ray particles constitute a significant radiation hazard to the astronauts.

Plastic track detectors were carried aboard Gemini IV and Gemini VI[60], and later aboard Apollo missions[61,64] as well as Skylab[63,64] and Cosmos.[65] Typically, fluxes of a few high-LET particles per cm² per day were measured.[64] High LET in this context usually refers to particles losing more than about 100 keV per micrometre (μm) of trajectory, via secondary electrons of energy < 350 eV. Such particles form tracks directly in sensitive plastics such as cellulose nitrate. These detectors have been used to determine the number of particles as a function of LET as well as the charge spectrum of particles incident upon the astronauts. It has been found that the rate of incidence of these high-LET particles during different space missions correlates with the solar activity, as might be expected.

Henke *et al.*[66] have described an automated system for measuring the charge and energy spectra of cosmic ray particles. The sheets which comprise a stack of plastics are etched to remove the surface layers of the detector to a considerable depth. The tracks then have the appearance of large through-etched holes. The holes are of reasonably high contrast, and are readily identified (by employing a suitable selection of "grey-level" threshold) by a computer-controlled optical digitizer. The area of the track hole is measured, and the position recorded. By following a track through several sheets, and measuring the hole area at each sheet, the charge of the particle can be estimated.

7.3.2 *Microdosimetry of negative-pion beams*

It has been proposed[67,68] that beams of negative pions (π^-) could provide valuable tools in radiotherapy. These pions lose energy by ionization as they slow down in a stopping medium; but they may also undergo strong interactions with atomic nuclei, leading to the liberation of large amounts of energy and to the disruption of the nucleus. This energy appears as the kinetic energy of the nuclear fragments, and is thus deposited with high LET in the region of the π^--nucleus interaction. Since these reactions take place most frequently close to the end of the pion trajectory, where the particle is moving more slowly, a considerable fraction of the pion-beam energy is deposited in the volume of material near to the end of the pion range; this region is known as the peak region of the beam. The preceding portion of the particle trajectory, extending from the surface of the absorber up to the onset of the peak, is known as the plateau region. Typically, the peak-to-plateau dose ratio is found to be about 1.6. The high-LET fragments are of considerable importance in radiotherapy because of their localized energy deposition and the enhanced biological effects that this produces.

Studies of the distribution of high-LET events in a π^- beam have been made by Knowles et al.,[69] Benton et al.,[70] and Bull.[71] In the last-mentioned experiment, several "sandwiches", each consisting of cellulose nitrate and Makrofol polycarbonate detectors, were irradiated in water by a beam of π^- mesons of momentum 146.4 MeV/c produced by the interaction of 7 GeV/c protons on a 10 cm tungsten target. The track detectors were deployed at four positions in the beam and the total doses absorbed at these positions were measured using an air-equivalent ionization chamber.

The results of both this study and that of Knowles et al.[69] clearly indicate that the peak to plateau ratio of high-LET events is far higher (~ 10) than the total dose ratio measured by the ionization chamber. The threshold LET values for the cellulose nitrate and Makrofol are about 100 keV/μm and 300 keV/μm, respectively.

Because of their compactness, plastic track detectors could be used to measure high-LET dose-distributions around the body of the patient during radiotherapy.[69,36a] Also, they have the advantage that their composition is approximately tissue-equivalent.

7.4   SSNTD Instrumentation: Automatic Evaluation and Methods of Track Image Enhancement

So far, we have concerned ourselves with the physical processes involved in the detection of various types of radiation. We now consider the means by which information stored in track detectors may be easily extracted. This is an extremely important factor in personnel dosimetry. Whereas in many research fields involving track detectors the analysis of a large number of tracks under a high-power microscope may be a viable proposition, in dosimetry it will generally be the case that a large number of detectors must be dealt with on a routine basis. If, therefore, track detectors are to provide a realistic means of radiation monitoring, some methods of rapid evaluation of these detectors must be devised. The SSNTD instruments described below have, of course, applications in many other fields of track research as well, e.g. elemental distribution studies, nuclear and reactor physics, geology and archaeology.

7.4.1   The spark counter

The spark counter technique was first devised by Cross and Tommasino,[72] and has since undergone several improvements.[73,73a,b] A typical circuit and electrode design for a spark counter are shown in Fig. 7.10. A detector sheet, etched in such a way that through holes have been made in it along (at least the steepest) tracks, is placed on top of the electrode and covered with a thin plastic foil aluminized on its lower surface (or with a thin Al foil supported by a plastic backing) as shown. When the switch is closed, the top plate of the capacitor C_1 is connected to the earth via resistance R_1, and there is no voltage across C_1 or across the electrodes. When the switch is opened, C_1 is raised in potential towards E_0, and a voltage appears across the electrodes and hence across the etched detector. Eventually, a discharge takes place between the cathode and the anode (via the aluminized surface of the Mylar* foil) across the etched track whose hole represents the easiest path for the current; C_1 is discharged and a current pulse flows. A voltage pulse appears across R_1, which may be processed and counted by the external circuitry, which usually includes a scaler. The flow of current, i.e. the spark, "burns out" a hole in the thin (typically $\sim 1 \ \mu m$) Al coating above the track end. This breaks the conducting path for the current through that particular track. The short-circuit ceases, and the voltage across C_1 again rises towards E_0. The first track is prevented from allowing a subsequent spark to take place through it because of the hole in the aluminium; and so the anode voltage rises until a discharge occurs across another track hole. The processes described

---

* Mylar is the registered trade mark of Dupont Chemicals.

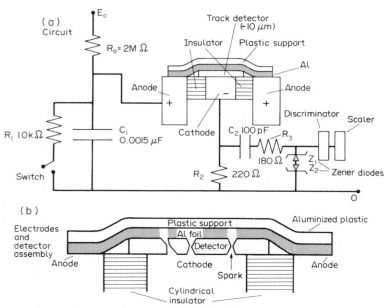

FIG. 7.10a,b   *Details of the circuit (*a*), and electrode design and detector assembly (*b*) for a spark counter. The anode and the cathode of the detector consist of two coaxial cylindrical conductors separated by an insulator. The irradiated plastic detector foil (~ 10–20 μm thick), etched so as to produce through-holes, is placed on the cathode, and covered by another plastic foil ~ 100 μm thick (essentially, for support as a backing), which is thinly aluminized on the lower side to offer a conducting path. When the switch is opened, the capacitor C_1 is raised in potential towards the applied voltage E_0, and a voltage appears across the electrodes and hence across the etched detector. Eventually, a discharge takes place between the anode and the cathode across an etched track. Sparks jump through different holes in the detector foil in random sequence: but only once per through-hole, since each spark destroys the conducting Al element in its vicinity. The sparks are counted by a scaler via a discriminator. After each spark the capacitor C_1 needs tobe recharged by the applied voltage E_0 to provide sufficient potential for the next spark. The aluminized surface of the plastic (usually Mylar) retains an easily visible replica of the distribution of tracks (e.g. the fission tracks from a uranium-bearing mineral sample, thus displaying the spatial distribution of uranium on, and just below, the sample surface). (Figure after Malik and Durrani,*[73a] *with modifications.)*

above are repeated and a second voltage pulse is recorded. This sequence continues until all track holes have been counted.

The method is quick, and the circuitry relatively simple. Also, the Al foil shows, at the end of counting, a number of holes punched through it (typically ~ 100 μm in diameter), which are equal to the number of tracks counted and which correspond exactly to the spatial distribution of the original tracks in the plastic detector. These holes may be counted easily by using a low-power microscope, and a permanent record of the number and distribution of tracks is thus obtained.

Incidentally, the spark counting technique has many other applications, besides dosimetry. Thus, the distribution of uranium or boron (which have high cross-sections for (n, f) and (n, α) reactions, respectively) in a sample can be accurately replicated on the Al foil by this method after irradiating a smooth section of the sample with thermal neutrons, while it is held in contact with the thin plastic detector-foil (see e.g., references 73a, 73c, 74, and also Chapter 9, for further details).

The voltages usually employed for counting are about 500 V. It is usual for the detector foil (typically 10–20 μm thick) to be given a pre-counting run at a higher voltage (say $\sim$ 700–900 V) in order to punch out holes which are originally not quite etched through (e.g. those from oblique or "sub-surface" tracks). This results in subsequent track counts which are more reproducible than they would otherwise be. After this "pre-sparking", the detector foils may be spark-counted repeatedly if desired, provided that a fresh area of aluminized foils is placed over the tracks on each occasion.

By sparking track detectors at several voltages it may be seen (cf. Fig. 7.11) that a plot of track count against sparking voltage yields a plateau region. The central region of the plateau will indicate the optimum counting voltage.[73a]

The requirement that track holes should extend nearly through the plastic foil limits the application to plastics of $\sim$ 10–20 μm thick for most ordinarily encountered charged particles. Fission-fragment tracks are especially suited to being counted in this way. Many of the plastics which may be obtained in very thin sheets (e.g. Makrofol KG*) are not sensitive to α-particles. Also, α-particles do not generally result in tracks of such small cone angle or long etchable range as do fission fragments. Thus the application of spark counting to tracks of more lightly ionizing particles, e.g. α's and neutron recoils, has been more difficult. Recently, however,[74,75] success has been achieved in spark-counting thin foils of α-sensitive plastic. The top thin red layer ($\sim$ 12–14 μm thick in the case of "type II" films, and $\sim$ 6–8 μm in "type I" films of the strippable LR 115 cellulose nitrate made by Kodak-Pathé of France), which can be floated off its thick ($\sim$ 100 μm) inert backing by soaking in warm water—usually after irradiation—, may thus be used for α-particle spark-counting. An alphagenic converter screen may be used in contact with the top layer of the detector to record (n, α) reactions from thermal (or fast) neutrons.

* Manufactured by Bayer AG of Leverkusen, W. Germany.

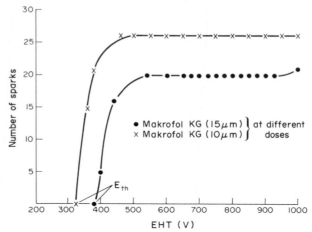

FIG. 7.11 *Operational characteristics of a spark counter for the final counting of through-holes. The thin (10 μm and 15 μm) detector foils, irradiated with two different fluences of normally incident fission fragments from a ^{252}Cf source, had previously been etched in 6.3 M NaOH at 60°C for 45 min, and had then been "pre-sparked" at 1000 V to produce (and enlarge) through-holes. The figure shows the variation of spark counts with the applied voltage for the final counting of the fission track holes using a fresh aluminized Mylar sheet. E_{th} is the threshold voltage, for the two thicknesses, at which the spark discharge sets in for these enlarged holes, and is soon thereafter followed by a broad plateau. A counting voltage well-ensconced within the plateau (e.g. 500 or 600 V in the above case) is then used for track counting and replication purposes. Typical counting efficiencies of 0.95 ± 0.05 are obtained, as normalized to visual counting with an optical microscope. (Figure from Malik and Durrani.[73c])*

The technique, however, is not without disadvantages. The first problem is the large size of the discharge spots formed in the Al layer. Since these are often a few hundred μm across, overlap becomes serious at track densities above $\sim 10^2 - 10^3$ tracks cm^{-2}. This may not be too serious a limitation in personnel dosimetry, where the low neutron doses generally encountered lead to a paucity, rather than an overabundance, of tracks produced either through ^{235}U fission or through alphagenic reactions. For radon dosimetry, too, the large discharge spots are usually no constraint. In the case of elemental analysis, the thermal neutron fluence may be chosen judiciously to yield a comfortable density of track spots on the aluminized foil.

The second problem is presented by tracks (in 2π-geometry) which are incident on the detector at low angles and hence may stop far short of the lower surface of the detector. Thus Balcázar-García[76] has shown that the efficiency of the spark counter decreases rapidly as the angle of incidence of the track with the detector surface decreases. This may result in a considerable angular dependence in the response of fast-neutron dosimeters based on neutron-recoil detection. (This will be less serious where (n, f) reactions are used, since the direction of the fission fragments will not be strongly dependent, if at all, on the neutron direction.) The problem for shallow tracks encountered in 2π-exposure to α-particles or fission fragments (say in fission-track dating), however, does remain—though it is reduced by employing thinner detector foils.

7.4.2 Other electrical-breakdown devices

A new detection method has recently been devised by Tommasino and co-workers[77] which also relies on breakdown phenomena. This detector has the advantage that it operates in real time, i.e. counts are recorded as the particles (specifically, fission fragments) penetrate the detector. The detector does not utilize etched tracks, and so it is somewhat outside the mainstream of the contents of this book. Nevertheless, being conceptually related to other devices discussed in this section, it seems worthy of a brief description.

The device consists (see Fig. 7.12) essentially of a commercially available thin-film capacitor of the metal–insulator–metal or metal–insulator–silicon type. The upper electrode is connected to a voltage supply V via a resistor. The passage of a fission fragment through the capacitor results in the breakdown of the insulator at an electric field strength (and thus applied voltage) which, in the set-up used by the authors,[77] was considerably less than the ~ 8 MV cm^{-1} (and the corresponding voltage) required for the breakdown of the undamaged insulator. Typically, applied voltages in the region of 30 V are used, and the insulating film—and the electrodes—are a few tens of nanometres thick. Breakdown spots are produced in the electrode, which, as with the spark counter, terminate the breakdown process for each individual fission fragment. Weak spots in the capacitor also produce localized breakdowns, and these must be cleared by running the device at a suitable voltage prior to irradiation. The change in the capacitance of the device resulting from prolonged bombardment with fission fragments may be monitored; this is related to the decrease in the effective area of the detector due to breakdown-spot formation. This factor may thus be allowed for in counts obtained late in the life of the detector. Only heavily ionizing particles (e.g. fission fragments) seem to be detectable by this technique; α-particles from ^{241}Am failed to produce the necessary breakdowns.

7.4.3 Scintillator-filled etch pit counting

Harvey and Weeks[77a] have recently described an elegantly simple technique for the rapid assessment of etched tracks, using the method of scintillation

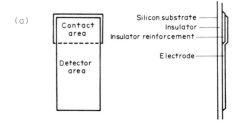

(a)

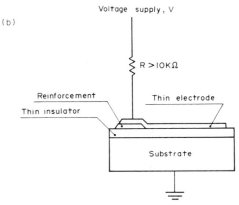

(b)

FIG. 7.12a,b *Schematic diagram of the apparatus used as a breakdown detector. The detector responds, in real time, to the passage of heavily ionizing particles (e.g. fission fragments from a nearby ^{252}Cf source) through the thin-film capacitor. The capacitor shown here (see (a)) is of the metalinsulator–silicon substrate type (e.g. Al-SiO$_2$–silicon). The upper electrode is connected, via a resistor R, to a voltage supply V (see (b)), which produces a high electrical field across the thin insulator (e.g. 30 V across a 40 nm film creates a 7.5 MV cm^{-1} field). The passage of a fission fragment through the capacitor results in the breakdown of the insulator at the applied electric field, producing spots—typically ~ 10 μm across—on the electrode (as in the case of a spark counter, Fig 7.10) by vaporizing a submicron-size hole through the insulator. The burning of a hole in the electrode terminates the breakdown process for each individual fission fragment. The applied voltage then recharges the capacitor in readiness for the registration of the next fission fragment passing through the detector assembly. (Figure after Tommasino et al.[77])*

counting. In essence, etch pits formed by charged particles (α's, fission fragments) in CR-39 plastic foils are filled with a suitable scintillator paste, and the excess wiped out. An intense source of α-particles, placed close to the scintillator-filled etch pits in a light-tight assembly, is then used to produce scintillations which are picked up by a photomultiplier (PM), and the output

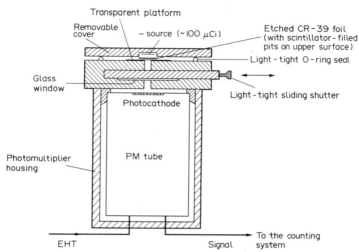

FIG. 7.13 *Scintillator-filled etch-pit counting system, using a photomultiplier. The detector to be counted (e.g. a ∼ 200 μm thick CR-39 foil) is chemically etched to produce sufficiently deep etch pits, which are then filled with a fine-grain (< 2 μm) scintillator (e.g. ZnS:Ag), made into a firm paste with ethylene glycol, etc.: the excess paste being carefully wiped off. The detector is placed in a light-tight assembly above the photocathode of a photo-multiplier (PM) tube, with the etch pits facing upwards. A strong α-source (e.g. 100 μCi* ²⁴¹*Am in the form of a strip) produces scintillations, which, when the shutter is opened, are observed through the body of the foil by the photocathode. A steel washer (I.D. ∼ 6 mm; not shown) is interposed above the CR-39 foil to define the area from which the scintillations are being observed. The signal, after suitable amplification, is fed, in a photon-counting mode, to a scaler (with an appropriate discriminator) or to a multi-channel analyser. One must ensure that a linear relationship exists between the microscopically counted track density and the photon counts. The method is very much quicker than visual counting, and is now routinely used by, for instance, the National Radiological Protection Board, Chilton, England, for radon measurements. Figure after Harvey and Weeks.*[77a]

fed to a scaler (see Fig. 7.13). The number and size distribution of the scaled pulses are found to be related to the number and size distribution of the etch pits.

These authors have used a fine-grain (< 2 μm) scintillator, ZnS:Ag, made into a firm paste with ethylene glycol (commercially available car-antifreeze fluid). CR-39 foils exposed to α-particles were chemically etched for 16 h in 6.25 M NaOH at 70 °C to make sufficiently deep etch pits. (A light pre-etching step helped reduce surface defects and background tracks prior to irradia-tion.) The CR-39 discs coated with the paste were subjected to a poorish vacuum to remove any air trapped in the etch pits, and excess scintillator mixture removed by careful wiping.

For readout, the discs, placed on a transparent platform above the photo-cathode of the PM tube, were exposed to an ^{241}Am α-source, ~ 100 μCi in strength, inside the light-tight assembly shown in Fig. 7.13, with the etch-pit-containing side up, i.e. facing the α-particles (which would not go through the ~ 200 μm-thick CR-39 foil). The scintillations produced in the etch pits were transmitted through the body of the CR-39 foil and detected by the photomultiplier (e.g. an EMI 6097 tube, run at 800 V). The output of the PM, in the form of electrical pulses resulting from individual photons incident on the photocathode, was, after amplification, fed to a scaler with an appropriate discriminator (or, alternatively, to a multichannel analyser (MCA)). A linear relation was found to exist between the exposure time (or the track density as measured by an optical microscope) and the counts recorded by the scaler or the MCA.

The above technique provides a simple and rapid method of counting etch pits, and is especially suited to situations where a large number of detector foils need to be evaluated by automatic means, and where the track densities are not too high, e.g. in the measurement of environmental α-activity such as radon emanation in dwellings, or even low fluences of thermal neutrons (by way of (n, α) reactions).[77b] Thus Pedraza et al.[77c,d] have successfully used the technique to measure radium concentrations in building materials such as "breeze-blocks" (lightweight aggregate blocks made by adding sand and cement to the residue from coal-burning at power plants). The technique has been recently extended to the measurement of "intrinsic" tracks made by recoiling H, C, and O nuclei in the CR-39 foils exposed to fast neutrons;[77e] other heavy-ion tracks (in addition to α- and fission-fragment tracks) may also be similarly detected. Further work, however, needs to be done to ensure linearity of response, elimination of background and spurious counts, optimization of etching conditions and source-strength matching, etc. (see also ref. 77f).

7.4.4 Electrochemical etching (ECE)

This is a method, not of automatic counting as such, as in the preceding sub-sections, but of enlarging the etched tracks to sizes, typically a few hundred micrometres across, such that counting of low track densities becomes a relatively easy matter, requiring only low magnification.

The method was first suggested by Tommasino,[78] and has since been developed by him as well as by a number of other workers.[79-85] A recent variant of the apparatus is shown in Fig. 7.14. In its simplest form, a plastic detector foil is used to divide a cell containing a suitable etchant, e.g. NaOH, KOH, etc. A platinum (or stainless steel) electrode is placed in each half of

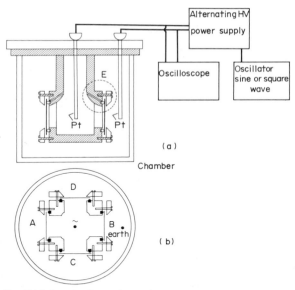

FIG. 7.14a,b   *An electrochemical etching (ECE) cell and its associated electronics (a). The cell shown is a multidetector device,
allowing four plastic detectors to be electrochemically etched
simultaneously. These detectors (A, B, C, D), each typically ∼ 100
to 500 μm thick, separate the inner and the outer parts of a container (b) filled with the etchant (e.g. 6 M NaOH at 60°C). Alter-
nating voltages (typically, several hundred to 1000 V) at high
frequencies (from ∼ 1 kHz to 100 kHz) are applied through a
pair of platinum electrodes (the rods marked Pt in (a)); the supply
is coupled to an oscillator and an oscilloscope. Although the
earthed electrode is placed opposite detector B in the figure (see
(b)), ECE spots appear with equal efficiency (and are of equal
diameters) on all four detectors. Alternatively, an electrode may
be placed against each detector, to minimize malfunctioning. (The
fine slanting channel shown within circle E (in (a)) allows the air
trapped in this and the other three O-ring grooves to escape.)
Typical ECE times for producing good etch spots are several
hours, sometimes preceded by chemical pre-etching in the same
chamber (without applying the electrical field, which in the ECE
phase may amount to several tens of kV cm⁻¹ peak-to-peak, as
measured on the oscilloscope). The ECE spots (typically
∼ 100 μm across) are normally produced on each face of the
detector foil. One-sided ECE may also be performed, if the inner
container is filled with a very weak etchant (e.g. 0.01 M NaOH).*
(*Figure from Al-Najjar et al.*[86])

the cell, and these are connected to a high-frequency oscillating voltage
supply. Typically, voltages are such that fields of ∼ 30–50 kV cm⁻¹ peak-to-
peak (or, in some applications, ∼ 15–40 kV cm⁻¹ rms) are produced across
the plastic detector. Frequencies in the range of several kHz to several tens

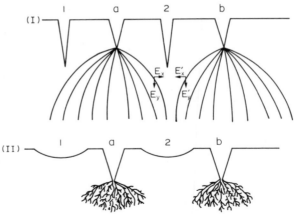

FIG. 7.15   *Stages in the production of electrochemical etch spots,
and other related phenomena. (I) Conical tracks are scooped out
by the etchant, and high electric fields ($\sim MV\ cm^{-1}$) develop at
the cone tips filled with the conducting fluid (the etchant), e.g.
at* a *and* b. *If, as here, tracks are in close proximity, some are
"shielded" from the electric field—e.g., in the figure, the horizontal components* E_x *and* E'_x *of the electric field, being in opposite directions, nearly or completely cancel each other out, leaving
an area of little or no field strength in between. (II) Electrical
breakdown ("treeing") of the insulator occurs at positions* a *and* b,
*where high electric fields had developed. Figure 7.15(II) suggests
a second state of affairs when some of the tracks may fail to
produce ECE etch spots: this is when some of the tracks may
have become over-etched (e.g. those marked 1 and 2 in II). This
may be done deliberately: say, by introducing a chemical pre-
etching step, which over-etches tracks due to particles with a
higher track etch rate* V_T, *e.g. fission fragments, while tracks due
to low-V_T particles such as α's produce conical etch pits (*a, b*) over
the same etching time. Such a stratagem allows one to discriminate
either between particles of different species (e.g. fission fragments
and α-particles from a* ^{252}Cf *source) or between particles of the
same charge but different energies (e.g. a mixture of α-particles)
by pre-etching the detector foils for different lengths of time to
produce underetched or overetched tracks of different types before
performing the ECE step. (Figure after Al-Najjar et al.[85])*

of kHz are employed (though devices employing mains a.c. voltages with
frequencies of ~ 50 Hz have also been constructed).

The theory of the electrical phenomena leading to the breakdown of the
dielectric and the consequent formation of the etch spots is not fully understood; but it is envisaged that the conducting etchant begins to etch out a
cone along the track path as shown in Fig. 7.15. The field lines become
concentrated at the cone tip, and high electric fields are thus produced in a
manner analogous to that familiar in electrostatics, in which high electric
fields are created in the vicinity of sharp points or corners on a conducting

material raised to a high potential. As the field oscillates, a repeated series of dielectric breakdowns will occur at each track tip. This apparently leads to severe damage over a considerable volume around the track tip, producing a large spot close to the etched track as shown in Fig. 7.16. The damage in the plastic foil is, of course, three-dimensional; the etch spots seen in Fig. 7.16 represent sections through the damaged region.

The situation may be interpreted in terms of Mason's equation[85a], which is applicable to "point and plane electrodes" embedded in a dielectric medium. Thus, if high voltage is applied between the tip of a needle (inserted from the "top" surface of a sheet of plastic) and a plane electrode attached to the "bottom" surface, Mason's equation gives

$$\frac{E_{tip}}{E_{av}} = \frac{2d}{r \ln\left(1 + \dfrac{4d}{r}\right)} \qquad \text{(for } d > 10r\text{)}$$

$$\simeq \frac{2d}{r \ln (4d/r)} \qquad \text{(for } d \gg r\text{)}$$

where E_{tip} is the electric field strength (or the "effective stress") at the tip of the needle (the etchant-filled track in our case), E_{av} is the average field strength *in the plane of the tip*, which is at a height d above the bottom surface, and r is the radius of the needle tip. If E_{tip} exceeds a critical value (say, E_{crit}), viz. the fundamental (or intrinsic) electric strength of the dielectric medium at a given temperature, "treeing" breakdown takes place. Since $E_{av} = V/d$, where V is the applied voltage, the above equation yields a value of $E_{tip} \simeq 2V[r \ln (4d/r)]^{-1}$ (for $d \gg r$). It is clear from the last expression that, for fixed values of V and the tip radius r, E_{tip} will vary inversely as $\ln(4d/r)$, i.e. the field strength at the tip of a long track originating from the top of the detector and reaching close to the bottom surface (so that d is small) will be higher than at the tip of a short track (d larger) of the same radius. This throws some light on the sequence or probability of treeing of different etched tracks; e.g. the track of a particle with a high V_T and sufficiently long range will, under the same ECE conditions, attain E_{crit} at its tip earlier or more readily (for similar r) than will a shorter track or one corresponding to a low-V_T particle. This fact allows particle discrimination to be achieved by adopting judicious combinations of chemical and electrochemical etching.[85]

In a practical case, with $V = 1000$ volt, $d = 100$ μm and $r = 1$ μm, Mason's equation gives the value of $E_{tip} \simeq 3.3$ MV cm^{-1}, which may exceed the fundamental electric strength of the plastic detector under the given ECE conditions (temperature, frequency of applied voltage, etc.). The radius of an etched track in the "conical phase" (§4.2.2) is, ideally, zero; in practice it may be assumed to be ~ 0.5–1 μm. Over-etching (say, by pre-etching) will blunt the end of the track and make treeing more difficult or impossible. It would

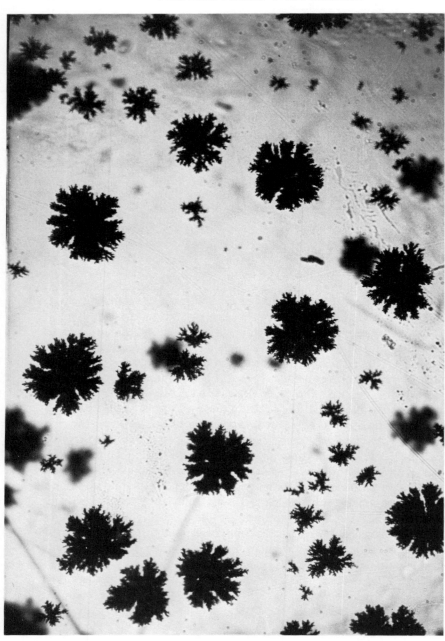

FIG. 7.16 ECE track-spots in CR-39 (485 μm thick) from the radon daughter ^{214}Po, which had been collected on a filter paper; pre-etching in 30wt% KOH at 60°C for 4 h; ECE for 3 h at a field strength of 15 kV cm^{-1} pk–pk and a frequency of 40 kHz. Some of the ECE spots have failed to develop fully because of local electrical shielding effects mentioned in the caption to Fig. 7.15. The damage in the plastic foil is three-dimensional (cf. Lichtenberg figures produced by electric

be desirable if the Mason equation could be re-expressed in terms of track-etching parameters, such as V_B, V_T, and the length of the etched-cone.* The dependence of the onset of ECE, and the growth of the size and number density of etch spots, on the frequency of the applied voltage[85b] and other etching conditions also needs further investigation.

A considerable amount of work has gone into finding the optimum electrochemical etching (ECE) parameters, viz. electric field strength, frequency, etching conditions such as the temperature and concentration of the etchant, and the duration of any ordinary pre-etching (see Somogyi,[80] Al-Najjar *et al.*,[85] Sohrabi[79]). Also, ECE cell design is important, as it is necessary to ensure that bubbles do not collect on the detector surface,[83,86] as these can lead to unwanted dielectric breakdowns. Figure 7.17a,b gives useful circuit diagrams for a variable voltage and frequency HT supply for ECE equipment, incorporating two different high-voltage amplifiers capable of a wide frequency response.[86a] Parts of this circuit may be replaced by semiconductors and integrated circuits. A suggestion for future development of the design would be to drive a specifically designed transformer with an integrated audio amplifier (as done by Thorngate *et al.*[86b]).

Recent designs of ECE chambers[85,86] allow several detectors to be simultaneously etched by arranging a number of cells around a central live electrode, with each cell separated from the surrounding electrolyte by a detector window facing its own earthed electrode placed nearby. Griffith[87] describes the system in use at Lawrence Livermore Laboratory (LLL) in which a large array of etch chambers is operated simultaneously for multiple-sample etching, but where each etch chamber holds an individual sample. The system has been designed in such a way that if one of the replicate samples sparks through or shorts out, it can be easily removed from the etching array and replaced by a blank plastic foil, in contrast to the earlier LLL system using a multiple-sample chamber when such a short would require the whole chamber to be disassembled. The new design also uses a much smaller volume of the electrolyte.

At first, electrochemical etching was applied mainly to fission fragment tracks; but later Somogyi[80] and Sohrabi[79,88] showed that alpha and neutron-recoil tracks may also act as sites for ECE spots. Some plastics respond better to ECE development than others. Lexan and most other polycarbonates are suitable, cellulose nitrate much less so. Fortunately the (from the track-etching point of view) newly discovered, highly sensitive CR-39 plastic responds excellently to electrochemical etching,[28,85,89] and even allows ECE development of proton tracks.[21,85,89] Wong and Tommasino[89a] have attempted to use the diameter of ECE spots as an indicator of the α-particle energy for spectrometric purposes.

---

* *Note added in proof.* For some recent developments along these lines see ref. 103.

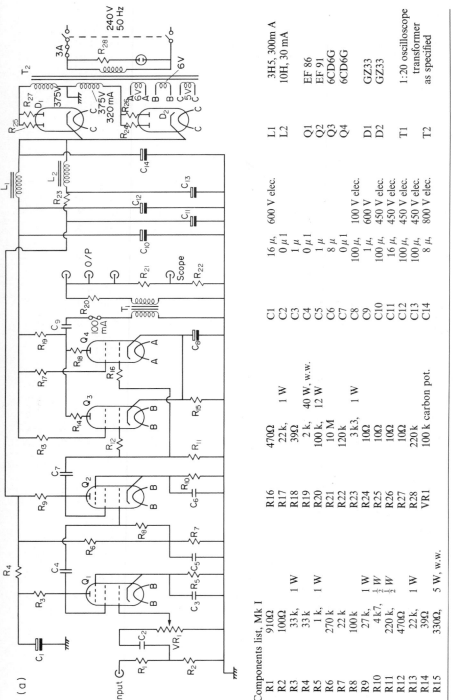

(a)

Components list, Mk I

R1	910Ω		R16	470Ω
R2	100Ω		R17	22 k, 1 W
R3	33 k, 1 W		R18	39 k
R4	33 k		R19	2 k, 40 W. w.w.
R5	1 k, 1 W		R20	100 k, 12 W
R6	270 k		R21	10 M
R7	22 k		R22	120 k
R8	100 k		R23	3 k3, 1 W
R9	27 k, 1 W		R24	10Ω
R10	4 k7, ½ W		R25	1Ω, 600 V
R11	220 k, ½ W		R26	10Ω
R12	470Ω		R27	10Ω
R13	22 k, 1 W		R28	220 k
R14	39Ω		VR1	100 k carbon pot.
R15	330Ω, 5 W, w.w.			

C1	16 μ, 600 V elec.	
C2	0 μ1	
C3	1 μ	
C4	0 μ1	
C5	1 μ	
C6	8 μ	
C7	0 μ1	
C8	100 μ, 100 V elec.	
C9	1 μ, 600 V	
C10	100 μ, 450 V elec.	
C11	16 μ, 450 V elec.	
C12	100 μ, 450 V elec.	
C13	100 μ, 450 V elec.	
C14	8 μ, 800 V elec.	

L1	3H5, 300m A
L2	10H, 30 mA
Q1	EF 86
Q2	EF 91
Q3	6CD6G
Q4	6CD6G
D1	GZ33
D2	GZ33
T1	1:20 oscilloscope transformer
T2	as specified

FIG. 7.17a,b   *Circuit details of two versions ((a) Mk I, and (b) Mk II) of a variable-voltage and variable-frequency amplifier and its HT supply for ECE. These high-voltage amplifiers are capable of a wide frequency response. In Mk I (Fig. 7.17a), two high-current output valves Q_3 and Q_4 are used in parallel. With this arrangement the output is 300 V pk-pk in the range 1–100 kHz; the voltage is stepped up 20:1 by transformer T_1 to make it sufficient for ECE purposes. In Mk II amplifier (Fig. 7.17b), a high-voltage valve Q_3, on the output, works off a 1.2 kV power supply. The amplifier is capable of delivering up to 1 kV pk-pk in the range 4–100 kHz. This design is capable of driving relatively low loads. Note that the following notation and abbreviations have been*

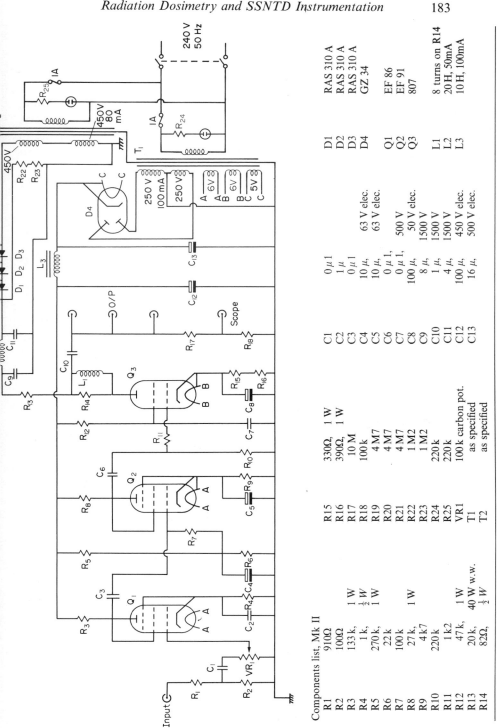

Fig. 7.17 *(continued.). For legend see caption on preceding page.*

Components list, Mk II

R1	910Ω	R15	330Ω, 1 W	C1	0 μ1	
R2	100Ω	R16	390Ω, 1 W	C2	1 μ	
R3	133k, 1W	R17	10 M	C3	0 μ1	
R4	1k, ½W	R18	100k	C4	10 μ,	63 V elec.
R5	270k, 1W	R19	4 M7	C5	10 μ,	63 V elec.
R6	22k	R20	4 M7	C6	0 μ1,	
R7	100k	R21	4 M7	C7	0 μ1,	
R8	27k, 1W	R22	1 M2	C8	100 μ,	500 V
R9	4k7	R23	1 M2	C9	8 μ,	50 V elec.
R10	220k	R24	220k	C10	1 μ,	1500 V
R11	1k2	R25	220k	C11	4 μ,	1500 V
R12	47k, 1W	VR1	100k carbon pot.	C12	100 μ,	1500 V
R13	20k, 40 W w.w.	T1	as specified	C13	16 μ,	450 V elec.
R14	82Ω, ½W	T2	as specified			500 V elec.

D1	RAS 310 A	
D2	RAS 310 A	
D3	RAS 310 A	
D4	GZ 34	
Q1	EF 86	
Q2	EF 91	
Q3	807	
L1	8 turns on R14	
L2	20H, 50mA	
L3	10H, 100mA	

Another important discovery was made by Somogyi,[80] who found that if tracks were pre-etched until rounded cone tips were formed, their subsequent ECE spot formation could be suppressed (since, presumably, high fields do not appear at "blunted" as opposed to sharp tips; cf. Fig. 7.15). This technique can be applied to discriminate between classes of tracks with different values of the ratio, V, of track etch rate to bulk etch rate (e.g. α-tracks versus fission tracks[85]). This method may also be used to reduce the background spot level in unirradiated plastics.[90,91] Background levels may be limited to values of the order of a few track spots per cm^2 (corresponding, in fast-neutron dosimetry, to a dose equivalent of a few tens of mrem[12]).

Once etched to produce ECE spots, a detector foil can be counted quickly under a low-power microscope, or on a microfiche reader. Also, the formation of large, optically dense regions around each track allows the technique to be adapted to various optical absorption methods of foil evaluation (e.g. microdensitometry[86]).

A disadvantage of the ECE technique, in common with spark counting, is that it imposes a rather low upper limit on the track density ($\sim 10^3$ cm^{-2}). Beyond this level, ECE spots tend to contract in size (presumably owing to reduction in the field strength developing at tracks in close proximity to each other[85]). Some tracks on the foil seem not to develop breakdown spots at all, the treeing process being inhibited, apparently, by a partial shielding of the electrical field. This, again, happens more frequently when the track density is high; thereby reducing the corresponding ECE "sensitivity" value (spot density/particle fluence) of the detector.[85,91a] These effects, however, do not prove to be significant in normal dosimetric work.

7.4.5 *Automatic and semi-automatic image-analysis systems*

None of the commercially available systems for automatic image analysis were originally designed specifically for the analysis of etched-track foils, although several "home-made" systems[92-94] have been developed or adapted for track work (see also the section on "Automation Systems" in the Proceedings of the 12th International Conference on Solid State Nuclear Track Detectors, *Nucl. Tracks* **8** (1984), 187-266).

The principle involved in image analysis may be briefly described as follows. The first step is to scan the etched plastic foil, usually by a closed-circuit television camera. In most cases, the camera takes the form of a conventional 625-line vidicon tube. The exception to this is the Quantimet 720 series* image analysers, which use a specially designed 700-line camera for improved image resolution. The next step in the analysis is to distinguish features of interest from the general surroundings. Two forms of feature discrimination are usually employed in the analysis of etched-track foils. The

---

* Manufactured by Imanco, Image Analysing Co., Melbourn, Herts, England.

first is the selection of a "grey level" such that only those regions of the image where the blackness exceeds this predefined lower limit are considered. This enhances the contrast of the image. The second form of feature discrimination is introduced by considering the shape of the features. The purpose here is to remove unwanted features that could not be excluded purely on the basis of the grey level setting.

Shape discrimination is not a very easy task to perform instrumentally. In favourable cases, however, e.g. when normally-incident tracks with circular openings are to be distinguished from other background features, shape discrimination based on the circularity of features can be applied quite successfully. Here, circularity may be defined as the perimeter to area ratio, which is a minimum for round features, and may be handled by the analyser. Other forms of shape discrimination are, however, more difficult to apply.

Ideally, an automatic image analyser should provide the operator with a variety of shape-discrimination techniques. In the older, hard-wired image analysers, e.g. the Quantimet 720, various plug-in modules allow a number of parameter determinations to be made, e.g. those of area, chord length, perimeter, etc. Combinations of these parameters can be used to provide a limited form of shape discrimination. The modular design of the Quantimet 720 makes it approach an application-specific analyser. Hand-held electronic devices (e.g. the "light pen") to demarcate areas of interest or selected features which can then be analysed automatically, are also available with the Quantimet system. However, the number of modules that may be required for a number of shape discrimination methods, and other forms of control, can make the design unattractive and very expensive. More recently, advances in microprocessor technology have resulted in the development of more sophisticated, and less expensive, analysers, such as the "Omnicon Pattern Analysis System,"* which are based on shape-discrimination techniques.

An improvement in the design of large automatic image analysers came about with the advent of powerful microcomputers. Examples of modern automatic image analysers are, amongst others, the Quantimet 900,[†] Magiscan 2,[‡] Leitz T. A. S.,[§] and Reichert-Jung IBAS.[¶] The basic principles in the operation of these image analysers are not very different from those of the older systems. The major departure is in the extensive use of the microcomputer to control the necessary steps in the analysis of the image. The previous grey-level discrimination can now be taken to an advanced stage, whereby

---

* Bausch and Lomb, Analytical Systems Division, Rochester, NY14625, USA.
† Cambridge Instruments, Cambridge, England.
‡ Joyce-Loebl, Team Valley, Gateshead, England.
§ Ernst Leitz GmBH, Wetzlar, W. Germany.
¶ Reichert-Jung UK, Slough, England. (The IBAS system is manufactured by Kontron Bildanalyse GmBH, Munich, W. Germany).

the grey level of each picture-element ("pixel") is digitized, typically to 8 bits (which allows 256 discrete grey levels), and then stored in the memory. The highest resolution possible with an ordinary 625-line TV camera is 512 × 512 pixels. Thus, the digital image is stored in 262 144 independently addressable memory locations. Grey-level selection can now be carried out at various discrete values; and, if done under suitable software control, can offer a much better discrimination than just a single pre-set value.

Similarly, with the newer systems, shape discrimination can be carried out under software control without the requirement of additional plug-in modules. A pixel can be independently accessed, and its relation to adjacent pixels can be determined in order to collect information on continuity, area, perimeter, diameter, etc. A very refined shape discrimination can be performed, which is limited only by the complexity of programming. Complex shape-discrimination routines which need repeated access to a pixel can take a very long time in analysing the image fully. For this reason, most large automatic image analysers offer an alternative "pipeline" mode, which increases the speed of analysis where complex shape discriminations are not required. Despite all these operations that are necessary in the carrying out of analysis, the minicomputer has enough reserve potential to control other related items, such as stage movement, focusing, and illumination. These large automatic image analysers, with their associated control peripherals, are also very expensive; but, when adequately programmed, they can perform repeated operations with little manual intervention.

Considerably cheaper image analysers are now commercially available, which are suitable for the analysis of etched-track foils. An example is the Optomax III,* which in its basic form is generally used as a colony counter based on grey-level discrimination. It can, for example, be interfaced with the Apple II,[†] Commodore PET,[‡] Hewlett-Packard HP85,[§] or BBC[¶] microcomputers for simple shape discrimination and sizing. Thus, for a modest outlay, the Optomax III can be used for counting tracks as well as for providing some indication of their size.

A completely different image analysing system has been developed in Hungary, called Vidimet.[‖94a,b] In it, the theory of geometrical topology is applied to recognize tracks of any shape, count their numbers, and measure the required parameters, such as area, perimeter, and minor and major axes. The system utilizes Euler's equation to relate quantities such as the number of

---

* Micromeasurements Ltd., Saffron Walden, Essex, England.
[†] Made by Apple Computer Inc., Cupertino, CA95014, USA.
[‡] Commodore Business Machines Inc., Santa Clara, CA95050, USA.
[§] Hewlett-Packard Ltd., Slough, England.
[¶] BBC computers are manufactured by Acorn Computers Ltd., Cambridge, England.
[‖] Central Research Institute for Physics and the Research Institute for Ferrous Materials, PO Box 49, H-1525 Budapest, Hungary.

pixels (*p*), adjoining corners (*c*) and edges (*e*): $G = \sum p + \sum c - \sum e$, where *G* is the Euler number giving the number of tracks in one image field. Further topological principles (viz. Hillard's, Buffon's and Saltykov's) and image-transformation techniques (erosion, dilation, rotation, copy, and subtraction) are used to obtain the above-mentioned parameters (i.e. perimeter, area, etc.).

This analyser has solved the problem by using hardware; but the data collected can be transferred through parallel or serial interface links to any type of (micro-)computer for statistical analysis. In this way, the Vidimet can be considered as the "main processor unit", as shown in the block scheme of another type of track analysing system—the Digitrack.

The microprocessor-supported system called Digitrack was developed at the Institute of Nuclear Research, Debrecen, Hungary,* specially for the automatic evaluation of nuclear track detectors[94c]. While this instrument is more or less analogous in the performance of its hardware and software to the commercially available image analysing systems (e.g. Quantimet 800 and 900, Magiscan 2, etc.), its automatic picture analysis is based on a new type of videoreceiver. This is a photosensitive semiconductor device called CCD (charge-coupled device), which is mounted on a Leitz Ortholux micro-scope. The use of large-scale integrated circuits and microprocessors has made it possible to build up a relatively inexpensive instrument which is very good for the evaluation of two-dimensional coordinates, sizes, areas and spatial densities of etched tracks, and their statistical distributions. The Digitrack system is able to recognize tracks with a contrast above a pre-set grey level, suppressing the unrequired background noises. As the system works at relatively low microscopic magnifications ($2-10 \times$), the track evaluation can be performed at high speed on large areas of plastic track detectors possessing reasonable track contrast (e.g. Kodak LR 115 type II, or CR-39).

The Digitrack consists of two main parts (see Fig. 7.18): the system receiving and recognizing picture elements (i.e. the image processor unit); and the data-analysing microcomputer equipped with INTEL 8080 microprocessors (i.e. the main processor unit). The videoreceiver is a 2.5 cm long line imager charge-coupled device of type Fairchild CCD 121HC, which converts one row of the picture to be analysed into 1728 binary signals by a thresholding logic. The grey levels 0 and 1 correspond, respectively, to the background and to the track signals. By this simplification, the picture memory necessary is greatly reduced.

For one cycle of evaluation the picture memory stores all the data collected by 3×8 lines of the CCD. The digitized video signals of 8 lines are fed, on-line, into one band of the picture memory; simultaneously, the other

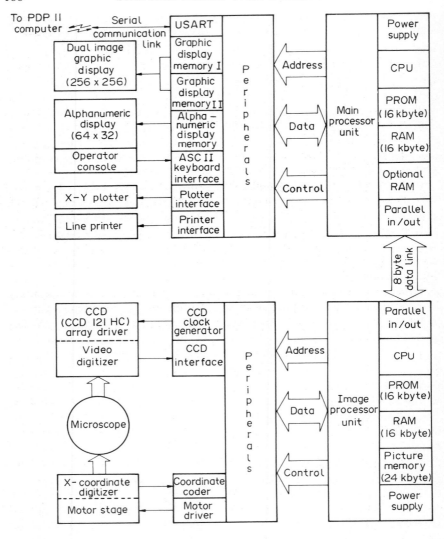

FIG. 7.18 *Block diagram of the Digitrack microprocessor-controlled system for the automatic counting and evaluation of nuclear tracks in plastic foils. It consists of two main parts: the image processor unit (bottom) and the main processor unit (top). The microscope image signals are received in the former unit by a charge-coupled device (CCD). The main processor unit handles the data with the aid of software, and displays the output as graphics, or prints/plots it with the help of other peripherals. See text for discussion. (Figure from Molnár et al.[94c])*

two bands are used for running the pattern-recognition program. The software procedure of pattern recognition examines the picture elements related to each nuclear track, separates the overlapping tracks, and determines their area and centre-of-mass co-ordinates. This set of data is then arranged in a suitable tabular form and transmitted to the main microprocessor unit for further evaluation. To store a larger quantity of track data collected by the system and by additional software for special track evaluations, the microprocessor is connected to a PDP 11/40 computer via a serial communication link.

The Digitrack nuclear track analysing system is equipped with an alphanumeric videoterminal, a dual-image graphic display of 256×256 pixel resolution, an $X-Y$ plotter, and a line printer. With the help of the videoterminal, one can set the measuring conditions at the start: the speed and movement limits of the stage-driver motor, the details of the scanning, evaluation and display procedures, etc. With the aid of the dual-image graphic display one can compare the digitized and the original microscopic images; and by using a special signal one can mark the nuclear tracks recognized by the equipment.

Other inexpensive image analysers are used solely for semi-automatic measurements. These require operator identification of features and movement of reference markers for measurements. Examples of these are the Videoplan* and VIDS.[†] These are based on a digitizing tablet connected to a microprocessor, and can therefore provide a high degree of flexibility in the analysis of measured parameters.

Digitizing tablets and video digitizing units are commercially available as independent units. These constitute the cheapest form of image analyser for track detectors. An example of this is the Video Position Analyser VPA-1000.[‡] With an appropriate hardware design, it is possible to link these units to a microprocessor and thus use them for rapid measurements of track parameters. Figure 7.19 shows such a system in use in the authors' laboratory.

Figure 7.20[31a,94d] gives the relevant block diagram for the system shown in the preceding figure. At the heart of the system is a Nascom 2 microprocessor,[§] supporting 48 kbytes of random access memory (RAM), with twin floppy-disk drives, and six parallel and one serial input/output (i/o) ports. The microprocessor runs on a disk-based operating system, which is the industry-standard Control Program/Microprocessor (CP/M).[¶] The serial i/o port is programmed to handle communication with either a printer (which

---

* Reichert-Jung UK, Slough, England.
[†] Micromeasurements Limited, Saffron Walden, Essex, England.
[‡] FOR.A. Co. Ltd., Japan.
[§] Lucas Logic Ltd., Warwick, England.
[¶] Copyright of Digital Research, Pacific Grove, CA 93950, USA.

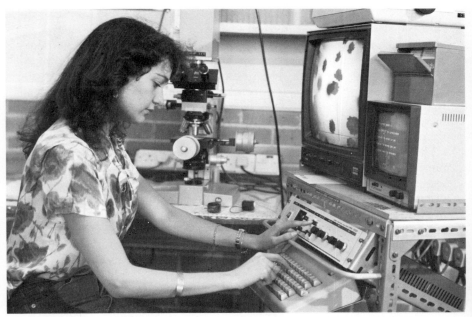

Fig. 7.19   *A typical microprocessor-based set-up for the semi-automatic measurement of etched tracks in plastics or minerals. A closed-circuit television (CCTV) camera, placed at the top of a Leitz Ortholux microscope on the left, looks at the etched tracks (in this case, electrochemical etch spots), and the picture is displayed on a TV screen (the "image monitor"; middle ground). The operator can, for example, set the X and Y cross-wires on the desired positions (e.g. tangentially at the opposite edges of an ECE spot, in sequence) with the help of a video position analyser (VPA), visible in the picture just above the keyboard. On the right is shown the (smaller) console monitor, which is used as a communication device for the operator. The microprocessor and disk drives for the computer (not shown) reside below the keyboard. A number of sophisticated measurements and calculations can be made with the microcomputer, using software and hardware, and displayed, printed, or drawn as histograms, etc., with the help of peripherals. See Fig. 7.20 for a block diagram showing the sequence and connections of the various components of this image measurement system.*

has a graphics facility) or an "intelligent" plotter. Two of the six parallel i/o ports are devoted to the twin floppy disks used for the mass storage of experimental data. Two are linked on an 8-bit analogue-to-digital converter (ADC), which is used to digitize the vertical (i.e. the Z-axis) movement of the microscope stage via the Z-attachment on the focusing knob. The final two parallel i/o ports are connected to a multiplexor/interface linked to a Video Position Analyser (VPA), which is the $X-Y$ digitizing unit.

The system incorporates two television monitors. The first, called the console monitor, is linked to the microprocessor. This provides visual communication to the experimenter. Interactive inputs required by the microprocessor are signified by appropriate "prompts" on the console monitor; and entry of

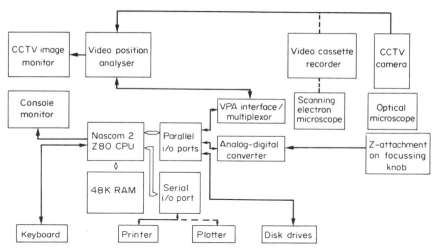

FIG. 7.20 *A typical microprocessor-based image measurement system (see also Fig. 7.19). The block diagram shows the fully linked set-up for the measurement of tracks observed microscopically in three dimensions as well as for the analysis of accumulated data. Alternative units are shown connected by dotted lines. For discussion, see text. (Diagram from A. G. Ramli.*[31a]*)*

numerical or other input parameters, other than the co-ordinates, is achieved via the keyboard. The second monitor (i.e. the CCTV image monitor) is linked to the VPA, and displays the image of the microscope field of view as seen by the closed-circuit television (CCTV) camera (mounted above the optical microscope) together with a set of axes and a movable cross-point generated by the VPA for the measurement of image features (e.g. track diameters). The cross-point or the axes are moved by knobs on the VPA controlling the X and Y movements separately.

The microprocessor can be programmed to carry out sophisticated measurements, e.g. the areas of ECE etch-spots, or to calculate range–energy relations for different ions in a variety of media (including the detectors), or to produce graphics of the evolution of track shapes with etching time for particles of different charges and energies. The system can also handle scanning electron microscope films or signals from a video cassette recorder (VCR). Further details about the performance and programming of the system will be found in references 31a and 94d.

The development of automatic systems for track counting and for the evaluation of track etch parameters is undoubtedly a very important growth area of great current interest. Such methods are aimed at removing the tedium traditionally associated with the manual and ocular measurements of SSNTDs and nuclear emulsions, and at greatly enhancing the speed of evaluation. These advances encompass the development of both hardware and software components of automatic scanning. The importance of such approaches is underscored, for instance, by the recent addition of a Software Section in the journal *Nuclear Tracks and Radiation Measurements* (published by Pergamon Press, Oxford).

We take this opportunity to mention that we give, in Appendix 1, a program (by Henke and Benton[94e]), which is widely used by many workers for the calculation of range–energy relations and the rate of energy loss (dE/dx and restricted energy loss (REL)) values for different ions of a wide range of energies in many media of interest (e.g. track-detecting plastics). Such programs can be used either with large computers or with microcomputers of the types mentioned above, which now exist in many nuclear track laboratories.

7.4.6   Other methods of measurement

A number of simple methods can be used for semi-automatic evaluation of track etch foils. For example, if the track density is high, the amount of light scattered from a detector surface will be proportional to the track density.[95,96] Also, foils in which the track holes penetrate a heavily dyed layer can be analysed using optical densitometry.[97] Hassib[98] has used microdensitometry for normally etched neutron-recoil tracks, while Al-Najjar *et al.*[86] have used a double-beam microdensitometer to count track spots produced in plastic foils by electrochemical etching. Other methods include the use of an α-particle source; here a scintillation counter or a surface barrier detector is employed to detect those α-particles that penetrate the etched foil.[99,100] Measurement of the dielectric constant of a track etch foil placed between two electrodes has also been used for track counting.[101]

Valuable earlier reviews of the application of track detectors to radiation dosimetry may be found in Fleischer, Price, and Walker[59] and in texts by Becker.[101,102]

References

1. ICRP (1977) Recommendations of the International Commission on Radiological Protection. *ICRP Publ.* **26.** Pergamon, Oxford.
2. R. M. Walker, P. B. Price & R. L. Fleischer (1963) A versatile disposable dosimeter for slow and fast neutrons. *Appl. Phys. Lett.* **3,** 28–9.
2a. C. S. Sims & G. G. Killough (1983) Neutron fluence-to-dose equivalent conversion factors: A comparison of data sets and interpolation methods. *Radiat. Prot. Dosim.* **5,** 45–8.
3. S. Glasstone & A. Sesonske (1967) *Nuclear Reactor Engineering.* Van Nostrand, New York.
4. G. Burger, F. Grünauer & H. Paretzke (1970) The applicability of track detectors in neutron dosimetry. In: *Proc. Symp. Adv. Rad. Detectors.* International Atomic Energy Agency, Vienna, paper SM-143.17.
5. Y. Nishiwaki, T. Tsuruta & K. Yamazaki (1971) Detection of fast neutrons by etch-pit method of nuclear track registration in plastics. *J. Nucl. Sci. Technol.* **8,** 162–6.
6. S. Prêtre (1970) Measurement of personnel neutron doses by fission fragment track etching in certain plastics. *Radiat. Effects* **5,** 103–10.

7. W. V. Baumgartner & L. W. Brackenbush (1966) Neutron dosimetry using the fission fragment damage principle. *Battelle-Northwest, Richland, USA, Report* BNWL-332.
8. W. V. Baumgartner, L. W. Brackenbush & C. M. Unruh (1966) A new neutron and high energy particle dosimeter for medical dosimetry applications. In: *Symp. on Solid State and Chemical Radiation Dosimetry in Med. and Biol.* International Atomic Energy Agency, Vienna.
9. M. Sohrabi & K. Becker (1972) Fast neutron personnel monitoring by fission fragment registration from ^{237}Np. *Nucl. Instrum. Meth.* **104**, 409–11.
10. P. Rago, N. Goldstein & E. Tochilin (1970) Reactor neutron measurements with fission-foil Lexan detectors. *Nucl. Appl.* **8**, 302–9.
11. K. G. Harrison (1978) A neutron dosimeter based on ^{237}Np fission. *Nucl. Instrum. Meth.* **157**, 169–78.
12. R. V. Griffith, D. E. Hankins, R. B. Gammage, L. Tommasino & R. V. Wheeler (1979) Recent developments in personnel neutron dosimeters—A review. *Health Phys.* **36**, 235–60.
13. H. Tatsuta & K. Bingo (1970) Evaluation of dose equivalent by fission fragment detectors. In: *Proc. 2nd Int. Cong. IRPA*, Brighton, paper 122.
14. D. E. Hankins (1973) Design of albedo neutron dosimeters. In: *Proc. Symp. on Neutron Monitoring for Radiation Protection Purposes*, Vienna, International Atomic Energy Agency, Vienna, Vol. I. pp. 15–29.
14a. K. G. Harrison, A. J. Taylor & J. A. B. Gibson (1982) The neutron response of a neptunium dosimeter worn on the body. *Radiat. Prot. Dosim.* **2**, 95–104.
15. M. Sohrabi (1979) A new dual response albedo neutron personnel dosimeter. *Nucl. Instrum. Meth.* **165**, 135–8.
16. M. A. Gomaa, S. A. Kasim & M. A. Abu-El Ela (1979) Development of albedo track dosimeter. *Nucl. Instrum. Meth.* **164**, 389–94.
17. M. A. Gomaa & S. A. Kasim (1980) Neutron personnel dosimetry using a CR-39 carbonate plastic detector. *Nucl. Instrum. Meth.* **176**, 579–82.
18. M. A. Gomaa, A. M. Eid, R. V. Griffith & K. J. Davidson (1981) CR-39 carbonate plastic as a neutron albedo and threshold dosimeter. *Nucl. Tracks* **5**, 279–84.
19. M. A. Kenawy, M. A. Gomaa, A. M. Ahmed & R. V. Griffith (1982) Neutron dosimetry using a CR 39-^{6}LiF albedo system. In: *Proc. 11th Int. Conf. Solid State Nucl. Track Detectors*, Bristol, and Suppl. 3, *Nucl. Tracks*. Pergamon, Oxford, pp. 469–79.
20. R. V. Griffith, J. C. Fischer, L. Tommasino & G. Zapparoli (1980) Development of a personnel neutron dosimeter/spectrometer. *Lawrence Livermore Laboratory, University of California Report* UCRL-82657 (and *Proc. 5th Int. Congress of IRPA*. Jerusalem, 1980).
21. E. V. Benton, A. L. Frank, R. A. Oswald & R. V. Wheeler (1980) Proton recording neutron dosimeter for personnel monitoring. In: *Proc. 10th Int. Conf. Solid State Nucl. Track Detectors*, Lyon, and Suppl. 2, *Nucl. Tracks*. Pergamon, Oxford, pp. 469–75.
21a. E. Piesch, J. Jasiak & M. Urban (1984) Makrofol and CR-39 recoil track etch detectors as a supplement of a universal albedo neutron dosimeter. *Nucl. Tracks* **8**, 323–6.
22. A. Dragu (1979) A SSNTD personnel neutron rem dosimeter. *Nucl. Tracks* **3**, 197–204.
22a. L. C. Northcliffe & R. F. Schilling (1970) Range and stopping power for heavy ions. *Nuclear Data Tables*, **A7** 233–63.
22b. D. Garber, C. Dunford & S. Pearlstein (1975) Data formats and procedures for the evaluated nuclear data file ENDF. *Brookhaven National Laboratory Report* BNL-NCS-50496.
22c. D. J. Hughes & R. B. Schwartz (1958) *Neutron Cross Sections (Brookhaven National Laboratory Report* BNL-325, 2nd ed.). Associated Universities Inc., Upton, New York.
22d. F. L. Ribe (1963) Neutron-induced reactions. In: *Fast Neutron Physics* (J. B. Marion & J. L. Fowler, eds.). Interscience, New York, Part II, pp. 1775–1864.
23. A. L. Frank & E. V. Benton (1970) High energy neutron flux detection with dielectric plastics. *Radiat. Effects* **3**, 33–7.
24. K. Becker (1969) Direct fast neutron interactions with polymers. *Oak Ridge National Laboratory Report* 4446, 226.
25. K. Becker (1969) Alpha particle registration in plastics and its applications for radon and neutron personnel dosimetry. *Health Phys.* **16**, 113–223.
26. J. W. N. Tuyn & J. J. Broerse (1970) On the use of Makrofol polycarbonate foils for the

measurement of the fast neutron dose distribution inside a human phantom. In: *Proc. 7th Int. Colloq. Corpuscular Photography and Visual Solid Detectors*, Barcelona, pp. 527–31.

27. F. Spurný & K. Turek (1977) Neutron dosimetry and solid state nuclear track detectors. *Nucl. Track Detection* **1**, 189–97.

28. S. A. Durrani (1978) Use of plastic track-detectors for fast neutron dosimetry. In: *Seventh DOE Workshop on Personnel Neutron Dosimetry. Battelle Pacific Northwest Laboratory, Richland, Washington, Report* PNL-2807/UC-48, pp. 49–55.

29. B. G. Cartwright, E. K. Shirk & P. B. Price (1978) A nuclear track recording polymer of unique sensitivity and resolution. *Nucl. Instrum. Meth.* **153**, 457–60.

30. R. M. Cassou & E. V. Benton (1978) Properties and applications of CR-39 polymeric nuclear track detector, *Nucl. Track Detection* **2**, 173–9.

31. D. L. Henshaw, N. Griffiths, O. A. L. Landen & E. V. Benton (1981) A method of producing thin CR-39 plastic nuclear track detectors and their application in nuclear science and technology. *Nucl. Instrum. Meth.* **180**, 65–77.

31a. A. G. Ramli (1983) The development of a dose-equivalent personnel fast-neutron dosimeter using CR-39. PhD thesis, University of Birmingham.

31b. A. G. Ramli & S. A. Durrani (1984) A flat dose-equivalent-response dosimeter for fast neutrons using CR-39. I: Experimental. *Nucl. Tracks* **8**, 327–30.

31c. A. G. Ramli & S. A. Durrani (1984) A flat dose-equivalent-response dosimeter for fast neutrons using CR-39. II: Mathematical model. *Nucl. Tracks* **8**, 331–4.

31d. S. A. Durrani & A. G. Ramli (1984) Modifying the response of electrochemically-etched CR-39 polymeric nuclear track detector for fast-neutron dosimetry. In: *Tenth DOE Workshop on Personnel Neutron Dosimetry. Battelle Pacific Northwest Laboratory, Richland, WA, Report* PNL-SA-12352, pp. 245–53.

31e. Matiullah & S. A. Durrani (1985) A fast-neutron dosimeter with nearly flat dose equivalent response using CR-39 detectors with various combinations of radiators. *Paper presented at the 13th Int. Conf. Solid State Nucl. Track Detectors*, Rome (and to be published in *Nucl. Tracks* **12**, 1986).

32. E. V. Benton, C. C. Preston, F. H. Ruddy, R. Gold & J. H. Roberts (1980) Proton and alpha particle response characteristics of CR-39 polymer for reactor and dosimetry applications. In: *Proc. 10th Int. Conf. Solid State Nucl. Track Detectors*, Lyon, and Suppl. 2, *Nucl. Tracks*. Pergamon, Oxford, pp. 459–68.

33. R. L. Fleischer, P. B. Price & R. M. Walker (1965) Solid state track detectors: applications to nuclear science and geophysics. *Ann Rev. Nucl. Sci.* **15**, 1–28.

34. M. Balcázar-García, R. K. Bull, I. D. Fall & S. A. Durrani (1979) Calibration of CA80-15 cellulose nitrate for fast neutron dosimetry. *Nucl. Instrum. Meth.* **161**, 91–5.

34a. M. Balcázar-García & S. A. Durrani (1980) High-energy neutron spectrometry with plastic SSNTDs. *Nucl. Instrum. Meth.* **173**, 131–5.

34b. S. A. R. Al-Najjar, A Abdel-Naby & S. A. Durrani (1985) Fast-neutron spectrometry using the triple-α reaction in the CR-39 detector. *Paper presented at the 13th Int. Conf. Solid State Nucl. Track Detectors*, Rome (and to be published in *Nucl. Tracks* **12**, 1986).

35. A. L. Frank & E. V. Benton (1972) Development of a high energy neutron detector. *Defense Nuclear Agency, USA, Report* 2918F.

36. J. W. N. Tuyn & J. J. Broerse (1970) Analysis of the etch pit size distribution in Makrofol polycarbonate foil after fast neutron exposure. In: *Proc. 7th Int. Colloq. Corpuscular Photography and Visual Solid Detectors*, Barcelona, pp. 521–6.

36a. N. Petoussi, S. A. Durrani & J. H. Fremlin (1984) A study of the propagation of a fast neutron beam in a water phantom using SSNTDs. *Nucl. Tracks* **8**, 349–54.

36b. N. Petoussi, S. A. Durrani, J. H. Fremlin & H. U. Mast (1986) Neutron-beam depth-dose dosimetry and spectrometry in water phantom using SSNTDs and multigroup transport calculations. *Nucl. Tracks* **11**, 59–66.

37. R. Gold, F. H. Ruddy & J. H. Roberts (1980) Applications of solid state track recorders in United States nuclear reactor energy programs. In: *Proc. 10th Int. Conf. Solid State Nucl, Track Detectors*, Lyon, and Suppl. 2, *Nucl. Tracks*. Pergamon, Oxford. pp. 533–47.

37a. P. F. Green, S. A. Durrani & J. Walker (1984) A passive environmental neutron spectrometer using SSNTDs. *Nucl. Tracks* **8**, 267–70.

37b. H. Ing & E. Piesch (Guest eds.) (1985) *Neutron Dosimetry in Radiation Protection*: (special issue) *Radiat. Prot. Dosim.* **10** (1–4), pp. 1–345.
38. M. Zamani & S. Charalambous (1981) Registration probability of alphas in cellulose nitrate. *Nucl. Tracks* **4**, 177–85.
38a. S. A. Durrani & P. F. Green (1984) The effect of etching conditions on the response of LR115. *Nucl. Tracks* **8**, 21–4.
39. J. Dutrannois (1971) Utilisation de détecteurs solides visuels en dosimétrie de rayonnement de haute énergie. In: *Proc. Int. Congr. on Protection Against Accelerator and Space Radiation*, CERN-71-16, Vol. 1., 271, CERN, Geneva.
40. J. H. Fremlin & F. Abu-Jarad (1980) Alpha-emitters in the environment. I: Natural sources. *Nucl. Instrum. Meth.* **173**, 197–200.
41. A. L. Frank & E. V. Benton (1977) Radon dosimetry using plastic nuclear track detectors. *Nucl. Track Detection* **1**, 149–79.
42. K. Becker (1968) Personnel radon dosimeter. *US Pat.* 3, 505, 523.
43. J. A. Auxier, K. Becker, E. M. Robinson, D. R. Johnson, R. H. Boyett & C. H. Abner (1971) A new radon progeny personnel dosimeter. *Health Phys.* **21**, 126–8.
44. A. M. Chapuis, D. Dajlevic, Ph. Duport & G. Soudain (1972) Dosimétrie du radon. In: *Proc. 8th Int. Conf. Nucl. Photography and Solid-State Track Det.*, Bucharest, pp. 319–28.
45. B. Haider & W. Jacobi (1972) Entwicklung von Verfahren und Geräten zur langzeitigen Radon-Überwachung in Bergbau. *Hahn-Meitner Inst.*, *Berlin, Report* BMBW-F.B 72-14.
46. A. L. Frank & E. V. Benton (1975) Active and passive radon-daughter dosimeters using track-etch detectors. *Dept. of Physics, Univ. of San Francisco, Tech. Report* 39.
47. D. B. Lovett (1967) Rifle mine-field test of track-etch dosimeters to measure radon daughters exposure. *General Electric Co., USA, Report* APED-5391.
48. D. B. Lovett (1969) Track-etch detectors for alpha exposure estimation. *Health Phys.* **16**, 623–8.
49. R. L. Rock (1968) *Memorandum Report on Experimentation with Track-Etch Films as Indicators of Cumulative Radon-Daughter Exposure.* Bureau of Mines, Denver Federal Center, Denver, CO, USA.
50. R. D. Evans (1969) Engineers' guide to the elementary behavior of radon daughters *Health Phys.* **17**, 229–52.
51. A. L. Frank & E. V. Benton (1975) Measurements with track-etch detectors on working-level exposures in houses near uranium tailings. *Dept. of Physics, University of San Francisco, Tech. Report* 40.
51a. T. Domański, A. Wojda, W. Chruścielewski & A. Żórawski (1984) Plate-out effects in the passive track detectors for radon and its daughters. *Nucl. Tracks* **9**, 1–14.
52. F. Abu-Jarad, J. H. Fremlin & R. K. Bull (1980) A study of radon emitted from building materials using plastic α-track detectors. *Phys. Med. Biol.* **25**, 683–94.
52a. A. V. Nero & W. M. Lowder (Guest eds.) (1983) *Indoor Radon*: (special issue) *Health Phys.* **45** (2), pp. 273–561.
52b. G. F. Clemente, H. Eriskat, M. C. O'Riordan & J. Sinnaeve (Proc. eds.) (1984) *Indoor Exposure to Natural Radiation and Associated Risk Assessment*: (special issue) *Radiat. Prot. Dosim.* **7** (1–4), pp. 1–439.
53. R. Weidenbaum, D. B. Lovett & H. D. Kosanke (1970) Flux monitor utilizing track-etch film for unattended safeguards application. *Trans. Amer. Nucl. Soc.* **13**, 524–6.
54. J. E. Gingrich & D. B. Lovett (1972) A track-etch technique for uranium exploration. *Trans. Amer. Nucl. Soc.* **15**, 118.
55. J. E. Gingrich (1973) Uranium exploration made easy. *Power Eng.* (Aug.), 48–50.
56. R. L. Fleischer & A. Mogro-Campero (1977) Mapping of integrated radon emanation for detection of long-distance migration of gases within the earth. Techniques and principles. *J. Geophys. Res.* **83**, 3539–49.
57. H. A. Khan, R. A. Akber, I. Ahmad, K. Nadeem & M. I. Beg. (1980) Field experience about the use of alpha-sensitive plastic films for uranium exploration. *Nucl. Instrum. Meth.* **173**, 191–6.
57a. R. L. Fleischer, W. R. Giard, A. Mogro-Campero, L. G. Turner, H. W. Alter & J. E. Gingrich (1980) Dosimetry of environmental radon: Methods and theory of low-dose, integrated measurements. *Health Phys.* **39**, 957–62.

57b. H. A. Khan, R. A. Akber, A. Waheed, M. Afzal, P. Chaudhary, S. Mubarakmand & F. I. Nagi (1978) The use of CA80-15 and LR115 cellulose nitrate track detectors for discrimination between radon and thoron. In: *Proc. 9th Int. Conf. Solid State Nucl. Track Detectors*, Munich, and Suppl. 1, *Nucl. Tracks*. Pergamon, Oxford, pp. 815–9.

57c. G. Somogyi, B. Paripás & Zs. Varga (1984) Measurement of radon, radon daughters and thoron concentrations by multidetector devices. *Nucl. Tracks* **8**, 423–8.

57d. P. C. Ghosh & M. Soundararajan (1984) A technique for discrimination of radon (^{222}Rn) and thoron (^{220}Rn) in soil-gas using solid state nuclear track detectors. *Nucl. Tracks* **9**, 23–7.

57e. E. Savvides, M. Manolopoulou, C. Papastefanou & S. Charalambous (1985) A simple device for measuring radon exhalation from the ground. *Int. J. Appl. Radiat. Isotop.* **36**, 79–81.

58. A. Mogro-Campero & R. L. Fleischer (1977) Subterrestrial fluid convection: a hypothesis for long-distance migration of radon within the earth. *Earth Planet. Sci. Lett.* **34**, 321–5.

58a. A. Birot, B. Adroguer & J. Fontan (1970) Vertical distribution of Radon 222 in the atmosphere and its use for study of exchange in the lower troposphere. *J. Geophys. Res.* **75**, 2373–83.

59. R. L. Fleischer, P. B. Price & R. M. Walker (1975) *Nuclear Tracks in Solids: Principles and Applications*. University of California Press, Berkeley.

60. E. V. Benton & M. M. Collver (1967) Registration of heavy ions during the flight of Gemini VI. *Health Phys.* **13**, 495–500.

61. E. V. Benton, R. P. Henke & J. V. Bailey (1974) Heavy cosmic-ray exposure of Apollo 17 astronauts. *Health Phys.* **27**, 79–85.

62. E. V. Benton, R. P. Henke & J. V. Bailey (1975) Heavy cosmic-ray exposure of Apollo astronauts. *Science* **187**, 263–5.

63. D. D. Peterson & E. V. Benton (1975) High-LET particle radiation inside Skylab (SL2) command module. *Health Phys.* **29**, 125–30.

64. E. V. Benton, R. P. Henke & D. D. Peterson (1977) Plastic nuclear track detector measurements of high-LET particle radiation on Apollo, Skylab, and ASTP space missions. *Nucl. Track Detection* **1**, 27–32.

65. D. Hasegan, V. E. Dudkin & A. M. Marenny (1980) Charge and LET distributions of cosmic heavy ions measured on Cosmos 690, 782 and 936. *Nucl. Tracks* **4**, 27–32.

66. R. P. Henke, E. V. Benton & R. M. Cassou (1980) A method of automated HZE-particle Z-spectra measurement in plastic nuclear track detectors In: *Proc. 10th Int. Conf. Solid State Nucl. Track Detectors*, Lyon, and Suppl. 2, *Nucl. Tracks*. Pergamon, Oxford, pp. 509–16.

67. P. H. Fowler & D. H. Perkins (1961) The possibility of therapeutic applications of beams of negative π mesons. *Nature* **189**, 524–8.

68. P. H. Fowler (1965) π mesons versus cancer. *Proc. Phys. Soc.* **85**, 1051–66.

69. H. B. Knowles, F. H. Ruddy, G. E. Tripard, G. M. West & M. M. Kligerman (1977) Status report: Direct track detector dosimetry in negative pion beams. *Nucl. Instrum. Meth.* **147**, 157–62.

70. E. V. Benton, S. B. Curtin, M. R. Raju & C. A. Tobias (1970) Studies of negative pion beams by means of plastic nuclear track detectors. In: *Proc. 7th Int. Colloq. Corpuscular Photography and Visual Solid Detectors*, Barcelona, pp. 423–8.

71. R. K. Bull (1977) A preliminary study of π^- interactions in plastic solid state track detectors. Unpublished report.

72. W. G. Cross & L. Tommasino (1968) Electrical detection of fission fragment tracks for fast neutron dosimetry. *Health Phys.* **15**, 196.

73. W. G. Cross & L. Tommasino (1970) A rapid reading technique for nuclear particle damage tracks in thin foils. *Radiat. Effects* **5**, 85–9.

73a. S. R. Malik, S. A. Durrani & J. H. Fremlin (1973) A comparative study of the spatial distribution of uranium and of TL-producing minerals in archaeological materials. *Archaeometry* **15**, 249–53.

73b. C. K. Wilson (1982) The jumping spark counter and its development as an aid to measuring alpha activity in biological and environmental samples, *Nucl. Tracks* **6**, 129–39.

73c. S. R. Malik & S. A. Durrani (1974) Spatial distribution of uranium in meteorites, tektites and other geological materials by spark counter. *Int. J. Appl. Radiat. Isotop.* **25,** 1–8.
74. G. Somogyi, I. Hunyadi & Zs. Varga (1978) Spark counting of α-radiograms recorded in LR-115 strippable cellulose nitrate film. *Nucl. Track. Detection* **2,** 191–7.
75. D. R. Johnson, R. H. Boyett & K. Becker (1970) Sensitive automatic counting of alpha particle tracks in polymers and its applications in dosimetry. *Health Phys.* **18,** 424–7.
76. M. Balcázar-García (1979) Applications of SSNTDs to nuclear physics and dosimetry. PhD thesis, Department of Physics, University of Birmingham.
77. L. Tommasino, N. Klein & P. Solomon (1977) Fission fragment detection by thin film capacitors—I: Breakdown counter. *Nucl. Track Detection* **1,** 63–70.
77a. J. R. Harvey & A. R. Weeks (1982) A simple system for rapid assessment of etch pits. *Nucl. Tracks* **6,** 201–6.
77b. S. A. Durrani, A. Abdel-Naby, H. Afarideh & S. A. R. Al-Najjar (1985) The scintillator-filled etch-pit method for neutron dosimetry: Part I—Thermal neutrons with (n, α) converters. *Paper presented at the 13th Int. Conf. Solid State Nucl. Track Detectors,* Rome (and to be published in *Nucl. Tracks* **12,** 1986).
77c. L. J. Pedraza, M. I. Edmonds & S. A. Durrani (1984) Radium concentration in building materials measured by nuclear track detectors. *Nucl. Tracks* **8,** 403–6.
77d. L. J. Pedraza C. (1983) Radium concentration in building materials measured by nuclear track detectors. M.Sc. dissertation, Department of Physics, University of Birmingham.
77e. S. A. Durrani, H. Afarideh, A. Abdel-Naby & S. A. R. Al-Najjar (1985) The scintillator-filled etch-pit method for neutron dosimetry: Part II—Fast neutrons, using intrinsic tracks. *Paper presented at the 13th Int. Conf. Solid State Nucl. Track Detectors,* Rome (and to be published in *Nucl. Tracks* **12,** 1986).
77f. S. A. R. Al-Najjar, A. Abdel-Naby, H. Afarideh & S. A. Durrani (1985) New methods of sample preparation and readout for the scintillator technique of etched-track counting of plastic detectors. *Paper presented at the 13th Int. Conf. Solid State Nucl. Track Detectors,* Rome (and to be published in *Nucl. Tracks* **12,** 1986).
78. L. Tommasino (1970) Electrochemical etching of damaged track detectors by H. V. pulse and sinusoidal waveform. *Internal Rept. Lab. Dosimetria e Standardizzazione, CNEN Casaccia, Rome.*
79. M. Sohrabi (1974) The amplification of recoil particle tracks in polymers and its application in fast neutron personnel dosimetry. *Health Phys.* **27,** 598–600.
80. G. Somogyi (1977) Processing of plastic track detectors. *Nucl. Track Detection* **1,** 3–18.
81. L. Tommasino & C. Armellini (1973) A new etching technique for damage track detectors. *Radiat. Effects.* **20,** 253–5.
82. L. Tommasino & G. Zapparoli (1978) Further investigations on electrochemical etching for personnel neutron dosimetry. In: *Seventh DOE Workshop on Personnel Neutron Dosimetry. Battelle Pacific Northwest Laboratory, Richland, Washington, Report* PNL-2807/UC-48, pp. 89–100.
83. L. Tommasino, G. Zapparoli & R. V. Griffith (1980) Electrochemical etching—I. *Nucl. Tracks* **4,** 191–6.
84. L. Tommasino, G. Zapparol, R. V. Griffith & A. Mattei (1980) Electrochemical etching—II. *Nucl. Tracks* **4,** 197–201.
85. S. A. R. Al-Najjar, R. K. Bull & S. A. Durrani (1979) Electrochemical etching of CR-39 plastic: Applications to radiation dosimetry. *Nucl. Tracks* **3,** 169–83.
85a. J. H. Mason (1959) Dielectric breakdown in solid insulation. In: *Progress in Dielectrics* (eds. J. B. Birks and J. H. Schulman). Heywood & Co. London, pp. 1–58.
85b. C. F. Wong & L. Tommasino (1982) The frequency response of electrochemical etching. *Nucl. Tracks* **6,** 25–34.
86. S. A. R. Al-Najjar, M. Balcázar-García & S. A. Durrani (1978) A multi-detector electrochemical etching and automatic scanning system. *Nucl. Track Detection* **2,** 215–20.
86a. A. G. Ramli, F. W. Lovick & S. A. Durrani (1981) A variable-voltage, variable-frequency HT supply for the electrochemical etching of plastic detectors. *Nucl. Tracks* **5,** 311–15.
86b. J. H. Thorngate, D. J. Christian & C. P. Littleton (1976) An apparatus for electrochemical etching of recoil particle tracks in plastic foils. *Nucl. Instrum. Meth.* **138,** 561–3.

87. R. V. Griffith (1978) Personnel neutron monitoring developments at LLL. In: *Seventh DOE Workshop on Personnel Neutron Dosimetry*. *Battelle Pacific Northwest Laboratory, Richland, Washington, Report* PNL-2807/UC-48, pp. 56–66.

88. M. Sohrabi (1975) The electrochemical etching amplification of low-LET recoil particle tracks in polymers for fast neutron dosimetry. PhD thesis, Georgia Institute of Technology, Atlanta, Ga.

89. S. A. Durrani & S. A. R. Al-Najjar (1980) Electrochemical etching studies of the CR-39 plastic. *Nucl. Instrum. Meth.* **173**, 97–102.

89a. C. F. Wong & L. Tommasino (1982) Energy discrimination of alpha particles by electrochemical etching of track detectors. *Nucl. Tracks* **6**, 17–24.

90. G. M. Hassib & E. Piesch (1978) Improvement of the electrochemical etching technique for fast neutron dosimetry. In: *Seventh DOE Workshop on Personnel Neutron Dosimetry*. *Battelle Pacific Northwest Laboratory, Richland, Washington, Report* PNL-2807/UC-48, pp. 71–8.

91. G. M. Hassib (1979) A pre-electrochemical etching treatment to improve neutron recoil track detection. *Nucl. Tracks* **3**, 45–52.

91a. S. A. R. Al-Najjar & S. A. Durrani (1984) Irradiation Fluence: An essential parameter in the optimization of ECE conditions. *Nucl. Tracks* **8**, 99–103.

92. S. Di Liberto & P. Ginobbi (1977) Automatic device for measurements of heavy ion tracks in plastics. *Nucl. Instrum. Meth.* **147**, 75–8.

93. W. Abmayr, P. Gais, H. G. Paretzke, K. Rodenacker & G. Schwarzkopf (1977) Read-time automatic evaluation of solid state nuclear track detectors with an on-line TV-device. *Nucl. Instrum. Meth.* **147**, 79–81.

94. J. U. Schott, E. Schopper & R. Staudte (1977) A high-precision video-electronic measuring system for use with solid state track detectors. *Nucl. Instrum. Meth.* **147**, 63–7.

94a. J. Pálfalvi, I. Eördögh & B. Verö (1980) Track density measurements using a VIDIMET-II A type image analyser. In: *Proc. 10th Int. Conf. Solid State Nucl. Track Detectors*, Lyon, and Suppl. 2, *Nucl. Tracks*. Pergamon, Oxford, pp. 503–7.

94b. I. Eördögh, B. Verö, M. Lányi & T. Réti (1981) The image analyser Vidimet-H. *Hungarian Machinery* **31** (2), 39–42; (also: J. Pálfalvi, private communication, 1984).

94c. J. Molnár, G. Somogyi, S. Szilágyi & K. Sepsi (1984) Development of a CCD based system called DIGITRACK for automatic track counting and evaluation. *Nucl. Tracks* **8**, 243–6; (also: G. Somogyi, private communication, 1982).

94d. A. G. Ramli, P. F. Green, S. Watt & S. A. Durrani (1984) A microprocessor-based track-measuring system. *Nucl. Tracks* **8**, 255–7.

94e. R. P. Henke & E. V. Benton (1967) A computer code for the computation of heavy-ion range-energy relationships in any stopping material. *US Naval Radiological Defense Laboratory, San Francisco, Report* TR-67-122.

95. J. W. N. Tuyn (1967) Solid-state nuclear track detectors in reactor physics experiments. *Nucl. Appl.* **3**, 372–4.

96. H. A. Khan (1971) Semi-automatic scanning of tracks in plastics. *Radiat. Effects* **8**, 135–8.

97. J. Barbier (1971) Some recent progress in neutron radiography and ionising particle dosimetry using cellulose nitrate as detecting material. *J. Photogr. Sci.* **19**, 108–11.

98. G. M. Hassib (1975) Microdensitometric measurements of neutron induced recoil tracks in plastic and their applications for fast neutron dosimetry. *Nucl. Instrum. Meth.* **131**, 125–8.

99. B. Dörschel (1969) Auswertung geätzter Festkörperspurdetektoren durch α-Absorptionsmessungen. *Kernenergie* **12**, 303–4.

100. H. A. Khan & S. A. Durrani (1972) Electronic counting and projection of etched tracks in solid-state nuclear track detectors. *Nucl. Instrum. Meth.* **101**, 583–7.

101. K. Becker (1973) *Solid State Dosimetry*. CRC Press, Cleveland, Ohio.

102. K. Becker (1972) Dosimetric applications of track etching. In: *Topics in Radiation Dosimetry*, Vol. 1 (ed. F. H. Attix). Academic Press, London.

103. S. A. R. Al-Najjar & S. A. Durrani (1985) Interpretation of Mason's equation in terms of measurable electrochemical-etching parameters governing the dielectric breakdown phenomenon. *Paper presented at the 13th Int. Conf. Solid State Nucl. Track Detectors*, Rome (and to be published in *Nucl. Tracks* **12**, 1986).

CHAPTER 8

Fission Track Dating

As has been mentioned on several occasions in this book, fission fragments from the spontaneous break-up of ^{238}U (or, for that matter, of any fissionable nuclide) are capable of registering chemically etchable tracks in most rock-forming minerals. This circumstance, added to the fact that uranium is a fairly ubiquitous trace element at the ~ 1 ppm ($= 10^{-6}$ g/g) to 1 ppb ($= 10^{-9}$ g/g) level in geological materials, opens up the possibility of a widely usable method of dating of rocks.[1]

In this chapter, the basic principles of this dating technique—fission track dating (FTD)—will be outlined. The emphasis here is on the methods rather than on a detailed list of the applications of the technique. However, the application of this method to the dating of meteorites is discussed at somewhat greater length, since this subject raises interesting problems of a more general nature.

First, by way of an introduction, a few general remarks about radioactive dating are presented.

8.1 Radioactive Dating

Consider atoms of a nuclide A which decays with a known decay constant λ_A into atoms of a chemically distinct nuclide B. If we denote the number of atoms of A present initially in a mineral by N_A^0, then after time t the surviving number of such atoms, N_A^t, is given by the familiar law of radioactive decay:

$$N_A^t = N_A^0 \exp\left(-\lambda_A t\right) \tag{8.1}$$

N_A^t is a measurable quantity, but N_A^0 is not. Since, however, in this simple case we know that all of the decaying atoms of type A have ended up as atoms of B (assuming the latter to be stable), a count of atoms B gives, in fact, the number of atoms A that have decayed. Adding this number of atoms of B to N_A^t, thus, immediately yields the value of N_A^0, which allows Eq. (8.1) to be easily solved for t.

In this procedure, either one knows that B is entirely radiogenic and has resulted only from the decay of A; or, if that is not the case, one makes an allowance for the non-radiogenic fraction of B. The latter can usually be computed from a knowledge of the natural abundances (undistorted by radio-activity) of isotopes of B, so that a measurement of one of the non-radiogenic isotopes in the sample yields the value of the "intrinsic" (i.e. the non-radiogenic) amount of nuclide B in it. The excess of B present in the sample is then the decay product of A, and it is this value that is then used in the analysis described above. The procedures adopted in FTD are put in quantitative terms, and explained in greater detail, in the next section.

Among the most important radioactive dating systems in common use are those based on the ^{87}Rb $\rightarrow$ ^{87}Sr and ^{40}K $\rightarrow$ ^{40}Ar decay schemes and on the U and Th decay chains. Of course there are many complications to the simple procedure outlined above, and the literature on such dating methods is now vast. Useful introductions to this subject may be found in references 2–4.

^{238}U decays by both α-emission and by spontaneous fission, with decay constants denoted by λ_α and λ_f, respectively. Alpha-emission is the dominant decay mode by far, and λ_α/λ_f is ~2×10^6. As with the other radioactive dating methods, in fission track dating we must measure the presently surviving abundance of the primary decaying species in the sample—in this case ^{238}U. However, rather than measuring, in addition, the abundance of daughter atoms, in the FTD method we measure the number of fission fragment tracks crossing unit area of the sample surface—for these represent the number of fission decays that have taken place since the "beginning" (see below).

8.2   The Fission Track Age Equation

Suppose that a mineral crystal contains N_8^0 atoms of ^{238}U per unit volume initially. The present number N_8^t of ^{238}U atoms is then given by

$$N_8^t = N_8^0 \exp\left(-\lambda_{D8}t\right) \tag{8.2}$$

where t is the age of the crystal and λ_{D8} is the total decay constant of ^{238}U. As was mentioned in Chapter 5, the fission track age of a sample measures the time which has elapsed since the mineral last cooled below the "closure temperature", i.e. that temperature at which quantitative track retention commences. (We shall return to the matter of the significance of fission track ages in §8.4). The total decay constant for ^{238}U is almost identical to the decay constant λ_α for α-emission.

The number of decays per unit volume which have taken place in time t is $N_8^0 - N_8^t$, and is therefore given by

$$N_8^t[\exp\left(\lambda_{D8}t\right) - 1]$$

Most of these decay events correspond to α-emission. The fraction of fission events is $\lambda_{f8}/\lambda_{D8}$; and so the number of fission events, N_f, per unit volume is given by

$$N_f = \frac{\lambda_{f8}}{\lambda_{D8}} N_8^t [\exp (\lambda_{D8} t) - 1] \tag{8.3}$$

The number density, ρ_s, of spontaneous fission tracks crossing a unit internal surface of the sample is related to the volume density N_f by

$$\rho_s = \eta R N_f \tag{8.4}$$

where R is the range of a *single* fission fragment, and η is a factor that takes into account the efficiency of revelation of these tracks. For tracks with constant $V (= V_T/V_B)$, it was shown in Chapter 4 that $\eta = \cos^2 \theta_c$, where θ_c is the critical angle for etching of such tracks, and where the track density is being measured on an "internal" surface (i.e. in "4π-geometry"). Upon combining Eqs. (8.3) and (8.4), we obtain

$$\rho_s = \frac{\lambda_{f8}}{\lambda_{D8}} N_8^t \eta R [\exp (\lambda_{D8} t) - 1] \tag{8.5}$$

Rather than measuring the ^{238}U content directly, it is more convenient to measure the ^{235}U content of the sample and to assume (as is almost always the case) that this stands in a constant ratio to the ^{238}U. (The determination of ^{238}U content by fast-neutron fission is avoided as these neutrons would produce fission also in the Th content of the sample.) If the sample is irradiated with a known fluence (i.e. the time-integrated flux) F of thermal neutrons, then the number of induced fissions per unit volume is

$$N_i = N_5^t \sigma F$$

where σ is the thermal-fission cross-section for ^{235}U, and N_5^t is the present number of ^{235}U atoms per unit volume. The number of induced-fission tracks per unit area, ρ_i, is then given by

$$\rho_i = \eta R N_5^t \sigma F \tag{8.6}$$

If we assume that the range R and the etching efficiency η for spontaneous and induced fission tracks are the same, then we can divide Eq. (8.5) by Eq. (8.6) and get

$$\frac{\rho_s}{\rho_i} = \frac{1}{\sigma F} \frac{\lambda_{f8}}{\lambda_{D8}} [\exp (\lambda_{D8} t - 1)] \frac{N_8^t}{N_5^t}$$

This yields

$$t = \frac{1}{\lambda_{D8}} \ln \left(\frac{\lambda_{D8}}{\lambda_{f8}} \frac{N_5^t}{N_8^t} \frac{\rho_s}{\rho_i} \sigma F + 1 \right) \qquad (8.7)$$

The present-day atomic ratio for uranium isotopes $I = N_5^t/N_8^t$ has been measured for a wide range of terrestrial samples, and is found to be constant at 7.26×10^{-3}. So we have

$$t = \frac{1}{\lambda_{D8}} \ln \left(\frac{\rho_s}{\rho_i} \frac{\lambda_{D8}}{\lambda_{f8}} I\sigma F + 1 \right) \qquad (8.8)$$

It may be mentioned here that, when $t \ll 1/\lambda_{D8}$, Eq. (8.5) can be simplified (since $\exp(\lambda_{D8}t) - 1 \simeq \lambda_{D8}t$) to

$$\rho_8 \simeq \lambda_{f8} N_8^t \eta R t \qquad (8.5')$$

Then, by dividing Eq. (8.5′) by Eq. (8.6), and rearranging, we obtain

$$t \simeq \frac{\rho_s}{\rho_i} \frac{I\sigma}{\lambda_{f8}} F \qquad (8.8')$$

The ratio $I\sigma/\lambda_{f8}$ (which consists entirely of natural constants) has a value $\simeq 6 \times 10^{-8}$ or 5×10^{-8} (yr cm^2), depending on whether we use $\lambda_{f8} \simeq 7 \times 10^{-17}$ yr^{-1} or 8.4×10^{-17} yr^{-1}, respectively, with $\sigma \simeq 5.8 \times 10^{-22}$ cm^2 for thermal-neutron fission. (The discrepancies between the various measurements of λ_{f8} are discussed in §8.5.) This yields the simple relation:

$$t \simeq 6 \text{ (or 5)} \times 10^{-8} \frac{\rho_s}{\rho_i} F \text{ years} \qquad (8.8'')$$

Equation (8.8) (or its simpler version (8.8′) in the case of young samples of ages $\ll 4.5 \times 10^9$ years) is then the basic fission-track age equation, and involves essentially the measurable quantities ρ_s, ρ_i, and the thermal-neutron fluence F. The measurement of F can be made by using the thermal-neutron activation of Au or Co foils, which are inserted into the irradiation facility immediately adjacent to the samples; However, the method used routinely in fission track dating, in general, is to irradiate a piece of glass with known uranium content (such as the reference glasses produced by the US National Bureau of Standards) in a position adjacent to the sample. Such glasses are often treated as secondary standards, their induced fission track densities ρ_{ig} having been measured for several thermal-neutron fluences F as standardized by (usually) one of the activation techniques. This calibration of the glass enables the constant B in the formula (cf. Eq. (7.4) in §7.1)

$$\rho_{ig} = BF \qquad (8.9)$$

to be established. This equation is identical in form to Eq. (8.6), with B replacing $\eta R N_5^i \sigma$. Then, in a dating experiment, ρ_{ig} may be measured and F calculated from Eq. (8.9). In practice, the induced fission track density relating to the monitor would usually be measured in an external detector, such as a plastic foil or, preferably, a sheet of mica placed in close contact with a flat surface of the reference glass during the irradiation.

The age determination thus reduces to the measurement of three track densities (ρ_s, ρ_i, and ρ_{ig}). Neutron dosimetry in fission track dating is discussed at greater length in §8.5.

8.3 Practical Steps in Obtaining a Fission Track Age

The essentials of the fission track dating method are perhaps best illustrated by considering typical steps that would be carried out, starting with a consideration of a feasible rock sample and proceeding towards the determination of a fission track age.

Although most minerals contain a little uranium, yet in order to measure ages in the region of a few Myr it is desirable, if not essential, that the minerals concerned have U contents of the order of at least 1 ppm. Minerals that usually contain relatively high U concentrations include apatite, zircon, and sphene. These are widely encountered accessory minerals in many rock types.

In order to isolate such accessory minerals out of a bulk rock sample, a number of standard mineralogical techniques may be applied. Most useful is the heavy-liquid separation, in which, to start with, a rock is crushed down to grain sizes that correspond roughly to the dimensions of typical accessory-mineral crystals within the rock (usually $\lesssim 200\ \mu m$). Monomineralic grains are then segregated into different fractions according to their density, i.e. their ability to float or sink in liquids of various densities.

For example, in order to isolate apatite and zircon from quartz and feldspar (bulk minerals in granite), and to separate them from each other, one first uses bromoform (density, 2.69 g cm^{-3}). The bulk minerals will float in this liquid, while apatite and zircon will both sink in it. Having removed the latter accessory minerals, one then tips them into a container of pure di-iodomethane (density, 3.325 g cm^{-3}). In this heavy liquid, apatite floats and zircon sinks; the two can thus be separated. These methods are described in many mineralogical textbooks (see, for example, reference 5).

Let us, by way of illustration, examine the fate of two fairly clean mineral separates, say apatite and zircon.

The apatite crystals may be mounted on glass slides, in batches of several tens or hundreds of grains, using conventional epoxy resins. Since zircon, on the other hand, requires a highly caustic etchant (as we shall see shortly), epoxies and glass backing are, therefore, usually not adequate—as they are not sufficiently resistant to the etchant. For this reason, zircons are usually mounted in a Teflon wafer in the following manner.[6] A glass slide is placed on a hotplate maintained at about 290°C, Zircon crystals are sprinkled onto the slide, and a Teflon wafer placed on top of them. A second glass slide, pre-heated to about 290 °C, is then placed on top of the assembly; and, upon apply-ing a slight pressure, the Teflon "flows" around the grains. The mount is then removed from the hotplate and, on cooling, the two glass slides may be easily detached from the Teflon—which now serves as a backing, contain-ing a layer of zircon crystals embedded in it.

After appropriate mounting as described above, both apatites and zircons are ground to produce flat crystal surfaces. This is carried out either by using carborundum powder spread on a glass plate and lubricated with water, or by using "wet and dry" emery paper. It must be stressed that, in the case of zircon crystals, the grinding should be limited to the minimum necessary, so as to prevent the removal of the constraining Teflon. The next step is the pre-liminary polishing of the crystals, achieved by using diamond paste spread on polishing pads, with a suitable lubricant. Lastly, fine polishing to a 1 μm, or preferably a $\frac{1}{4}$ μm, finish is performed, using the finest diamond paste (or alumina powder). Both the polishing and the grinding steps can be auto-mated, and many samples thus prepared simultaneously.

Such mounting and polishing procedures are, however, not always required. Cleaved sheets of mica, for example, require neither mounting nor polishing.

Polished crystals are now ready for etching. Apatites are treated with dilute nitric acid at room temperature. A typical etching treatment might be 1 minute in 5% aqueous solution of HNO_3 at room temperature; but, of course, the exact conditions required will vary somewhat for apatites from different sources. Etched tracks in apatite are shown in Fig. 8.1.

Zircons require more exacting etching conditions. One treatment that has been found to be effective[7] utilizes a molten eutectic mixture of NaOH and KOH (that is, a mixture of NaOH and KOH in such proportions that the melting point of the mixture is minimized; this corresponds to 11.5 parts KOH:8 parts NaOH). Wafers containing the polished zircons are then floated onto this mixture at a temperature of about 220 °C. Etching is performed in platinum crucibles, and etching times vary from a few hours to several days.

In instances such as above, where a highly caustic etch is used, it is particu-larly important that the etching treatment is followed by thorough washing of the samples (possibly in an ultrasonic cleaning bath), and by an equally thorough drying: it is important to ensure that no caustic traces remain whose fumes could attack the microscope objective lens and, in time, cause severe corrosion. This is a particularly important precaution in the case of etching with hydrofluoric acid. Some commonly employed etching conditions for minerals of general interest have been given in Table 4.1.

Finally, the etched tracks may be counted under an optical microscope using an overall magnification of ~ 500–$1000 \times$.

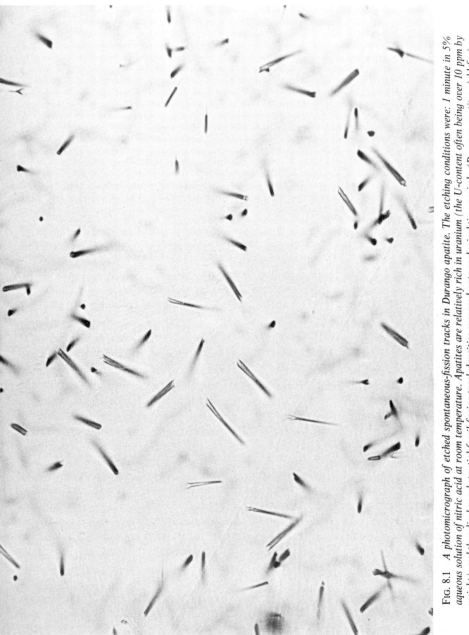

FIG. 8.1 *A photomicrograph of etched spontaneous-fission tracks in Durango apatite. The etching conditions were: 1 minute in 5% aqueous solution of nitric acid at room temperature. Apatites are relatively rich in uranium (the U-content often being over 10 ppm by weight), and thus display substantial fossil-fission track densities over moderate geological time-periods. (Durango apatites yield fission-track ages of ~30 Myr.)*

After describing the basic procedures, it is necessary at this juncture to consider several different routes by which a fission track age may be obtained. The relative merits of these alternative strategies have been reviewed by Gleadow.[8]

8.3.1 The population method

In this method[9] it is assumed that the uranium content of the material to be dated is fairly uniform. The mineral separate is divided into two aliquots, both being regarded as representative of the "population" of crystals in the mineral separate. The first batch is reserved for the measurement of the spontaneous fission track density. The second batch is annealed to remove the spontaneous fission tracks. It is then irradiated with thermal neutrons (along with an appropriate neutron fluence monitor, as described above). Finally, it is mounted, polished and etched along with the unannealed set. The induced fission track density is measured in the annealed crystals, and the spontaneous fission track density in the unannealed aliquot.

The population method will be unreliable if the uranium concentrations are highly variable—particularly if small aliquots of crystals are used. Also, erroneous age values will be obtained if the annealing process alters the track etching properties of the crystals. In the case of both sphene and zircon,[8,10] alpha-recoil damage accumulated over geological timescales through the decay of their U and Th content renders the etching process isotropic in its action; whereas in young sphenes or zircons (and of course in the annealed crystals, whose α-recoil damage has been healed), the etching process is anisotropic (see Chapter 4). Thus the efficiency for track revelation is higher in highly α-recoil damaged (natural) crystals than in the annealed ones, and the ratio of spontaneous to induced fission track densities obtained by the population method will give rise to incorrect age determination unless this effect can be allowed for.

This problem can be overcome by eliminating the annealing step. In such a case, spontaneous track densities are measured in one aliquot, and spontaneous plus induced track densities are measured in the other. The induced track density in the second aliquot is then obtained by subtraction. The statistical errors introduced by this subtraction may be high, unless the induced track density is very much higher than the spontaneous track density.

In using the population method it is also necessary to watch out for the effect of the prolonged-etching factor $f(t)$ (§4.2.3.1 and reference 10a), which can introduce serious errors in obtaining the fission track age. To avoid these errors, it is best to etch the two aliquots containing the spontaneous- and the induced-fission tracks under the same etching conditions (and, preferably, simultaneously). Otherwise one may have to obtain the values of $\rho(0)$ for a "zero" etching time by extrapolating backwards the track-density values $\rho(t)$ as a function of etching time t (see §4.2.3.1). This precaution is especially important in the case of glasses (e.g. tektites).

8.3.2 The external-detector method

In the external detector method[11] the aim is to determine both the spontaneous and the induced fission track density for each individual crystal; thus the problems associated with wide variations in uranium contents, from crystal to crystal, can be overcome. To apply this method a number of grains are mounted and polished as described above. They are then etched to reveal the fossil (i.e. spontaneous fission) tracks. The grain mount is then covered with a sheet of a suitable detector material such as Lexan or muscovite mica (see Fig. 8.2), and irradiated with thermal neutrons. The induced fission tracks emanating from the crystals are then recorded in the detector; and, upon etching the detector sheet, the induced track density due to each crystal can be measured. A number of precautions must be observed if reliable results are to be obtained. First, the uranium content of the external detector itself must be low compared to those of the crystals. The uranium contents of micas are somewhat variable, and some preliminary work may be necessary in order that a mica of low U content (a few parts U in 10^9 is acceptable) be selected; suitable synthetic mica foils are also available. Lexan contains very little U ($\ll 1$ part in 10^9), but is unsuitable for use with very high neutron fluences ($\gtrsim 10^{17}$ n cm^{-2}) as it tends to become "fogged" with neutron recoil tracks produced in it by the fast component (usually $\sim 1-10\%$) which is normally present in the thermal-neutron fluxes available from nuclear reactors (whether in a beam tube or at thermal columns); moreover, most plastics become brittle and difficult to handle after exposure to very high neutron fluences. Partial fading of induced fission tracks may also take place in plastics, unlike in mica, especially if the detector is used *inside* a reactor (e.g. in a beam tube). Mica is also preferable because it has track registration properties which are closer to those of minerals than has Lexan. Secondly, conditions of great cleanliness must be maintained during the preparation of the samples for neutron irradiation; otherwise U-bearing contaminations may lead to the production of interfering induced tracks on the detector foil. Finally, since one needs to measure the spontaneous and induced track densities corresponding to identically the *same* individual area of a particular crystal in calculating its age, it is most important to ensure that one is able to identify the induced-track "image" produced by each crystal in the external detector (mica, etc.). This can be achieved by making a careful map of the crystal locations before irradiation and by inscribing any necessary fiducial marks on the crystal mounting as well as the detector placed in firm contact with it. A better

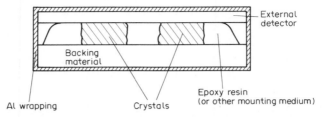

FIG. 8.2   *Arrangement for the measurement of induced-fission track density by thermal neutron irradiation at a reactor, using the external detector method. The sample—consisting, say, of a number of crystals—is set in an epoxy resin, etc., on a rigid backing material (e.g., glass or thick plastic), and given some degree of polish. A suitable external detector, e.g. a sheet of muscovite mica or fused silica—both of which are low in, or virtually devoid of, uranium content themselves—is placed rigidly in position in close contact with the sample surface. (Lexan polycarbonate, which would not record the (n, p) recoils from the fast-neutron content of the reactor beam, is a possible external detector; but it tends to become brittle, and its etching properties deteriorate, when exposed to high fluences of neutrons and γ-rays in a reactor.) Usually some fiducial marks are scratched on the detector surface and on the same sample mount to allow subsequent identification to be made of the exact areas corresponding to different individual crystals. (On etching the detector, one also gets a fairly good replica of the individual crystals imprinted on the detector surface, provided that a sufficient number of induced-fission tracks have been produced.) The whole mount is usually covered with and secured by an aluminium foil. Sometimes a monitoring foil or wire (e.g. [197]Au or [59]Co), or an "age standard", is placed close to the sample to monitor the thermal neutron fluence imparted during the irradiation. Fission fragments emanating from each crystal are recorded in the external detector, at corresponding positions, and the fission tracks in the detector, revealed by subsequent etching, are counted. The method can also be used for producing an exact replica of the spatial distribution of uranium in the sample and to determine the U-content.*

method is to make a photographic map of the etched external detector and the corresponding crystal mount.

Since the induced tracks are measured from an external surface, whereas the spontaneous tracks are measured on an internal surface (revealed by polishing off the equivalent of at least one fission fragment range, say ~ 10–20 μm, from the top of the crystal before etching), the latter track density (ρ_s) must be multiplied by a factor of 0.5 prior to insertion into the age equation, Eq. (8.8) or (8.8'). This is because material on only one side of an external-detector surface contributes to the induced track density ρ_i, whereas material from both sides of a surface originally interior to the crystal will have contributed to the fossil track density ρ_s. This geometrical factor may differ[12] from the "ideal" value of 0.5 if the etching efficiencies of the external detector and the crystal (or glass) to be dated differ. Glasses will generally have etching efficiencies well below 1, owing to their low ratio of track to bulk etching velocities, whereas mica and Lexan have detection efficiencies close to 1 (see Chapter 4). Green and Durrani[12] have pointed out that if observational techniques impose a lower limit on the detectable tracks, such that the *minimum* observable track length is significantly greater than 0.5 μm, it becomes necessary to determine the applicable geometry factor by calibration, when using external detectors.

Anisotropies in bulk etching rates can lead to problems with crystals, particularly zircon and sphene.[7,8,10] If the bulk etch rate normal to the crystal surface is high, then the etching efficiency will be reduced owing to the same effect as mentioned above in the case of glass (viz., $\eta = \cos^2 \theta_c$ is small, where $\theta_c = \sin^{-1} V_B/V_T$; see §4.2.3). On the other hand, if there are widely different bulk etching rates operating in two directions parallel to the crystal surface, then tracks in some directions will be revealed more slowly than in others, and a sufficiently long etching time must be allowed if the efficiency is to reach its maximum value. [13,8] Crystals that have high bulk etch rate normal to the surface are usually distinguishable upon visual inspection of the etched grains[10], and can be eliminated from the analysis.

Finally, the effect of the prolonged-etching factor (cf. §4.2.3.1) must be borne in mind, especially in the case of glasses. Since only the internally produced track densities (viz., those in the sample) are subject to increase by prolonged etching, and not those in the external detector, it may become necessary for accurate age determination to obtain the "zero" etching time value $\rho(0)$ in the glass sample (e.g. a tektite). Alternatively, calibration factors based on age standards (§8.5.3) would need to be used.

The population and external-detector methods are the two most widely adopted schemes in routine fission track dating, although other approaches are possible. For example, a set of grains can be etched to reveal spontaneous fission tracks and then irradiated. After the irradiation, the grains are re-polished and re-etched to reveal the spontaneous plus induced fission tracks in them. A discussion of several alternative strategies in fission track dating is given by Gleadow.[8] The external-detector method is almost invariably used in the case of sphene and zircon crystals, since problems with the annealing of α-recoil damage and non-uniformity of U-content are thereby avoided.

8.4 The Interpretation of Fission Track Ages

We have so far used the term "fission track age" rather loosely, without carefully defining exactly what event the fission track age refers to. It is clear that any radiometric age relates to some specific event in the history of the rock. Thus K–Ar ages[14] represent the amount of time that has elapsed since the rock was last held at a high enough temperature for Ar to diffuse out of it at a significant rate. Rb–Sr ages usually measure the time since the chemical fractionation of Rb and Sr within different mineral phases of the rock.

The discussion of annealing in Chapter 5 showed that, as a mineral cools down, fission tracks become stable for ever longer periods of time. Eventually a temperature T_c is reached for a given mineral in a monotonically cooling rock when, although tracks subsequently formed will fade somewhat, this fading is balanced by the presence of some surviving tracks from earlier stages in the cooling process. This T_c is the "closure temperature", at which quantitative track retention effectively begins; and it is the time period reckoned from the juncture when this temperature was reached that is measured by a fission track age. Only if a rock cooled and crystallized on a timescale that was short compared with the total fission track age, will this fission track age approximate to the true crystallization age of the rock.[15] This mode of cooling and the corresponding interpretation of the age, however, frequently correspond to the real situation, particularly in the case of volcanic rocks, for example.

As we saw in Chapter 5, different minerals have different closure (or closing) temperatures. Suppose we measure fission track ages for, say, apatite and zircon from a rock that cooled slowly. They should yield different track retention ages, t_a and t_z, respectively, with $t_a < t_z$ since zircon is the more retentive mineral. If we know the closure temperatures $T_{c,z}$ and $T_{c,a}$ for zircon and apatite, respectively, from laboratory annealing data, we can estimate a cooling rate for the rock, given by $(T_{c,z} - T_{c,a})/(t_z - t_a)$, provided that we assume that the rock cooled monotonically between $T_{c,z}$ and $T_{c,a}$. Ideally, since closure temperatures are themselves weakly dependent on cooling rate, an iterative approach may be adopted. The cooling rate obtained by making a first guess at the closure temperatures would then be used to select better values for the closing temperatures. Such fine corrections, however, are rarely warranted by the accuracy of the data.*

Apatite, in view of its relatively greater temperature sensitivity, has been particularly extensively used in the study of the thermal history of rocks. For example, by measuring the fission track ages of apatites as a function of topographic elevation in the Central Alps, Wagner and co-workers[16,17] have determined the uplift rates for this region. Cooling of rocks is frequently caused by uplift.

So far, we have concerned ourselves with monotonically cooling rock systems. Sometimes, however, rocks are subject to reheating. In terrestrial rocks

* As a rough guide, the following values of T_c may be useful:- apatites: $105 \pm 20°C$; merrillite: $135 \pm 15°C$; epidotes: $225 \pm 25°C$; sphenes: $330 \pm 20°C$; zircons: $365 \pm 50°C$. The exact values would depend on the chemical composition and the cooling rate assumed. See also Table 8.2.

complex cooling histories may be caused by igneous intrusive heating or metamorphism.[18] Lunar rocks may be heated during some meteoritic impact event.

Wagner[15,18] has presented a careful discussion of the geological interpretation of fission track ages when the geological systems have experienced different types of thermal histories. If, for example, the reheating brings the rock to very high temperatures, all minerals will be cleared of fission tracks, and they will subsequently record different stages of cooling from this high-temperature event. If, however, a lower degree of reheating is experienced, it may happen that tracks in the apatite, for example, are erased but those in zircon remain because of the latter mineral's higher closure temperature. This fission track age of the zircon would thus measure the time since the primary cooling of the rock, and that of the apatite would measure the time of occurrence of the reheating (see Fig. 8.3). Also, some limits could be placed on the peak temperature reached during this latter event.

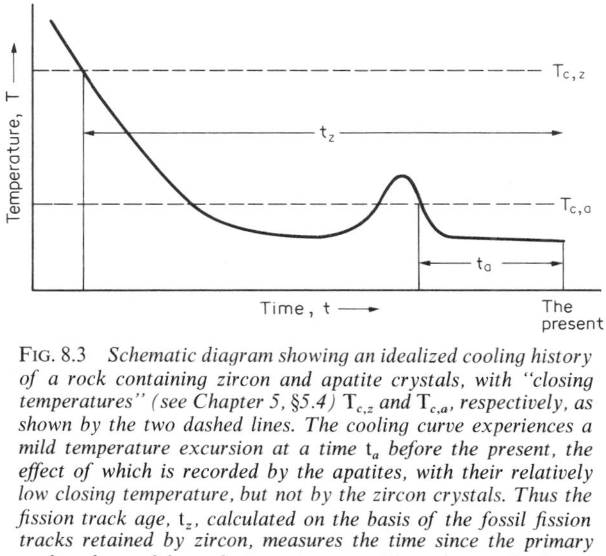

FIG. 8.3 *Schematic diagram showing an idealized cooling history of a rock containing zircon and apatite crystals, with "closing temperatures" (see Chapter 5, §5.4)* $T_{c,z}$ *and* $T_{c,a}$*, respectively, as shown by the two dashed lines. The cooling curve experiences a mild temperature excursion at a time* t_a *before the present, the effect of which is recorded by the apatites, with their relatively low closing temperature, but not by the zircon crystals. Thus the fission track age,* t_z*, calculated on the basis of the fossil fission tracks retained by zircon, measures the time since the primary cooling down of the rock to a temperature* $T_{c,z}$*, whereas the fission track age of the apatite,* t_a*, measures the time since the end of the reheating event (which erased the pre-existing tracks in apatite) or, more precisely, since the temperature once again fell below the value* $T_{c,a}$*. The two ages,* t_z *and* t_a*, can thus tell us something about the cooling-down history of the rock, and also allow us to place some limits on the peak temperatures reached during the secondary reheating episode (viz. that the temperature must have gone above* $T_{c,a}$ *but not above* $T_{c,z}$*).*

Even if the reheating is insufficient to clear the tracks from a mineral, many of the pre-existing tracks may be shortened. The tracks formed subsequently to this heating will be of the normal length (barring any further thermal excursions). By measuring the numbers of both thermally affected and thermally unaffected tracks it is possible to make some deductions about the time at which the reheating occurred.[19]

The fission track dating technique therefore offers a powerful method for the investigation of the thermal histories of geological samples, the full potential of which has yet to be exploited. The reader is referred to Wagner[18] for a useful review of the geological interpretation of fission track ages and discussion of the potentialities of the method.

8.5 Neutron Dosimetry, Fission Decay Constant of ^{238}U, and Age Standards

8.5.1 *Neutron fluence measurements*

Neutron fluence measurements during the irradiation of samples for fission track dating are usually made either by activating Au or Co foils using the reactions $^{197}Au(n, \gamma)^{198}Au$ and $^{59}Co(n, \gamma)^{60}Co$, or alternatively by utilizing the reaction ^{235}U (n, fission), where the uranium has been incorporated into reference glasses. Glasses with well defined trace element contents are manufactured by the US National Bureau of Standards, and a range of U contents is available. (Note that the uranium in these glasses has a non-standard isotopic composition, being greatly depleted in ^{235}U.) Uranium concentrations ranging *nominally* from ~ 0.02 ppm to 500 ppm are available.* A reference glass with U content roughly matching that expected for the unknown is usually the most suitable, because the same thermal-neutron fluence will then induce reasonable amounts of fission in both of them. Tracks are usually counted in an external detector (cf. §8.3.2) such as mica or Lexan placed in contact with a flat, polished surface of the glass. Usually the glass is treated as a secondary standard and is calibrated against an activation foil.

Attention must be paid to a number of points if accurate neutron dosimetry is to be achieved.

(1) Most irradiation facilities (e.g. nuclear reactors) exhibit neutron flux gradients. These may exist either along the axis of the reactor or at right angles to it, although an axial gradient is more usual. For this reason it is important to take care over the layout of samples within the reactor. Hurford and Gleadow[20] studied the J1 irradiation facility at the Herald

---

* The glasses have reference numbers SRM 610 to SRM 618 (wafer thickness, 1–3 mm; diameter 12–14 mm) and may be purchased from: Office of Standard Reference Materials, National Bureau of Standards, Washington, DC 20234, USA. Their certificates will give the exact values of the U contents as well as the other trace elements added to these glasses.

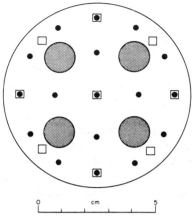

FIG. 8.4 *Composite diagram showing the irradiation geometry used by Hurford and Gleadow[20] in three different experiments for measuring the flux distribution in the J1 neutron-irradiation facility at the Herald reactor, AWRE, Aldermaston, England. The Al holder, a disk 8 cm in diameter, carried a number of monitors indicated as follows: (a) 17 Au monitors, used in the first experiment, shown by solid black circles; (b) 9 pairs of Au and Co monitors, used in the second experiment, shown by open squares; and (c) 4 uranium-bearing standard glasses in the third irradiation experiment, shown by large, hatched circles. The experiments indicated that there was no significant variations in the neutron flux over a plane tangential to the reactor core in the J1 facility of the Herald reactor.*

reactor, AWRE, Aldermaston (England), in some detail, and found a gradient of about 5–10% per centimetre outwards from the reactor core. They attached samples and monitors to high-purity aluminium disks, as shown in Fig. 8.4, and found that no significant flux variations occurred radially across these disks.

In view of the possibility of flux gradients it is always advisable to have the samples and flux monitors in close proximity to each other.

(2) If a large number of samples of U-rich material (or of material bearing other nuclides with large neutron-capture cross-sections) are placed too close together in the irradiation facility, the innermost samples within an assemblage may suffer from flux depression. This is best avoided by adequate spacing out of samples within the facility. Batches of specimens, placed within an irradiation can, may be separated by the use of a suitable inert filler material, and each batch provided with its own fluence monitor.

(3) It is advantageous to use an irradiation facility with a low component of fast and epithermal neutrons. Wagner *et al.*[21] recommend a thermal-to-epithermal ratio of at least 100:1. This is because the age equation

(Eq. (8.8)) involves the product of the thermal-neutron fluence F with the thermal-neutron fission cross-section σ. Clearly, erroneous results will be obtained if a significant number of induced fissions are caused by neutrons in the non-thermal domain. A significant fast-neutron component will also result in contributions to the induced track density made by the fission of ^{238}U and ^{232}Th; this will require correction if an accurate determination of the fission track age is to be made.

Neutron dosimetry is one of the most problematic areas in fission track dating at present. Hurford and Green,[22] for example, have compared the induced fission track density ρ_D (analogous to ρ_{ig} in Eq. (8.9)) in micas adjacent to reference glasses, with measurements of thermal-neutron fluences, F_{Co}, obtained from Co-wire activation monitors, in more than fifty irradiations carried out in the J1 facility at the Herald reactor at Aldermaston, over a period of more than five years. The measured ratios ρ_D/F_{Co} (corresponding to B of Eq. (8.9)), obtained in these irradiations, show variations far in excess of those expected from the experimental uncertainties. Much work remains to be done on this subject. (See references 22–24 for a useful discussion of this topic.)

8.5.2 The fission decay constant λ_{f8}

A further possible source of systematic errors in fission track dating is the uncertainty in the value of the fission decay constant of ^{238}U. For minerals which are young compared to the overall half-life of ^{238}U ($=4.47 \times 10^9$ yr),

TABLE 8.1* *Some determinations of the spontaneous fission decay constant λ_{f8} of ^{238}U*

Authors	Value $(yr)^{-1}$	Method	Reference
Flerov and Petrzhak (1940)	$(0.7 \text{ to } 7) \times 10^{-17}$	Ion chamber	25
Spadavecchia and Hahn (1967)	8.42×10^{-17}	Rotating bubble chamber	26
Gerling *et al.* (1959)	1.19×10^{-16}	Xe analysis	27
Fleischer and Price (1964)	6.85×10^{-17}	Mica-U sandwich + minerals of known age	28
Roberts *et al.* (1968)	7.03×10^{-17}	Mica-U sandwich	29
Khan and Durrani (1973)	6.82×10^{-17}	Mica-U sandwich	30
Storzer (1970)	8.49×10^{-17}	Fission tracks in glass of known age	31
Thiel and Herr (1976)	8.57×10^{-17}	Fission tracks in glass of known age	32
Hurford and Gleadow (1977)	7.0×10^{-17}	Fission tracks in minerals of known age	20
Hadler *et al.* (1981)	8.6×10^{-17}	Mica-U sandwich	33

* A more complete list of determinations of λ_{f8} is given by Bigazzi.[23]

the fission track age is inversely proportional to λ_{f8} (see Eq. (8.8′)). A large number of measurements on this parameter have been reported over the years, and a wide spread in the results obtained is clearly apparent (see Table 8.1). Track detectors have been widely used in recent determinations of λ_{f8}. The method used involves counting spontaneous fission tracks either in samples of known ages (e.g. those determined by other radiometric methods) or in detectors held adjacent to uranium foils for measured lengths of time.

It may be seen from Table 8.1 that there is a large variation in the values obtained for λ_{f8}, and there is an apparent clustering of values around $\sim 7 \times 10^{-17}$ yr^{-1} and 8.5×10^{-17} yr^{-1} (cf. Fig. 8.5, from Bigazzi[23]). The track detector methods for determining λ_{f8} nearly always involve the measurement

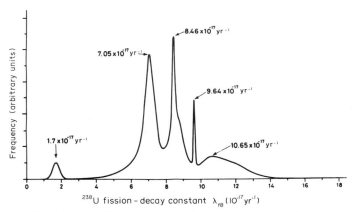

FIG. 8.5 *Plot showing the relative frequency with which different values for the* ^{238}U *fission-decay constant,* λ_{f8}, *have been obtained by various observers using a large variety of measurement techniques. (A few early or imprecise measurements have been ignored.) The curve is the addition of standard-area Gaussian curves whose width has been determined by the experimental errors stated by the given workers. The diagram shows two major peaks, well-separated, which can be correlated with two values of the decay constant:* 7.05×10^{-17} yr^{-1} *and* 8.46×10^{-17} yr^{-1} *in addition to three minor peaks (two of which—at* $1.7 \times$ *and* 9.64×10^{-17} yr^{-1}—*represent single measurements only). The author (Bigazzi[23]) concludes that such big differences are probably due to systematic differences inherent in the approaches made. Most track workers use either of the two main values,* $\sim 7 \times 10^{-17}$ yr^{-1} *and* $\sim 8.4 \times 10^{-17}$ yr^{-1}, *with some preference for the former— which is close to the value obtained by a number of groups using SSNTD methods. Various ways of circumventing this uncertainty have been devised (see text for discussion). If, however,* λ_{f8} *has been explicitly used in fission track dating, the value assumed must be stated in reporting results, so that valid comparisons can be made with other age determinations.*

of neutron fluences. The measurement of λ_{f8} is therefore bound up with problems in neutron dosimetry.[21-23] Systematic differences have, for example, been found between fluences measured simultaneously by Au and Co monitors in the same reactor (see references 22 and 30). It may be noted that whatever method of neutron dosimetry is used to measure F in a given fission track dating laboratory, the value of F so determined may be inserted into Eq. (8.8) or (8.8′) for a sample of *known* age, and a value of λ_{f8} so obtained. This value of λ_{f8} will allow consistent ages of other samples to be obtained by the laboratory in question, so long as the same method of fluence determination is used. However, as Hurford and Green[22] emphasize, problems can arise if attempts are made to recalculate fission track ages by using a different combination of dosimetry method and decay constant than the one that gave the mutually compatible values. In other words, the λ_{f8} value and the dosimetric procedure are interrelated, or interdependent, in the context of FTD, and ought not to be separated without further calibration or adjustment.

8.5.3   Age standards

The explicit dependence of fission track ages on λ_{f8} and F can be removed by the use of age standards. The age equation (Eq. (8.8)) for the unknown sample is:

$$t_{un} = \frac{1}{\lambda_{D8}} \ln \left[\left(\frac{\rho_s}{\rho_i} \right)_{un} \frac{\lambda_{D8}}{\lambda_{f8}} I\sigma F + 1 \right]$$

If spontaneous and induced fission tracks are measured in a sample of well-known age t_{st}, then we may write

$$t_{st} = \frac{1}{\lambda_{D8}} \ln \left[\left(\frac{\rho_s}{\rho_i} \right)_{st} \frac{\lambda_{D8}}{\lambda_{f8}} I\sigma F + 1 \right]$$

where we have assumed that both the standard and the unknown have been irradiated together in the same thermal-neutron fluence F and that both contain uranium of the same isotopic ratio I. These two equations can be combined to give, upon some rearrangement:

$$t_{un} = \frac{1}{\lambda_{D8}} \ln \left\{ \left[\left(\frac{\rho_s}{\rho_i} \right)_{un} \middle/ \left(\frac{\rho_s}{\rho_i} \right)_{st} \right] [\exp(\lambda_{D8} t_{st}) - 1] + 1 \right\} \qquad (8.10)$$

Thus the age is written in terms of the four measured quantities $(\rho_s, \rho_i)_{un}$, $(\rho_s, \rho_i)_{st}$; a well known constant of nature λ_{D8}; and a well-established age t_{st}.

A suitable age standard should have a number of concordant, well-determined ages obtained by independent techniques, such as K–Ar, Rb–Sr, and from stratigraphy.[24] Also, the sample should be homogeneous, and should not contain material of different ages. Furthermore, tracks in the age standard should show no signs of shortening due to annealing. Apatites, in particular, frequently show evidence of annealing effects and are therefore undesirable if this is found to be the case; zircons are probably better suited as age standards. Ideally, several such age standards should be used, along with the unknown sample.

A somewhat different approach in the use of age standards is taken by Fleischer and Hart[34] (see also Fleischer, Price, and Walker[35]). If we take the age equation, Eq. (8.8), and replace F by using Eq. (8.9) (where F is related to the induced track density ρ_{ig} in or from a standard reference glass), then we have:

$$t = \frac{1}{\lambda_{D8}} \ln \left[\left(\frac{\rho_s}{\rho_i} \right) \frac{\lambda_{D8}}{\lambda_{f8}} \frac{I\sigma\rho_{ig}}{B} + 1 \right]$$

Upon collecting together the constants, we obtain

$$t = \frac{1}{\lambda_{D8}} \ln \left[\left(\frac{\rho_s}{\rho_i} \right) \lambda_{D8} \rho_{ig} \zeta + 1 \right] \tag{8.11}$$

where

$$\zeta = \frac{I\sigma}{\lambda_{f8} B} \tag{8.12}$$

with all the constants of nature as defined in §8.2, and B ("tracks per neutron") being the calibration constant in Eq. (8.9). Now if we have a sample of known age t_{st}, we can rearrange Eq. (8.11) to obtain an expression for ζ:

$$\zeta = \frac{\exp(\lambda_{D8} t_{st}) - 1}{\left(\frac{\rho_s}{\rho_i} \right) \rho_{ig} \lambda_{D8}} \tag{8.12'}$$

Thus, by using a series of samples of known age, a value for the constant ζ is obtained (ζ will, of course, be constant only if the same reference glass is always used). In this approach, the age standards are, in effect, used to calibrate the reference glass; and the fluence F and fission decay constant λ_{f8} do not appear explicitly in the relevant age equation, Eq. (8.11). The reference glass, so calibrated, is used therefore as a secondary standard, and there is no need to include primary age standards in each subsequent irradiation. Only three measured quantities are now required. These are ρ_s and ρ_i in the sample under investigation, as before, and the induced track density ρ_{ig}, usually measured in a detector adjacent to the reference glass.

8.6 Annealing Corrections

It was noted in Chapter 5 that as latent-damage trails are progressively annealed, they gradually shorten. In the case of the two "prongs" of a fission event, shortening takes place from the end of each prong, which is the site of the least intense radiation damage. For internally distributed tracks, the observed track density at a surface depends on the etchable range R of a single fission fragment (see Eq. 8.4); and so this length reduction is accompanied by a track density reduction despite the fact that no tracks may actually be totally removed. In deriving the age equation earlier in the present chapter (§8.2), it was assumed that the spontaneous and induced fission tracks both had the same etchable range. If, however, the fossil (i.e. spontaneous fission) tracks have been shortened by some heating event in nature, then the "apparent" fission track age is lower than the true age. It may be noted that apatites and glasses commonly show the effects of fossil-track fading, whereas zircons and sphenes generally do not.

Suppose that a rock is subjected, for a certain length of time, to some constant, rather high, temperature at which a moderate amount of track shortening takes place. If tracks are shortened by an amount δR, yielding a track density reduction $\delta \rho$, this track density reduction will, in turn, lead to a reduction δt in the measured fission track age. The important point is that this apparent age, $t - \delta t$, corresponds to no "significant" physical event (because the moderate heating could have taken place at any time during the track retention age of the rock), unlike the case of severe annealing events in which all tracks are eliminated, so that the fission track "clock" is thereby reset.

Such partial annealing effects must therefore be allowed for if a meaningful fission track age is to be obtained.

Several methods have been devised to allow for partial annealing, none of which have met with universal acceptance. Nevertheless, two of the most important of these will be described below.

8.6.1 The track size correction method

As we have seen above, moderate amounts of annealing are manifested as reductions in the length of the (internal) tracks, and this reduction in the etchable range is one reason (but not the only one) for reductions in the track density encountered at any given surface. Moreover, by partially healing up the radiation damage along the track, annealing reduces the track etch velocity, and hence the diameter of the track opening—which is the most readily measured parameter for tracks in glass. It would seem reasonable, then, to use track size reductions as a diagnostic tool for the investigation of annealing effects. Bigazzi[36] and Mehta and Rama[37] were among the first to report reductions in the lengths of spontaneous fission tracks in mica compared with induced fission tracks. Storzer and Wagner[38] have produced a plot showing the correlation between the mean diameter of fission tracks in australite glass and the observed track density in an annealing experiment. Many authors have since produced similar plots and used them to correct fission track ages for annealing (see, e.g., Durrani and Khan[39]). There does, however, exist some variation in the form of these correction curves. There is controversy, for example, as to whether a reduction of $\sim 10-15\%$ in the length of fission tracks in apatites does[39a] or does not[39b,c] result in a reduction in the observed track density (see also reference 39d).

The usual method is to take a batch of crystals or glasses from which fossil tracks have been removed by annealing, and to irradiate them so as to induce a fresh generation of fission tracks (the annealing step may be unnecessary if the induced tracks are expected to be much more numerous than the spontaneous tracks). Samples are then subjected to various annealing treatments (whether "isothermal" or "isochronal") prior to etching, and the consequent lengths (or diameters) of the induced tracks are plotted as a function of the surviving track density. These quantities are usually normalized to the size and density of tracks in an unannealed control sample. It is then possible to construct graphs of l/l_0 (or D/D_0) versus ρ/ρ_0 (cf. Fig. 8.6), where l, D, and ρ are, respectively, the track length, diameter, and number density measured after some annealing treatment, and l_0, D_0, and ρ_0 are the unannealed values of these parameters. By measuring l (or D) for the fossil tracks in an unknown sample, one can compute l/l_0 or D/D_0 and then obtain ρ/ρ_0 from a plot for the material concerned, such as that shown in Fig. 8.6. Since ρ is, of course, measurable, ρ_0—the track density corrected for thermal lowering—can then be obtained quite easily.

Several model calculations of the lowering of track density with decreasing track size have been performed,[34,40,41] and these generally give good agreement with experimentally derived curves. It seems[34] that in glasses the reduction in V_T (and hence an increase in the cone semi-angle θ_c) along the track is primarily responsible for decreases in the observed etched-track density (the track revelation efficiency, for constant V_T, being proportional to $\cos^2 \theta_c$ on an internal surface). The work of Hashemi-Nezhad and Durrani[41] has shown that the effect of prolonged etching factor[10a] assumes importance in the revelation of annealed tracks, especially in glass, as θ_c increases, and thus needs to be taken into account in calculating the correction curves. (The dependence of the prolonged-etching factor $f(t)$ on the critical angle θ_c has been shown explicitly in §4.2.3.1.)

The measurement of either track length or track diameter for both spontaneous and induced fission tracks is an important part of any fission track dating experiment in order to ascertain whether, and how much, correction needs to be applied for any possible natural annealing of the fossil tracks.

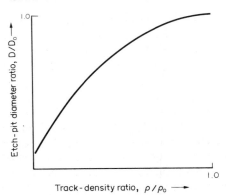

FIG. 8.6   *Schematic representation of curve used in the correction of fission track ages of natural glasses (e.g. tektites). The plot shows the variation in the ratio of etch pit diameters D of annealed samples to unannealed etch pit diameters D_0 as a function of the ratio of the annealed to the unannealed track density, ρ/ρ_0. With the help of such calibration curves based on laboratory experiments, it would be possible, by determining the diameter ratio D/D_0 between the etch pits (under the same etching conditions) of existing fossil fission tracks and freshly-induced fission tracks, respectively, to estimate the reduction in the fossil track density due to thermal fading in nature. This would allow a correction to be made—if laboratory results can be applied to fading over geological time periods—to the fission track age of the sample, which utilizes the ratio ρ_s/ρ_i of the fossil to the induced fission track densities in its calculation (cf. Eqs. (8.8) and (8.8')). Calibration curves for annealed-length ratio l/l_0, as a function of ρ/ρ_0, in the case of crystals lead to similar correction methods—through effects of crystallographic anisotropy on the length and track density reductions and controversy as to some initial length reduction without a corresponding track density reduction in some crystals lead to complications (see text for discussion).*

8.6.2   *The plateau correction method*

This method is somewhat analogous to the stepwise heating approach to K–Ar dating (Turner[42]) in that it involves the application of increasingly severe heating steps to the samples and relies on the presence of a relatively stable core within both spontaneous and induced fission tracks. The method was first proposed by Storzer et al.[43] and by Burchart et al.,[44,44a] and is now being used by a number of workers.[45–48]

The procedure for the "isochronal plateau technique" is as follows.[43,46] Two aliquots of a sample are taken, one containing spontaneous fission tracks and the other freshly induced fission tracks. Each aliquot is heated for the same time (typically one hour) at a given temperature, and then the surviving track density in each is measured. The two samples are then further annealed at a higher temperature for the same length of time; and after repolishing and etching, the track density in each sample is again measured. At first the induced track density falls more rapidly than does the spontaneous track density; but eventually the ratio of spontaneous to induced track densities reaches a constant value (see Fig. 8.7a)—the fission track age derived from this ratio is termed the "plateau age".

The principle underlying this approach is that the damage along a fission track has a spectrum of activation energies for removal by thermal annealing (see Chapter 5). In the case of thermally affected fossil tracks, only the most resistant part of the track has survived. Thus, for a given number of fission events, the surface density of naturally annealed (and hence shortened) fossil fission tracks is lower than for the freshly induced (full-length) fission tracks. This argument, which is based on track lengths, is similar to that presented earlier in connection with Eq. (8.4). After a sufficiently severe laboratory annealing step, only the highly resistant portions of *both* induced and fossil tracks remain, and the ratio of fossil-to-induced track densities is equal to the ratio of volume densities of fission events for the two track populations. The plateau method should, therefore, yield an age corrected for any naturally caused thermal reduction of the fossil track density.

This technique has been applied to both glasses and crystals, and an accuracy of $\pm 3\%$ in the plateau age has been claimed.[44,46]

The principle of the "isothermal plateau method" is the same as that of the isochronal plateau method, except that a constant (relatively high) temperature is now applied[45] for ever-increasing lengths of time. Once again, a plateau is eventually reached, after which time the ratio of the surviving fresh and fossil tracks remains unchanged (see Fig. 8.7b).

Since the track-size correction as well as the plateau-correction methods are essentially population methods of age correction, all the problems and limitations involved in such procedures (see §8.3.1) must be kept in mind. Uncertainties introduced by the inhomogeneity of uranium, and the poor statistical accuracy appertaining to the low fossil track densities ρ_s generally encountered, beset all age-correction methods (see §8.10). The plateau-age corrections have been successfully applied, in general, only to glasses and apatites (see, e.g. references 45–48), but not to sphenes and zircons.

The importance of the effect of anisotropy on the annealing of fission tracks in minerals such as apatite must not be neglected in this context[47a] (whereby the healing of etchable damage of latent tracks depends on their crystallographic orientation–resulting in differences in their observability after etching; see §§5.2 and 4.2.5). The problem of the change in the track-revelation efficiency caused by laboratory annealing, especially in the case of sphenes and zircons,[10.7] has already been alluded to in §8.3.1.*

---

* *Note added in proof.* For recent advances in this and other fields of fission track dating see reference 47b.

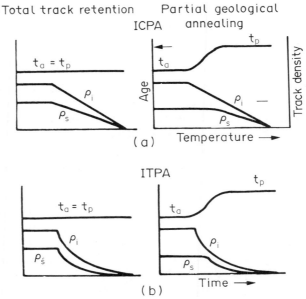

FIG. 8.7a,b   *Plateau-age correction methods. The figure illustrates the concepts of (*a*) the "isochronal plateau age" (ICPA),
and (*b*) the "isothermal plateau age" (ITPA). Here* t_a *stands for
the apparent age;* t_p *for the plateau age;* ρ_s *for the spontaneous
(or fossil) fission track density; and* ρ_i *for the induced fission track
density. In all four plots the upper part of the ordinate represents
the ages: and the lower part of the ordinate, the track densities.
The result of applying isochronal or isothermal annealing is as
follows. If no fossil track loss has taken place in nature (i.e. there
has been total track retention: left-hand plots in* a *and* b*), both
populations of tracks (*ρ_s *and* ρ_i*) behave similarly, and the calculated ages, which depend on the ratio* ρ_s/ρ_i*, remain constant with
the temperature or time variable, so that* $t_a = t_p$ *(flat lines). In the
case, however, of partial geological annealing (right-hand plots in*
a *and* b*), the behaviour of the two track populations is different.
As the fossil tracks have already undergone annealing, only the
hardier, or more resistant, part of the tracks has survived. The
application of the laboratory heating, therefore, makes no impact
on* ρ_s *at first: only when the severity of annealing begins to surpass
the equivalent geological conditioning, does the fossil track density
* ρ_s *begin to decline. The freshly induced fission tracks, however,
begin to react to the laboratory-annealing treatment much earlier,
and begin to decline at a lower value of temperature or time. The
net effect is that one gets a false, lower, age* t_a *from the earlier
parts of the curves. When, however, the severity of annealing has
caught up with the geological conditioning of the fossil tracks,
the two populations begin to react to further annealing in the same
way. The ratio of* ρ_s *to* ρ_i *thus begins to rise and eventually reaches
a constant value. This yields a higher, and true, plateau-age value
* t_p *in both graphs* a *and* b*. (Figure after Carpena et al.[45])*

8.7   Fission Track Dating of Lunar Samples and Meteorites

Rock samples from meteorites and from the lunar surface are, as yet, the only macroscopic quantities of extraterrestrial material to have been studied directly in the laboratory. Their study has yielded important information on the early history of solar-system bodies; and we discuss below some of the special problems which confront workers attempting fission track dating of meteorites and lunar rocks. One special feature common to all rocks older than about 4×10^9 yr, namely the presence of ^{244}Pu spontaneous fission tracks in them, has such a drastic effect on the interpretation of fission track measurements that the following section of this chapter (§8.8) will be devoted to that topic. (No terrestrial rocks authentically older than $\sim 3.8 \times 10^9$ yr have, to date, been unearthed; so that the ^{244}Pu fission tracks have not had to be considered in them—as practically all ^{244}Pu would have decayed by that time.) In the present section, however, we focus attention on some other problems posed by extraterrestrial samples. For a recent summary of the use of nuclear track methods to study meteorites, see reference 48a.

In the foregoing treatment of fission track dating of terrestrial samples, it has been assumed that there are no sources of tracks in ancient samples other than the spontaneous fission of ^{238}U (the contribution from the spontaneous fission of ^{235}U—whose natural abundance is $\sim 0.7\%$ of all U— amounts to only $\sim 0.04\%$ generally). Only in one special case—the Oklo prehistoric natural fission reactor[49]—was a high density of neutron-induced fission tracks produced in the ^{235}U component of terrestrial rocks;[50,51] but this need not be further considered here.

In the case of rocks from relatively unshielded locations on the moon, or in meteorites (or indeed on the surface of any atmosphereless planetary body without a significant magnetic field), tracks from a number of sources other than the fission of ^{238}U may be present, and these are briefly described here. The principal sources of tracks in meteorites were first discussed in two classic papers by Fleischer *et al.*[52,52a]

8.7.1   Heavy cosmic-ray primaries

Interstellar space is pervaded by the galactic cosmic rays—energetic nuclear particles, whose origin and modes of transport and acceleration are still controversial, but which consist mostly of protons and helium nuclei, and whose containment within the galaxy is believed to be caused by the action of a galactic magnetic field. The energy spectrum of these primary particles (i.e. number of particles $(m^2 \ s \ sr \ MeV)^{-1}$ versus energy) falls steeply with increasing energy (see Fig. 8.8, curve a); but the solar magnetic field acts to exclude many of the lower-energy particles from the inner solar system. This results in a modulated spectrum, as shown in Fig. 8.8 (curve b), peaking at an energy rather below ~ 1 GeV/nucleon. Passage of these particles to the surface of the earth is further hindered by our atmosphere and by the geomagnetic field. However, these latter restrictions are almost entirely absent in the case of the moon and the meteorites.

A small fraction ($\sim 10^{-3}$) of the primary cosmic ray particles consists of heavy nuclei ($Z \gtrsim 20$), which are capable of registering tracks in meteoritic minerals such as olivine and pyroxene at energies of ~ 1 MeV/nucleon. Take a typical Fe nucleus possessing an energy of ~ 1 GeV/nucleon on striking a meteorite, which slows down via ionization losses. Then, provided that it escapes a nuclear collision, it will register an etchable track over the last

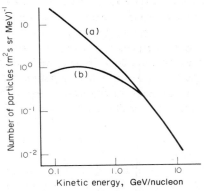

FIG. 8.8   *The low-energy portion of the energy spectrum of cosmic ray protons (of kinetic energy < 10 GeV). Curve b shows the spectrum as measured close to the earth by balloon-flight and satellite experiments, and peaks at an energy rather below ~ 1 GeV/ nucleon. Curve a shows an estimate of the original primary spectrum corrected for the modulation effect of the solar magnetic field, which has acted to exclude the lower-energy particles from the inner solar system (including the moon and the earth). The geomagnetic field, and the atmosphere of the earth, further remove the very-low-energy particles (including some of solar origin) at the start of curve b (not shown in the graph), some of which are trapped in the Van Allen belts surrounding the earth, and whose intensity varies through the solar cycle. The VH ($20 \leq Z \lesssim 28$) primary flux follows the same pattern, and would have similar intensities as shown by these curves, provided that the ordinate scale were changed to differential flux per GeV (instead of MeV^{-1}). The differential spectra, I_d, follow relations of the form $I_d = KE^{-(\gamma+1)}$, where K is a constant, of dimensions indicated by the ordinate, E is the energy per nucleon, and γ is the "slope" of the spectrum. (Figure after Wilson.[52b])*

~ 10–$20\ \mu m$ of its trajectory after passing through several centimeters of rock. Typical cosmic ray tracks in a meteorite crystal are shown in Fig. 8.9.

Because the energy spectrum of the cosmic rays is dominated by low-energy particles, and because of the effects of nuclear interactions (resulting in the production of cascades, spallations, etc.), the profile of cosmic ray track density versus depth within a meteorite is a steeply falling function of increasing depth (see Fig. 8.10). Typically, track densities for a meteorite with a cosmic ray exposure age (i.e. age since the meteorite progenitor was reduced by collision, etc., to fragments ~ 1 m across, which were thus largely unshielded from cosmic rays) of 20 Myr, vary from $\sim 10^7$ cm^{-2} in the top centimetre or so, down to $\sim 10^5$ cm^{-2} at a depth of ~ 20 cm. These track densities are relatively weakly dependent on the nature of the mineral crystal.

To distinguish cosmic ray tracks from ^{238}U- (and ^{244}Pu-) fission tracks is not an easy matter. It may be best to count such tracks in U-poor crystals (e.g. olivines, pyroxenes, or feldspars). Differential annealing can also be used as an aid, since fission tracks are more resistant to heat than the VH ($20 \leq Z \lesssim 28$) tracks, so that those surviving after relatively high-temperature annealing may be attributed to fission. Finally, since in a moderately sized meteorite, samples near an outside surface (which may be distinguished by a fusion crust, provided that too much fragmentation has not taken place during the atmospheric passage) show a non-isotropic distribution of cosmic ray tracks—falling off in number density from outside inwards—one may be able to distinguish them from fission tracks, which are isotropic. This is usually done by plotting the number of tracks as a function of angle with respect to some arbitrary direction: the distribution in a U-rich phase would be fairly isotropic while that in a U-poor phase, exposed to a high, directional, cosmic-ray flux, would be significantly anisotropic (see, e.g., Green *et al.*[52d]). The length distributions of tracks (TINT, TINCLE, or surface tracks) in meteoritic crystals can also display differences indicative of their origin, viz. fission versus cosmic rays.[52a,d]

The sun emits its own low-energy "cosmic rays" (or solar particles). These "solar cosmic rays," or solar flare particles, exhibit a softer energy spectrum than the galactic cosmic rays, and are therefore severely attenuated by a few millimetres of rock. Since a meteorite typically loses at least a few centimetres from its outer layers as a result of ablation during atmospheric entry, the solar flare tracks are usually not observed in meteorites. However, a few meteorites—those possessing an abundance of trapped rare gases—have been found to contain some randomly distributed grains displaying very high track densities ($\sim 10^7$–10^9 cm^{-2}). These track-rich grains are believed to have been irradiated by solar flare particles while they were part of a soil layer on the surface of a meteorite parent body. During this period solar wind irradiation would have implanted rare-gas atoms into the crystals. The

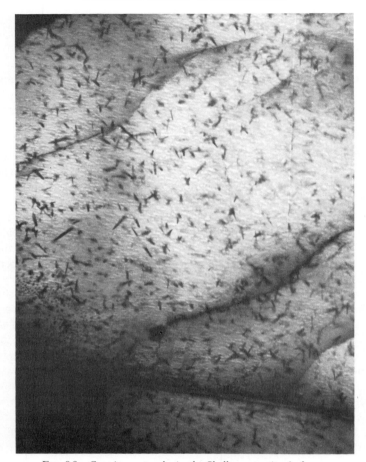

FIG. 8.9 *Cosmic ray tracks in the Shalka meteorite (a hyper-sthene achondrite which fell in W. Bengal, India, in 1850). The hypersthene (a variety of pyroxene, viz. Fe-rich orthopyroxene) crystals were mounted in epoxy resin, polished, and then etched in boiling 6:4 NaOH (i.e. 60% aqueous solution of NaOH), in a reflux system for periods of 1–2 h. Only the VH ($20 \leq Z \leq 28$ or 30) and VVH ($Z \gtrsim 30$) components of cosmic rays, when they have slowed down sufficiently, produce etchable tracks in silicate minerals—and these are usually spatially non-isotropic in a large meteorite, since the cosmic rays travel from outside to the inside of a meteorite. [In the case of the Shalka meteorite, the tracks were found to have been completely or partially annealed over depths of up to ∼ 1000 μm below the "fusion crust" owing to the high temperature generated during the atmospheric passage (Bull and Durrani[52c])]. Only a part of the track density is in sharp focus in this photomicrograph.*

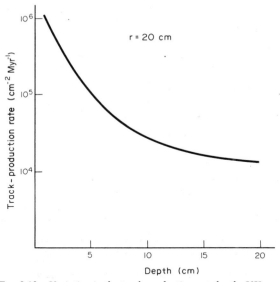

FIG. 8.10 *Variation in the track production rate by the VH component (20 ≤ Z ≲ 28) of cosmic ray primaries as a function of depth in a meteorite. The cosmic ray flux values have been based on near-earth measurements, but are assumed to apply, on average, to typical meteorite orbits over many millions of years. The curve has been computed for a spherical stony meteorite of radius 20 cm, and represents an average between curves for crystal-sample orientations parallel and perpendicular to the radius vector (or, alternatively, ⊥ and ‖ to the nearest surface) of the meteorite. The mean etchable range has been taken as 10 μm. The depths are below an outside surface of the meteorite; the curve has a symmetrical (i.e. a mirror-image) limb (not shown) rising to the opposite surface. The outer parts of a meteorite have almost always been lost in ablation (thin-film melting) during its atmospheric passage; but the steepness of the production-rate curve (such as the one shown here) gives an indication of the preatmospheric size of the meteorite, since over the same distance (say 1–2 cm) the curve falls more steeply close to the outer surface than further inside (because the "softer" component of the primary cosmic rays is attenuated more readily than the "harder" component penetrating further). Hence by fitting the rate of fall of the track density with distance from an arbitrary inner surface (usually a fusion crust on a fragment) on to a number of possible total radii of the initial (pre-atmospheric) meteorite, and making adjustments on the basis of several trial values of assumed ablated thickness outside the fusion crust, it is possible to discover the preatmospheric radius of the meteorite (usually assumed to have been spherical). It is that radius on which the observed track density values (as a function of depth) sit best, within experimental errors. Usually a cosmic ray exposure age is assumed, or is known from other sources; the track-density data thus yield the track production rates. (Figure after Fleischer et al.[52])*

soil grains, it is thought, were later compacted, along with an admixture of unirradiated material, into a rock which was ultimately fragmented to form a meteorite.

Lunar rocks and soils also show abundant evidence of galactic cosmic ray tracks as well as solar flare tracks (and implanted solar wind gases). The results obtained from the study of these tracks represent an important application of the etched-track technique, but their detailed consideration is beyond the scope of this book (for extensive reviews see, e.g., references 35 and 53). The pertinent question for the present is as to how we may allow for the presence of solar flare and cosmic ray tracks when attempting to obtain the fission track age of a lunar rock or meteorite.

The important point is that cosmic ray tracks (note that for the rest of this discussion the generic term "cosmic ray tracks" is used to denote tracks of both solar and galactic origin) occur independently of the U content of a crystal. Thus a U-rich crystal (such as whitlockite* or apatite) will contain approximately the same number of cosmic ray tracks as an adjacent U-poor phase (such as olivine or pyroxene). The track densities in these latter minerals can therefore be used to subtract the approximate cosmic ray contribution, from the total track density in the U-rich phases. In practice the situation is a little more difficult. Sensitivities of different minerals for recording cosmic ray tracks do vary somewhat, owing to differences in the etchable range of cosmic ray nuclei in these minerals. Thus olivine will probably exhibit fewer cosmic ray tracks per unit area than an adjacent phosphate (e.g. whitlockite or apatite). Such differences in sensitivity are, however, not known with great accuracy; so that the most reliable fission track ages will be derived when $\rho_{\text{fission}} \gg \rho_{\text{cosmic ray}}$.

8.7.2   Cosmic-ray-induced fission

Energetic protons in the primary cosmic ray flux (which constitute all but a few percent of that flux) are able to induce fission in the heavy-element impurities within the meteorite or the lunar rock. Furthermore, as the primary protons interact with the constituent nuclei of the meteoritic material, a cascade of secondary protons and fast (and eventually thermal) neutrons develops. These secondary particles are also capable of producing induced fission. These fissions will mostly take place in those minerals which have high U and Th contents—and which are precisely those that contain high spontaneous fission track densities.

To allow for these induced fission tracks is no simple matter, although fortunately they usually represent a fairly small fraction of the total track density.

Thiel et al.[54] have irradiated thick targets with high-energy protons. These targets were loaded with U-bearing glasses and the subsequent induced

* Also called merrillite.

fission track densities were measured. Durrani and Khan (1972; unpublished data) also irradiated a number of sandwich assemblies, containing U foils and other metals and reference glasses in contact with glass, mica, and plastic detectors, with 7 GeV/c protons—the assemblies being placed at various depths inside large concrete blocks (up to $127 \times 76 \times 66 \ cm^3$ in volume) to simulate cosmic ray bombardment of meteorites. Woolum and Burnett[55] irradiated a series of mica–uranium sandwiches at different depths below the lunar surface in an experiment deployed by the Apollo 17 astronauts. Their results were used to derive the thermal-neutron fission rate as a function of depth below the lunar surface.

Damm *et al.*[56] have calculated the fast-neutron- and proton-induced fission rates by using cross-sectional data in conjunction with nucleon fluxes as a function of depth in rocks of lunar composition as computed by Reedy and Arnold.[55] They have then combined these results with the thermal-neutron fission rate data of Woolum and Burnett[55] to obtain the total induced fission rate versus depth below the lunar surface (see Fig. 8.11).

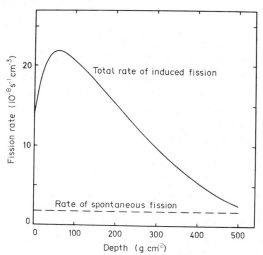

FIG. 8.11 *Calculated total induced fission rate versus depth below the lunar surface, in $1 \ cm^3$ of a sample of mass density $2.6 \ g \ cm^{-3}$ and containing heavy elements in the following concentrations: U_{nat}, 1 ppm (^{235}U, 7.2 ppb); ^{232}Th, 4 ppm; ^{209}Bi, 1 ppb; ^{197}Au, 1 ppb. These heavy elements are assumed to undergo induced fission by the influx of cosmic ray high-energy protons and neutrons (as computed by Reedy and Arnold[57]) and by thermalized neutrons (from the data of Woolum and Burnett[55]). The spontaneous fission rate of 1 ppm ^{238}U (assuming $\lambda_{f8} = 8.46 \times 10^{-17} \ yr^{-1}$) is given by the dashed line for comparison. (Figure after Damm et al.[56]).*

These corrections become important only for meteorites or lunar rocks having cosmic ray exposure ages that are an appreciable fraction of their fission track ages.

8.7.3 Spallation recoil tracks

When energetic protons and neutrons in the cosmic ray flux collide with target nuclei within the meteoritic or lunar sample, further nucleons are spalled off. The residual (target) nucleus recoils with considerable energy and is, on occasion, capable of forming a track. Fortunately, most of the bulk constituents of common meteoritic or lunar minerals are too light for etchable recoil tracks to result from their fragmentation in such collisions. Nevertheless, heavy-element impurities may contribute spallation recoil tracks to the total track density. It is found, however,[52] that spallation recoils are usually rather short—generally less than 2–3 μm—and, provided some minimum-length criterion is set for the counting of genuine fission tracks, they do not constitute a serious interference to the fission-track densities.

Other, more exotic, sources of tracks have also been considered,[52] but have not yet been shown to be important in nature—with one notable exception: that of spontaneous fission of ^{244}Pu (see §8.8 below).

Another variety of problems associated with fission track dating of extraterrestrial samples arise from the small sample sizes often available for study, and from the extreme paucity of U-rich phases in many extraterrestrial rocks, particularly meteorites. In most chondritic meteorites, the only U-rich phases are whitlockite* $[\beta\text{-}Ca_3(PO_4)_2]$ and chlorapatite $[Ca_5(PO_4)_3Cl]$.

Despite the attendant difficulties, fission track studies of meteorites and, to a lesser extent, of lunar rocks have provided a wealth of information on the thermal history of the early solar system.

8.8 ^{244}Pu Fission Tracks in Very Ancient Samples

It has been known since the early work of Fleischer, Price, and Walker[58,59] that meteoritic minerals contain more fission tracks than can be accounted for on the basis of spontaneous fission of ^{238}U over the age of the solar system. The mineral whitlockite was found to contain large excesses of fission tracks,[60] and these were reported by Wasserburg et al.[61] to correlate with the presence of fission-type Xe. It was surmised that this Xe, and the fission tracks, originated from the fission of the extinct ^{244}Pu content;[62] and this was confirmed when the isotopic spectrum of Xe from the fission of artificially

---

* This mineral is now usually referred to as "merrillite". The older name will, however, be retained throughout this discussion.

produced ^{244}Pu was found to match that of the meteoritic fission-Xe.[63] It is now well established that excesses of fission tracks in lunar samples and meteorites, over those expected from the decay of ^{238}U (provided that all other sources of tracks have been allowed for), are indeed due to the fission of ^{244}Pu.

8.8.1 ^{224}Pu fission track age equation

In conventional ^{238}U fission track dating we can measure the number of fission tracks that have been produced, determine (usually via the ^{235}U content) the number of ^{238}U atoms that still remain, and use these two measured quantities to determine two unknown quantities: the fission track age and the initial uranium content (cf. §§8.1, 8.2). In ^{244}Pu fission track dating, since any ^{244}Pu ($\tau_{1/2} \sim 82$ Myr) surviving now is present in essentially unmeasurable quantities, we have only one measured quantity—the fission track density. One way of getting some idea of the original ^{244}Pu content is to assume that Pu and U are chemically coherent, i.e. they maintain a constant ratio in a wide variety of materials (we shall see later that, in actual fact, this assumption is not generally true). Let us, then, assume that at some reference time t_0 years ago the ratio of ^{244}Pu to ^{238}U had some known value $(^{244}\text{Pu}/^{238}\text{U})_0$, and that Δt years later a given mineral began to record fission tracks. We can now proceed to derive an age equation.

If the initial number of ^{244}Pu atoms is designated by N_4^0, then the surviving number Δt years later is $N_4^0 \exp(-\lambda_{D4}\Delta t)$, where λ_{D4} is the total decay constant for ^{244}Pu. The number present now being $N_4^0 \exp(-\lambda_{D4}t_0)$, the number of ^{244}Pu atoms that have decayed by fission, between the time when track retention began and today, is given by:

$$\frac{\lambda_{f4}}{\lambda_{D4}} N_4^0 \left[\exp(-\lambda_{D4}\Delta t) - \exp(-\lambda_{D4}t_0)\right]$$

where λ_{f4} is the fission decay constant for ^{244}Pu. A similar expression can be written down for ^{238}U. The ratio of present-day fission track densities from ^{244}Pu and ^{238}U is therefore

$$\frac{\rho_4}{\rho_8} = \frac{\lambda_{f4}}{\lambda_{D4}} \frac{\lambda_{D8}}{\lambda_{f8}} \left(\frac{^{244}\text{Pu}}{^{238}\text{U}}\right)_0 \frac{\exp(-\lambda_{D4}\Delta t) - \exp(-\lambda_{D4}t_0)}{\exp(-\lambda_{D8}\Delta t) - \exp(-\lambda_{D8}t_0)} \quad (8.13)$$

where the subscripts 4 and 8 refer to ^{244}Pu and ^{238}U, respectively. Note that the symbol ρ_8, used in this section to denote the density of spontaneous fission tracks due to ^{238}U, is identical to the ρ_s employed for the same purpose earlier in the chapter (§8.2 *et seq.*).

If $(^{244}\text{Pu}/^{238}\text{U})_0$ and t_0 are known, i.e. if we know the ratio of ^{244}Pu to ^{238}U atoms at some reference time t_0 years ago, then we can use the track density ratio ρ_4/ρ_8 to obtain Δt, and thus find the fission track age (i.e. the track retention age) t from the relation $t = t_0 - \Delta t$. In fact, neither ρ_4 nor ρ_8 is separately measurable (the two species of tracks having much the same appearance): we measure only the *total* number of fission tracks ρ_f. By making a reasonable guess at the fission track age of the sample, we can estimate ρ_8 from a knowledge of the present-day uranium content. Then ρ_4 and ρ_4/ρ_8 corresponding to this preliminary estimate can be computed, and a better value of the fission track age found; this age can then be iteratively used to obtain an improved value of ρ_8 and hence of t. Since ρ_8 is a rather slowly varying function of t, this iterative procedure will rapidly converge to a constant track-retention age t ($= t_0 - \Delta t$, where t_0 is the "reference age" and Δt is the track non-retention interval immediately following the reference time).

The Pu-fission-generated Xe content of a meteorite can be used to derive an estimate of the ^{244}Pu content at the time of rare-gas retention. This time can be fixed fairly reliably by the ^{39}Ar–^{40}Ar age[42] of the meteorite. If the U content is also measured then we can fix the ratio ^{244}Pu/^{238}U at some known time in the past. Provided that we assume the chemical coherence of Pu and U, we can use Eq. (8.13) to obtain the fission track age of any mineral from a very ancient meteorite or lunar rock.

It has, however, become increasingly clear in recent years[64] that the ratio of Pu to U varies considerably from one mineral to another and from one meteorite to another. Fission track ages can still be obtained by using Eq. (8.13), provided that both the fission track and the fission-Xe measurements are made on a given mineral (such as whitlockite) from the same meteorite. Even in such circumstances, however, care must be exercised since, as emphasized by Pellas and Storzer,[65] considerable variations in Pu fission-track density occur between individual whitlockites (or apatites) from the same meteorite, even though the *average* Pu content of various chondrites may have been reasonably constant.[70]

It is important to realize here that we are really measuring a time interval Δt between the onsets of Xe retention and track retention. This interval can only be placed within an absolute timescale by fixing the gas retention time with the help of another chronometer, such as the ^{39}Ar–^{40}Ar system.

8.8.2　The "contact" track density method

A powerful extension of the ^{244}Pu track method has been developed by Pellas and Storzer.[65–69] They measure the track densities at the surface of crystals poor in uranium and plutonium, such as pyroxene and feldspar, which are contiguous to U-, Pu-rich phases such as whitlockite (see Fig. 8.12). These tracks arise from fissions occurring in the whitlockite (or apatite)—which phase thus acts as a common source for all tracks recorded in the surfaces

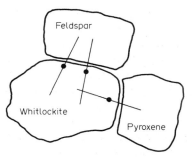

FIG. 8.12 *Schematic diagram showing the principle of the multi-crystal boundary- (or contact-) track method of Pellas and Storzer.*[65-69] *The source of the fission tracks is a U-, Pu-rich phase such as whitlockite ("merrillite"), as shown here (alternatively, it could have been apatite); the detector is a crystal poor in fissile elements (say a feldspar or a pyroxene or an olivine), which is contiguous to the former. The fission fragments from the break-up of U and Pu in the source crystal (whitlockite) leave tracks in the boundary layers (~ 10–15 μm deep) of the detector crystals. By measuring the fission track densities in two different types of detector crystals (feldspar and pyroxene in the present figure), both adjoining different surfaces of the same source crystal, but each having a different track non-retention time interval Δt, it is possible to calculate the cooling-down rate of the meteorite during the time that it took for the temperature to fall from the track-retention value for one detector crystal (or "phase") to that for the other phase. See text for further discussion.*

of adjacent crystals (whose own fossil fission is considered to be negligible). If we write down Eq. (8.13) for two different U- and Pu-poor phases, denoted by subscripts 1 and 2, both of which border whitlockite crystals, then we have

$$\left(\frac{\rho_4}{\rho_8}\right)_1 = \frac{\lambda_{f4}\lambda_{D8}}{\lambda_{D4}\lambda_{f8}} \left(\frac{^{224}\text{Pu}}{^{238}\text{U}}\right)_0 \frac{\exp\left(-\lambda_{D4}\Delta t_1\right) - \exp\left(-\lambda_{D4}t_0\right)}{\exp\left(-\lambda_{D8}\Delta t_1\right) - \exp\left(-\lambda_{D8}t_0\right)} \quad (8.13a)$$

and

$$\left(\frac{\rho_4}{\rho_8}\right)_2 = \frac{\lambda_{f4}\lambda_{D8}}{\lambda_{D4}\lambda_{f8}} \left(\frac{^{244}\text{Pu}}{^{238}\text{U}}\right)_0 \frac{\exp\left(-\lambda_{D4}\Delta t_2\right) - \exp\left(-\lambda_{D4}t_0\right)}{\exp\left(-\lambda_{D8}\Delta t_2\right) - \exp\left(-\lambda_{D8}t_0\right)} \quad (8.13b)$$

In the above equations we have assumed that, since both "detector minerals" have the same source for their fission tracks, therefore the values of $(^{244}\text{Pu}/^{238}\text{U})_0$ and t_0 are the same in each case; the two track non-retention intervals Δt have, however, been taken to be different for the two detector minerals.

On examining Eqs. (8.13a,b) (or Eq. (8.13) itself) more closely, it would be noticed that $\exp\left(-\lambda_{D4}\, t_0\right)$ is very close to zero since ^{244}Pu is, for all practical purposes, now extinct. In addition, since the number of ^{238}U decays is rather insensitive to small changes in t (because of the long half-life of that isotope), hence $[\exp(-\lambda_{D8}\Delta t_1) - \exp(-\lambda_{D8}t_0)] \simeq [\exp(-\lambda_{D8}\Delta t_2) - \exp(-\lambda_{D8}t_0)]$. Therefore, upon dividing Eq. (8.13a) by (8.13b), we obtain:

$$\left(\frac{\rho_4}{\rho_8}\right)_1 \bigg/ \left(\frac{\rho_4}{\rho_8}\right)_2 \approx \exp\left[-\lambda_{D4}(\Delta t_1 - \Delta t_2)\right]$$

or

$$\left(\frac{\rho_4}{\rho_8}\right)_1 \bigg/ \left(\frac{\rho_4}{\rho_8}\right)_2 \approx \exp\left(-\lambda_{D4}\Delta t'\right) \qquad (8.14)$$

where $\Delta t'$ is now the time interval between the onset of track retention in phase 1 and in phase 2, respectively. If it is assumed that the ^{244}Pu tracks are much more numerous than ^{238}U tracks (which is generally the case for minerals older than $\sim 4 \times 10^9$ yr (track retention age))*, and when use is made of the fact that the ^{238}U track density is insensitive to small changes in the fission track age (compared with the effect on the ^{244}Pu track density), then Eq. (8.14) may be further simplified to:

$$\rho_{f1}/\rho_{f2} \simeq \exp\left(-\lambda_{D4}\Delta t'\right) \qquad (8.15)$$

Here ρ_{f1}, ρ_{f2} are the total track densities measured on the surfaces of Pu-, U-poor phases 1 and 2 in contact with a Pu-, U-rich phase.

Thus, by measuring track densities in whitlockites or apatites as well as on the surfaces of crystals such as feldspars, pyroxenes, or olivines which are contiguous to these phases, it is possible to determine the times which separate cooling down through the closure temperatures for different detector minerals. If these closure temperatures are known, the corresponding cooling rate of the meteorite can be deduced (see Table 8.2 for a rough guide of the track-retention properties of some minerals of interest). It should be noted here that track densities measured internally *within* whitlockite or apatite should be divided by a factor of 2 prior to comparison with track densities in the contacting minerals, since these latter "detector" surfaces are irradiated in 2π-geometry in contrast to the 4π-irradiation within the "source" minerals.

This method of studying the cooling history of meteorities (where, in essence, the decay of ^{244}Pu, with its known half-life, is used as the "clock", and different track-recording crystals serve as a "palaeothermometer"), has been recently reviewed by Pellas and Storzer.[65] Typically, cooling rates of a few degrees K per million years are obtained for cooling through the temperature range $\sim 800-300$ K.[65] These cooling intervals can only be placed

* This results from the very large value ($\sim 1.5 \times 10^5$) of the ratio $\lambda_{f4}/\lambda_{f8}$, which on multiplication with the factor $\lambda_{D8}/\lambda_{D4}$ in Eq. (8.13) still gives a value $\sim 2.7 \times 10^3$. Reasonable values of $(^{244}Pu/^{238}U)_0$ (say ~ 0.017, $\sim 4.6 \times 10^9$ yr ago) lead to $\rho_4/\rho_8 \approx 100$ (see reference 48a).

TABLE 8.2   *Temperatures required for 50% reduction of fission track density in different minerals over 1 hour of isothermal heating**

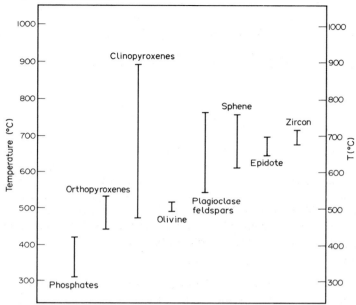

* The values shown are based on published experimental data and are meant to serve only as a rough guide; the actual values depend on the etching conditions employed. The large spread in the temperature values for 50% track retention (or loss) is a reflection of wide variations both in the chemical composition of the minerals and in experimental conditions (a small spread often indicates a paucity of data rather than tightness of distribution). Closure temperatures (T_c) depend on several factors (see §5.4) but may be expected to show the same broad trend from mineral to mineral as exhibited above (extrapolation of 50% track-density loss Arrhenius line to geological timescales of $\sim 10^6$–10^8 yr (corresponding to different cooling rates) yields an approximate value of T_c). See also Table 5.1 for 100% loss of fission tracks.

within an absolute time scale if a $^{244}Pu/^{238}U$ ratio at some reference time is assumed.

A number of experimental difficulties exist with this method. The Pu track densities in individual whitlockite and apatite crystals are extremely variable, and so a large number of contiguous track densities must be measured in order to obtain meaningful cooling-rate data. Also, problems can arise from the presence of shock-induced dislocations,[65] which may be mistakenly counted as tracks.

Although the ratio $^{244}Pu/^{238}U$ varies from one phosphate crystal to another, there is some evidence that the average initial Pu content of whitlockites was roughly the same in different chondritic meteorites,[70,70a] corresponding to ~ 20 parts in 10^9 at about 4.57×10^9 years ago. It has been suggested[71,72] that, instead of relying on the Pu/U ratios, it may be better to seek some other element to which the Pu content may have been better correlated—e.g. the rare-earth elements Nd or Sm. The accurate determination of the latter trace elements in the whitlockites and apatites is, however, no easy matter.[73]

8.9 Fission Track Dating in Archaeology

As an example of the application of the fission track method for the dating of very much younger samples than meteorites let us consider its use in archaeology. This field has been well covered by a concise review article by Wagner,[74] to which the reader is referred for further details (see also Fleischer *et al.*[35]). We shall give here only a brief overview of the subject area.

Since human civilization spans a mere ten thousand years or so, and the activities of the hominids not much more than $\sim 10^6$ years, the density of spontaneous fission tracks in samples of archaeological interest is, of necessity, relatively small (the fission-decay half-life of ^{238}U being $\sim 10^{16}$ years). Reference to Eq. 8.5' will show that in glass (with $\eta = \cos^2 \theta_c \simeq 0.75$; density $\simeq 2.5$ g cm^{-3}; R $\simeq 10^{-3}$ cm), a uranium content of 1 ppm will yield a mere ~ 3 tracks cm^{-2} over a period of 10^4 years. To obtain a statistically reliable result one has, therefore, to count tracks over large areas; this means, in practice, polishing off a large number of surface layers in succession, etching each layer in turn, and counting the revealed fission tracks cumulatively. This has been done, but is tedious, and entails considerable statistical error in the age obtained—especially for young samples, e.g. Roman glass ~ 1500–2000 years old. Man-made glasses containing high concentrations of uranium (a few per cent, to give them a pleasant yellowish coloration) are more amenable to fission-track dating; but these are rare.

In practice, the most useful archaeological material for fission track dating (FTD) is obsidian, or volcanic glass, which was used by early man as a cutting tool or a weapon as long as 30 000 to 100 000 years ago, and even travelled far-afield by way of trade.[75] Obsidian often has a uranium content ranging from several ppm up to 10 or 20 ppm by weight. One has, however, to select reasonably clear obsidian samples for FTD, for they are often either very dark in colour and/or patterned by multicolored "striae", or full of bubbles, crystallites (which, when etched by hydrofluoric acid, leave track-like pits) and other troublesome inclusions (see Fig. 8.13). Glass shards from tuff layers and pumice can also be used for FTD, but similar remarks apply to them. Flint, much used by prehistoric man, has unfortunately not proved amenable to being dated by this method, for it is not possible to develop properly etched tracks in this material—though the induced-fission method has been used (with an external detector) to determine the concentration and spatial distribution of uranium in flint,[76] which are important in the thermoluminescence method of dating.

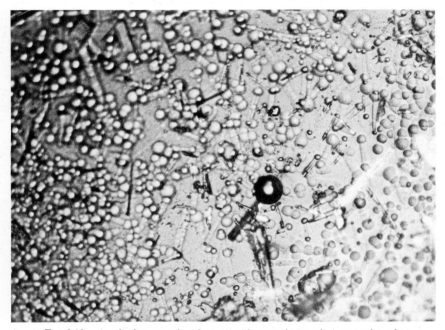

FIG. 8.13 *An obsidian sample (from Acigöl, central Anatolia), viewed in the reflected-light mode by an optical microscope, after etching with hydrofluoric acid. A genuine fission-track etch pit is seen on the lower right-hand side, surrounded by a sea of etched bubbles and other imperfections. The needle-like tracks are believed to be due to feldspar crystallites located close to the top surface, which have become etched. To facilitate fission track dating, it is advisable to select obsidian samples which have clear areas, not obscured by dark or multicoloured "striae" (furrows or bands) and "flow patterns"; microscopic viewing in the "epi-illumination" (or reflected-light) mode, rather than in the transmission mode, may also be helpful.*

When it is man's use of natural glasses, such as obsidians or tektites,* which is to be dated, and not the samples themselves, then they need to have been severely heated by him to reset the "fission-track clock". This could have been either by design or, more probably, by accident. Thus, Fleischer et al.[77] were able to date an obsidian knife-blade found by L. S. B. Leakey in Kenya in 1927. The knife had been severely heated ~3700 ± 900 years ago (so that, in 36 successive etchings, with a total scanned area of 5.6 cm², only 17 fossil-fission tracks were found—hence the large uncertainty in age). The date agrees rather well with the C-14 dating of a nearby neolithic crema-tion site. Another example of FTD is that of dating the activities of the Peking Man (*Sinanthropus*), who made fires. The Chinese scientists collected ~5 kg of the ancient ashes from his cave at Zhoukoudian (Chou-Kou-Tien), and dated the sphene crystals found amongst them by the FTD method. An age of 0.462 ± 0.045 Ma has been reported by Guo,[78] which is in good agree-ment with K–Ar and thermoluminescence (TL) methods of dating. Similarly, Zimmerman et al.[79] were able to establish the authenticity of a putatively ancient bronze horse by dating the zircon-crystal inclusions from the ceramic core of the statue—though the authentication in this case was done by the TL method, with the uranium mapping by the induced fission track tech-nique playing an essential part.

In cases where an archaeological object has not been heated severely enough to completely erase all pre-existing tracks but only to anneal them partially, the age-correction methods described earlier in this chapter (§§8.6.1, 8.6.2) must be applied. Thus Wagner[74] reports the use of such age-correction procedures in reconciling the age of a number of obsidian flakes found at the site of ancient obsidian workings at El Inga in Ecuador; for instance, one flake, which yielded an apparent age of 0.14×10^6 years, gave a corrected age (on the basis of track-size correction method) of 1.72×10^6 years (cf. the case of tektite age reduction attributed to bush fires.[38,39])

On occasion, the only feasible method of deciding between different pos-sible origins of archaeological artefacts has proved to be the FTD technique. Thus the identification of the source of obsidian artefacts and waste flakes found at the mesolithic cave settlements at Franchthi in southern Greece had proved impossible on the basis of "conventional" techniques such as optical spectroscopy, X-ray fluorescence, and neutron activation. There is no known source of obsidian on the mainland of Greece itself; possible sources were thought to be the Aegean islands, Anatolia in Turkey, and the Carpathian mountains in Hungary. Trace-element classification had failed to distinguish between obsidians from all these contending sources. The fission track anal-ysis proved to be the only method which could successfully pinpoint the

* Small globules of molten glass found in certain areas of the globe and believed to have been generated by meteoritic or cometary impacts with the earth some millions of years ago.

origin of the Franchthi cave obsidians. Thus Durrani *et al.*[80] were able to show, on the basis of eruption age of the obsidians as well as their U-content, that the only source which matched the Franchthi material in both these respects was the island of Melos in the Aegean sea, ~ 120 km south-east of Franchthi on the mainland. (These authors compared the fossil-track etch-pit diameters with those of the freshly induced ones, and concluded that no age correction was necessary.) The establishment of the Melian source for the Franchthi cave obsidian thus provided the earliest evidence of maritime travel by ancient man, and carried the history of seafaring back a thousand years or more. It seems that the hunters and fishers of southern Greece were already, by the beginning of the seventh millennium BC, and perhaps earlier, reaching the island of Melos by boat. This coincides with the beginning of the intensive exploitation of marine fish in the region.

One final example of the FTD method to archaeology would suffice: that of dating man-made glasses. It had been postulated that a certain ornamental ring of greenish uranium glass had been made in China during the eighteenth century, well before the earliest known production of uranium glass in Bohemia around 1840. However, fission track dating of the ring revealed that it had been manufactured only around the turn of the twentieth century.[74,81] Sometimes, even when the determination of an exact date of a man-made glass object is difficult, it may be sufficient to discriminate between two widely different possibilities by the FTD method. Thus, a fragment of a glass vessel found at an iron age archaeological site (Landberg Fort) on Fair Isle off the northern Scottish coast was suspected by archaeologists to be either of ancient Roman origin (in which case it would have been the most northerly example of imported Roman glass ~ 2000 years old) or a comparatively recent adjunct. In the event, the FTD method put its age at no more than a few hundred years.[82]

8.10   Errors in Fission Track Dating

Suppose that in determining a track density ρ, a number of tracks N have been counted within an area A of the crystal. Then

$$\rho = N/A$$

If the tracks have been measured in n crystals, then the average track density is given by

$$\rho = \left(\sum_{i=1}^{n} N_i \right) \bigg/ \left(\sum_{i=1}^{n} A_i \right) \tag{8.16}$$

where the subscript i refers to the value of the relevant parameter in the ith crystal.

It is usually assumed that track counts follow Poissonian statistics, so that the error $\sigma(\rho)$ in the track density is given by

$$\sigma(\rho) = \left(\sum_{i=1}^{n} N_i \right)^{1/2} \bigg/ \left(\sum_{i=1}^{n} A_i \right) \tag{8.17}$$

Errors in track counting obtained in this manner are then combined with errors assigned to other measurements, such as the neutron fluence F, in order to arrive at the uncertainty in the fission track age. This approach is probably valid if the spread in track densities in different crystals from the sample is due only to the statistical nature of radioactive decay. Where other sources of fluctuation in the track density are present (such as variations in U content, for example), a more detailed analysis may be required.

The subject of errors in fission track dating is an important one, but currently somewhat controversial. The reader is referred to a number of recent papers[83-87] on this topic (see also references 88–90, where certain statistical tests for the analysis of FTD results are described).

References

1. P. B. Price & R. M. Walker (1962) Observation of fossil particle tracks in natural micas. *Nature* **196**, 732–4.
2. G. Faure (1977) *Principles of Isotope Geology*. Wiley, New York.
3. E. I. Hamilton (1965) *Applied Geochronology*. Academic, London.
4. E. Jäger & J. C. Hunziker (eds.) (1979) *Lectures in Isotope Geology*. Springer-Verlag, Berlin.
5. L. D. Muller (1977) Laboratory methods of mineral separation. In: *Physical Methods in Determinative Mineralogy* (J. Zussman, ed.). Academic Press, London (2nd edition), pp. 1–34.
6. A. J. Hurford (1976) *The Birkbeck Laboratory Manual on Fission Track Dating*. Birkbeck College, University of London.
7. A. J. W. Gleadow, A. J. Hurford & R. D. Quaife (1976) Fission track dating of zircons: Improved etching techniques. *Earth Planet. Sci. Lett.* **33**, 273–6.
8. A. J. W. Gleadow (1981) Fission-track dating: What are the real alternatives? *Nucl. Tracks* **5**, 3–14.
9. C. W. Naeser (1967) The use of apatite and sphene for fission track age determination. *Bull. Geol. Soc. Am.* **78**, 1523–6.
10. A. J. W. Gleadow (1978) Anisotropic and variable track etching characteristics in natural sphenes. *Nucl. Track Detection* **2**, 105–18.
10a. H. A. Khan & S. A. Durrani (1972) Prolonged etching factor in solid state track detection and its applications. *Radiat. Effects* **13**, 257–66.
11. R. L. Fleischer, P. B. Price & R. M. Walker (1964) Fission track ages of zircons. *J. Geophys. Res.* **69**, 4885–8.
12. P. F. Green & S. A. Durrani (1978) A quantitative assessment of geometry factors for use in fission track studies. *Nucl. Track Detection* **2**, 207–13.
13. S. A. Durrani, H. A. Khan, R. K. Bull & G. W. Dorling (1974) Charged-particle and micrometeorite impacts on the lunar surface. In: *Proc. 5th Lunar Sci. Conf., Geochim. Cosmochim. Acta*, Suppl. 5. Pergamon, New York, pp. 2543–66.
14. C. W. Naeser, G. A. Izett & J. D. Obradovich (1980) Fission track and K–Ar ages of natural glasses. *US Geol. Surv. Bull.* **1489**, 1–31.
15. G. A. Wagner (1981) Fission-track ages and their geological interpretation. *Nucl. Tracks* **5**, 15–25.

16. G. A. Wagner & G. M. Reimer (1972) Fission track tectonics: The tectonic interpretation of fission track apatite ages. *Earth Planet. Sci. Lett.* **14**, 263–8.

17. G. A. Wagner, G. M. Reimer & E. Jäger (1977) Cooling ages derived by apatite fission track, mica Rb–Sr and K–Ar dating: The uplift and cooling history of the Central Alps. *Memoir. Instit. Geol. Min. Univ. Padova* **30**, 1–27.

18. G. A. Wagner (1979) Correction and interpretation of fission track ages. In: *Lectures in Isotope Geology* (E. Jäger and J. C. Hunziker, eds.). Springer-Verlag, Berlin, pp. 170–7.

19. D. Storzer (1970) Fission track dating of volcanic glasses and the thermal history of rocks. *Earth Planet. Sci. Lett.* **8**, 55–60.

20. A. J. Hurford & A. J. W. Gleadow (1977) Calibration of fission track dating parameters. *Nucl. Track Detection* **1**, 41–8.

21. G. A. Wagner, G. M. Reimer, B. S. Carpenter, H. Faul, R. van der Linden & R. Gijbels (1975) The spontaneous fission rate of U-238 and fission track dating. *Geochim. Cosmochim. Acta.* **39**, 1279–86.

22. A. J. Hurford & P. F. Green (1981) A reappraisal of neutron dosimetry and Uranium-238 λ_f values in fission-track dating. *Nucl. Tracks* **5**, 53–61.

23. G. Bigazzi (1981) The problem of the decay constant λ_f of ^{238}U. *Nucl. Tracks* **5**, 35–44.

24. A. J. Hurford & P. F. Green (1981) Standards, dosimetry and the Uranium-238 λ_f decay constant: A discussion. *Nucl. Tracks* **5**, 73–5.

25. G. N. Flerov & K. A. Petrzhak (1940) Spontaneous fission of uranium. *J. Phys. USSR* **3**, 275–80.

26. A. Spadavecchia & B. Hahn (1967) Die Rotationskammer und einige Anwendungen. *Helv. Phys. Acta.* **40**, 1063–79.

27. E. K. Gerling, Yu A. Shukolyukov & B. A. Makarochkin (1959) Determination of the half-life of the spontaneous uranium-238 disintegration from the xenon content of uranium minerals. *Radiokhimiya* **1**, 223–6.

28. R. L. Fleischer & P. B. Price (1964) Decay constant for spontaneous fission of U^{238}. *Phys. Rev.* **133**, B63–4.

29. J. H. Roberts, R. Gold & R. J. Armani (1968) Spontaneous-fission decay constant of ^{238}U. *Phys. Rev.* **174**, 1482–4.

30. H. A. Khan & S. A. Durrani (1973) Measurement of spontaneous-fission decay constant of ^{238}U with a mica solid state track detector. *Radiat. Effects* **17**, 133–5.

31. D. Storzer (1970) Spaltspuren des 238-Urans und ihre Bedeutung für die Geologische Geschichte natürlicher Gläser. PhD Thesis, University of Heidelberg.

32. K. Thiel & W. Herr (1976) The ^{238}U spontaneous fission decay constant redetermined by fission tracks. *Earth Planet. Sci. Lett.* **30**, 50–6.

33. J. C. Hadler, C. M. G. Lattes, A. Marques, M. D. D. Marques, D. A. B. Serra & G. Bigazzi (1981) Measurement of the spontaneous-fission disintegration constant of ^{238}U. *Nucl. Tracks* **5**, 45–52.

34. R. L. Fleischer & H. R. Hart., Jr. (1972) Fission track dating: Techniques and problems. In: *Calibration of Hominoid Evolution* (W. W. Bishop, D. A. Miller and S. Cole, eds.). Scottish Academic Press, Edinburgh, pp. 135–70.

35. R. L. Fleischer, P. B. Price & R. M. Walker (1975) *Nuclear Tracks in Solids: Principles and Applications*. University of California Press, Berkeley.

36. G. Bigazzi (1967) Length of fission tracks and age of muscovite samples. *Earth Planet. Sci. Lett.* **3**, 434–8.

37. P. P. Mehta & Rama (1967) Annealing effects in muscovite and their influence on dating by the fission track method. *Earth Planet. Sci. Lett.* **7**, 82–6.

38. D. Storzer & G. A. Wagner (1969) Correction of thermally lowered fission track ages of tektites. *Earth Planet. Sci. Lett.* **5**, 463–8.

39. S. A. Durrani & H. A. Khan (1970) Annealing of fission tracks in tektites: Corrected ages of bediasites. *Earth Planet. Sci. Lett.* **9**, 431–45.

39a. P. F. Green (1981) "Track-in-track" length measurements in annealed apatites. *Nucl. Tracks* **5**, 121–8.

39b. K. K. Nagpaul, P. P Mehta & M. L. Gupta (1974) Annealing studies on radiation damages in Biotite, Apatite and Sphene and corrections to fission track ages. *Pageoph.* **112**, 131–9.

39c. A. J. W. Gleadow & I. R. Duddy (1981) A natural long-term annealing experiment for *apatite*. *Nucl, Tracks* **5**, 169–74.

39d. G. M. Laslett, A. J. W. Gleadow & I. R. Duddy (1984) The relationship between fission track length and track density in apatite. *Nucl. Tracks* **9**, 29–38.

40. G. Somogyi & M. Nagy (1972) Remarks on fission track dating in dielectric solids. *Radiat. Effects* **16**, 223–31.

41. S. R. Hashemi-Nezhad & S. A. Durrani (1981) Correction of thermally-lowered fission-track ages of minerals and glasses: The importance of the prolonged etching factor. *Nucl. Tracks* **5**, 101–11.

42. G. Turner (1969) Thermal histories of meteorites by the ^{39}Ar–^{40}Ar method. In: *Meteorite Research* (P. M. Millman, ed.). D. Reidel, Dordrecht, pp. 407–17.

43. D. Storzer, G. Poupeau & M. J. Orcel (1973) Ages-plateaux de minéraux et verres par la méthode des traces de fission. *Compt. Rend.* **276**, 137–9.

44. J. Burchart, M. Dakowski & J. Gałazka (1973) A method for fission-track dating of minerals with very high track densities (Abstract). *3rd Europ. Colloq. Geochron. Cosmochron. & Isotope Geol.* (ECOG III), Oxford, p. 9.

44a. J. Burchart, M. Dakowski & J. Gałazka (1975) A technique to determine extremely high fission track densities. *Bull. Acad. Polon. Sci.* (Ser. Sci. de al Terre) **23**, 1–7.

45. J. Carpena, D. Mailhe, G. Poupeau & D. Vincent (1981) Model ages in fission track dating (Abstract). *Nucl. Tracks* **5**, 240–2.

46. D. S. Miller & G. A. Wagner (1981) Fission-track ages applied to obsidian artifacts from South America using the plateau-annealing and the track-size age correction techniques. *Nucl. Tracks* **5**, 147–55.

47. D. Chaillou & A. Chambaudet (1981) Isothermal plateau method for apatite fission-track dating. *Nucl. Tracks* **5**, 181–6.

47a. S. Watt, P. F. Green & S. A. Durrani (1984) Studies of annealing anisotropy of fission tracks in mineral apatite using track-in-track (TINT) length measurements. *Nucl. Tracks* **8**, 371–5.

47b. D. S. Miller, I. R. Duddy, R. L. Fleischer & T. M. Harrison (Proc. eds.) (1985) *Fission-Track Dating:* (special issue) *Nucl. Tracks* 10(3), pp. 291–442.

48. C. Arias, G. Bigazzi & F. P. Bonadonna (1981) Size corrections and plateau age in glass shards. *Nucl. Tracks* **5**, 129–36.

48a. S. A. Durrani (1981) Track record in meteorites. *Proc. Roy. Soc. Lond.* **A374**, 239–51.

49. R. Bodu, H. Bouzigues, N. Morin & J. P. Pfiffelman (1972) On the existence of anomalous isotopic abundances in uranium from Gabon. *Compt. Rend.* **275**, 1731–2.

50. S. A. Durrani, K. A. R. Khazal, S. R. Malik, J. H. Fremlin & G. L. Hendry (1975) Thermoluminescence and fission-track studies of the Oklo fossil reactor materials. In: *Le Phénomène d'Oklo*. International Atomic Energy Agency, Vienna, pp. 207–22.

51. J. C. Dran, M. Maurette, J. C. Petit, R. Drozd, C. Hohenberg, J. P. Duraud, C. Le Gressus & D. Massignon (1975) A multidisciplinary analysis of the Oklo uranium ores. In: *Le Phénomène d'Oklo*. International Atomic Energy Agency, Vienna, pp. 223–34.

52. R. L. Fleischer, P. B. Price, R. M. Walker & M. Maurette (1967) Origins of fossil charged-particle tracks in meteorites. *J. Geophys. Res.* **72**, 331–53.

52a. R. L. Fleischer, P. B. Price, R. M. Walker, M. Maurette & G. Morgan (1967) Tracks of heavy primary cosmic rays in meteorites. *J. Geophys. Res.* **72**, 355–66.

52b. J. G. Wilson (1976) *Cosmic Rays*. Wykeham Publications, London, pp. 58–65.

52c. R. K. Bull & S. A. Durrani (1976) Cosmic-ray tracks in Shalka meteorite. *Earth Planet. Sci. Lett.* **32**, 35–9.

52d. P. F. Green, R. K. Bull & S. A. Durrani (1978) The fission-track records of the Estherville, Nakhla and Odessa meteorites. *Geochim. Cosmochim. Acta* **42**, 1359–66.

53. D. Lal (1972) Hard-rock cosmic-ray archaeology. *Space Sci. Rev.* **9**, 623–50.

54. K. Thiel, G. Damm & W. Herr (1974) Simulated cosmic-ray induced fission tracks in artificial lunar soil and implications for the ^{238}U fission track dating of lunar surface samples. In: *Proc. 5th Lunar Sci. Conf., Geochim. Cosmochim. Acta*, Suppl 5. Pergamon, New York, pp. 2609–21.

55. D. S. Woolum & D. S. Burnett (1974) In situ measurement of the rate of ^{235}U fission induced by lunar neutrons. *Earth Planet. Sci. Lett.* **21**, 153–63.

56. G. Damm, K. Thiel & W. Herr (1978) Cosmic-ray induced fission of heavy nuclides: Possible influence on apparent ^{238}U fission track ages of extraterrestrial samples. *Earth Planet. Sci. Lett.* **40**, 439–44.

57. R. C. Reedy & J. R. Arnold (1972) Interaction of solar and galactic cosmic-ray particles with the Moon. *J. Geophys. Res.* **77**, 537–55.

58. R. L. Fleischer, P. B. Price & R. M. Walker (1965) Spontaneous fission tracks from extinct Pu^{244} in meteorites and the early history of the solar system. *J. Geophys. Res.* **70**, 2703–7.

59. R. L. Fleischer, P. B. Price & R. M. Walker (1968) Identification of Pu^{244} fission tracks and the cooling of the parent body of the Toluca meteorite. *Geochim. Cosmochim. Acta* **32**, 21–31.

60. Y. Cantelaube, M. Maurette & P. Pellas (1967) Traces d'ions lourds dans les minéraux de la chondrite de Saint-Séverin. In: *Radioactive Dating and Methods of Low Level Counting.* International Atomic Energy Agency, Vienna, pp. 215–29.

61. G. J. Wasserburg, J. C. Huneke & D. S. Burnett (1969) Correlation between fission tracks and fission-type xenon in meteoritic whitlockite. *J. Geophys. Res.* **74**, 4221–32.

62. C. M. Hohenberg, M. N. Munk & J. H. Reynolds (1967) Spallation and fissiogenic xenon and krypton from stepwise heating of the Pasamonte achondrite. The case for extinct ^{244}Pu in meteorites. Relative ages of chondrites and achondrities. *J. Geophys. Res.* **72**, 3139–77.

63. E. C. Alexander, R. S. Lewis, J. H. Reynolds & M. C. Michel (1971) Plutonium-244: Confirmation as an extinct radioactivity. *Science* **172**, 837–40.

64. T. Kirsten, J. Jordan, H. Richter, P. Pellas & D. Storzer (1977) Plutonium in phosphates from ordinary chondrites. *Meteoritics* **12**, 279–81.

65. P. Pellas & D. Storzer (1981) ^{244}Pu fission track thermometry and its application to stony meteorites. *Proc. Roy. Soc. Lond.* **A374**, 253–70.

66. P. Pellas & D. Storzer (1975) Mesures des taux de refroidissement des chondrites ordinaires a partir des traces de fission du Plutonium-244 enregistrées dans les cristaux détecteurs. *C. R. Acad. Sci.* **280D**, 225–8.

67. P. Pellas & D. Storzer (1975) Uranium and plutonium in chondritic phosphates. *Meteoritics* **10**, 471–3.

68. P. Pellas & D. Storzer (1977) On the early thermal history of chondritic asteroids derived by 244-Plutonium fission track thermometry. In: *Comets, Asteroids and Meteorites* (A. H. Delsemme, ed.). University of Toledo Press, Toledo, pp. 355–62.

69. P. Pellas & D. Storzer (1979) Differences in the early cooling histories of the chondritic asteroids. *Meteoritics* **14**, 513–15.

70. P. Mold, R. K. Bull & S. A. Durrani (1981) Constancy of ^{244}Pu distribution in chondritic whitlockite. *Nucl. Tracks* **5**, 27–31.

70a. P. Mold, R. K. Bull & S. A. Durrani (1982) Plutonium-244 concentrations in chondritic phosphates and their significance in fission-track dating of meteorites. In: *Proc. 11th Int. Conf. Solid State Nucl. Track Detectors,* Bristol, and Suppl. 3, *Nucl. Tracks.* Pergamon, Oxford, pp. 851–4.

71. K. Marti, G. W. Lugmair & N. B. Scheinin (1977). Sm–Nd–Pu systematics in the early solar system. *Lunar and Planes. Sci.* **VIII** (Abstracts). Lunar and Planetary Institute, Houston, pp. 619–21.

72. S. B. Jacobsen & G. J. Wasserburg (1980) Sm–Nd isotopic systematics of chondrites. *Lunar and Planet. Sci.* **XI** (Abstracts). Lunar and Planetary Institute, Houston, pp. 502–4.

73. P. Mold (1982) Fission-track studies and plutonium fractionation in meteoritic phosphates. PhD thesis, University of Birmingham.

74. G. A. Wagner (1978) Archaeological applications of fission-track dating. *Nucl. Tracks* **2**, 51–64.

75. J. E. Dixon, J. R. Cann & C. Renfrew (1968) Obsidian and the origins of trade. *Sci. Am.* **218** (3), 38–46.

76. S. R. Malik, S. A. Durrani & J. H. Fremlin (1973) A comparative study of the spatial distribution of uranium and of TL-producing minerals in archaeological materials. *Archaeometry* **15**, 249–53.

77. R. L. Fleischer, P. B. Price, R. M. Walker & L. S. B. Leakey (1965) Fission track dating of a mesolithic knife. *Nature* **205**, 1138.

78. S-L. Guo (1982) Some methods in fission track dating of Peking Man. In: *Proc. 11th Int. Conf. Solid State Nucl. Track Detectors*, Bristol, and Suppl. 3, *Nucl. Tracks*. Pergamon, Oxford, pp. 371–4.

79. D. W. Zimmerman, M. P. Yuhas & P. Meyers (1974) Thermoluminescence authenticity measurements on core material from the bronze horse of the New York Metropolitan Museum of Art. *Archaeometry* **16**, 19–30.

80. S. A. Durrani, H. A. Khan, M. A. Taj & C. Renfrew (1971) Obsidian source identification by fission track analysis. *Nature* **233**, 242–5.

81. R. H. Brill (1964) Applications of fission-track dating to historic and prehistoric glasses. *Archaeometry* **7**, 51–7.

82. S. A. Durrani (1972) Obsidian source identification by fission-track analysis (and dating of Roman and iron-age glasses). *MASCA Newsletter* (University of Pennsylvania) **8** (1), 2.

83. P. F. Green (1981) A new look at statistics in fission track dating. *Nucl. Tracks* **5**, 77–86.

84. J. Burchart (1981) Evaluation of uncertainties in fission-track dating: Some statistical and geochemical problems. *Nucl. Tracks* **5**, 87–92.

85. N. M. Johnson, V. E. McGee & C. W. Naeser (1979) A practical method of estimating standard error of age in the fission track dating method. *Nucl. Tracks* **3**, 93–99.

86. V. E. McGee & N. M. Johnson (1979) Statistical treatment of experimental errors in the fission track dating method. *Math. Geol.* **11**, 255–68.

87. S. A. Durrani (1982) The current status and problems of fission-track dating. In: *Proc. 11th Int. Conf. Solid State Nucl. Track Detectors*, Bristol, and Suppl. 3, *Nucl. Tracks*. Pergamon, Oxford, pp. 361–70.

88. R. F. Galbraith (1981) On statistical models for fission track counts. *Math. Geol.* **13**, 471–8.

89. R. F. Galbraith (1982) Statistical analysis of some fission-track counts and neutron-fluence measurements. *Nucl. Tracks* **6**, 99–107.

90. W. E. Bardsley (1984) Note on confidence bounds for fission-track dating with low track counts. *Nucl. Tracks* **9**, 69–70.

CHAPTER 9

Further Applications of Track Detectors and Some Directions for the Future

In the preceding chapters of this book an attempt has been made to outline the basic principles of solid state nuclear track detectors and to examine some of their main areas of application. These applications have, by and large, been limited to those uses which have evolved into areas of considerable research activity or even routine use, and which serve to illustrate the strengths and the versatility of track detectors. Such uses are exemplified by fission track dating, neutron and α-particle dosimetry, and particle identification. In this chapter a few other interesting areas of application are described. We conclude by indicating some prospects for the future of the technique.

9.1 Applications to Nuclear Physics

Applications of track detectors to nuclear physics involving particle identification have already been discussed in general terms in Chapter 6. Here, we describe a few diverse problems in nuclear physics which have been tackled with the help of track detectors.

9.1.1 Fission phenomena and related studies

Track etch detectors are particularly well suited to studies of fission reactions because they are capable of registering low fluxes of fission fragments while remaining unaffected by very large fluxes of lightly ionizing particles. Track detectors can be arranged to have a very large sensitive area and thus integrate very low fission-fragment fluxes. Two types of experimental set-up may be envisaged for nuclear physics experiments (see Figs. 9.1a,b). Either the detector is placed adjacent to a target foil, so that the number of reaction products is integrated over all angles of emission; or the detector (or a number of small detector foils) is placed at some distance from the target. In the latter case, information may be gathered on the angular distribution of reaction products by measuring track density as a function of detector position.

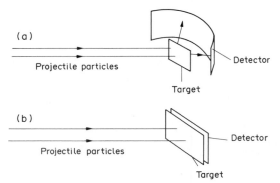

FIG. 9.1a,b *Schematic diagram showing two possible target–detector arrangements in nuclear physics experiments. In arrangement (a), the detector is placed some distance away from the target which is being bombarded by incident projectile particles. The detector may consist of a number of strips or elements, numbered beforehand; or it may be cut up into elements after the experiment. In either case, one is able to study the angular distribution of the product nuclei by measuring the track density as a function of the detector-element position. Other, more complex, geometries are also possible. In arrangement (b), the detector is placed adjacent to the target foil, so that the number of reaction products is integrated over all angles. The SSNT detectors have the advantage of integrating small fluxes over large areas and measuring times, which is especially important in the case of small reaction cross-sections or low projectile fluxes. Time-dependent counting can also be achieved by employing, say, a detector strip moving at a known time rate. By deliberately choosing a detector which is insensitive to (high) fluxes of lightly ionizing particles, one is able, moreover, to discriminate against interference by undesirable types of radiation, which is often difficult with conventional detectors. Alternatively, by successively employing detectors of different sensitivities—i.e. with different thresholds of charge sensitivity—one may discriminate between several different types of product nuclei.*

Muscovite mica or a variety of plastics can be used as detectors. The plastics will also register tracks due to the projectile ions of charge $Z \geq 2$ units (though even protons will be detected if the polymer CR-39 is used). To avoid this low-Z contribution to the track density, a mica detector can be used, since it will only record ions with $Z \geq 10$. Mica may, however, suffer from the disadvantage of having a background (albeit a low one) of spontaneous-fission tracks from the fission of ^{238}U impurities. The background can be reduced by pre-annealing of the mica, or by employing synthetic mica. Other possible detectors are quartz or fused silica, which usually have very low U contents ($\sim 10^{-9}$ g/g), or annealed glass.

Studies of the fission cross-section as a function of incident α-energy have been used[1] to examine fission barriers for ^{201}Tl.

Brandt and co-workers[2-6] have used mica detectors to examine ternary and quaternary fission in heavy elements, especially uranium, induced by very heavy ions such as ^{84}Kr, ^{136}Xe, or ^{238}U. Such heavy projectile ions register tracks of their own, even in mica; but since they are normally-incident on the mica, they are revealed as small dots, and the fragments resulting from fission can be readily picked out (see Fig. 9.2). A number of other workers are now conducting similar studies (see, for example, references 7–9). In a typical experiment of this kind,[6] mica sheets, covered with a layer of UF_4 of thickness 1 mg cm^{-2}, were irradiated with ^{208}Pb ions of total energy 1477 MeV and fluence approximately 10^6 cm^{-2}. Subsequently, after etching in HF, the mica sheets were scanned for multipronged events: 154 were discovered. These reactions have been interpreted in terms of sequential fission of excited reaction products following deep inelastic collisions.[6] The application of track detectors to fission physics has been reviewed by Brandt.[2] *

Track detectors are also finding use in the search for superheavy elements (SHE). Experiments designed to artificially create elements within the "island of stability" around $Z = 114$ have, for example, involved the bombardment of ^{248}Cm targets with ^{48}Ca projectiles. (Several other target–projectile combinations are, of course, also possible.) Any superheavy elements formed in the target are then concentrated by chemical means, and can be placed adjacent to mica foils to detect the spontaneous fission of such nuclei. A variety of research efforts on superheavy elements have been reviewed by Brandt.[10] An alternative approach for the detection of short-lived ($\tau_{1/2} \sim 1$ ns) superheavies is an arrangement described by Nitschke.[11] Here, any superheavy nuclei that may be produced recoil out of the target foil and pass down a tube lined with mica, which detects fragments from fission-in-flight of the heavy nuclei.

Track detectors have also been employed in the search for any naturally occurring superheavy elements *in nature*. If such elements still exist, they are likely to be detected as a result of the emission either of fission fragments and neutrons from their spontaneous fission, or of high-energy α-particles from their decay. Of course, surface-barrier detectors can be used to detect fission fragments; and they can be coupled with neutron detectors so that coincidences of fission-fragment and neutron emissions can be recorded. However, activities are very low, so that counting may need to be carried out over months or years. A passive detector, such as a plastic foil, which can be used to record fission events with a very high efficiency over long periods of time, is therefore a valuable alternative tool.

Malý and Walz[12] have used sheets of Mylar† to detect an excess of spontaneous fission events from radiogenic lead. By measuring the lengths of both

---

* *Note added in proof.* For more recent work in this field see references 105–108.
† Mylar is the registered trade mark of Dupont Chemicals.

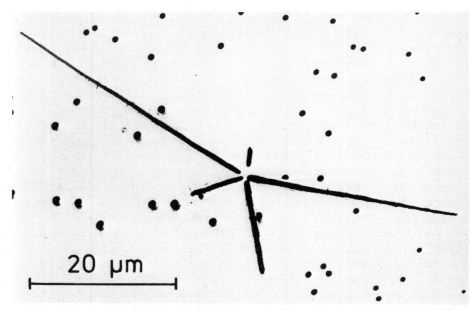

FIG 9.2   *Emission of five heavy (> 30 amu) reaction products in a high-energy fission (or spallation) event produced by the interaction of a 16.7 MeV/nucleon beam of* ^{238}U *ions incident on a natural-uranium target. The detector was a mica sheet used in 2π-geometry. The event, believed to be a "triple sequential fission", has been reconstructed from (~13) stepwise photographs taken at different focal depths; the longest track is ~80 μm. Brandt and co-workers*[2-6] *have also recorded tertiary and quaternary fission events in heavy elements (e.g. by high-energy U or Xe ions incident on Au, U, etc.). The heavy projectile ions register tracks of their own in the mica detector; but since they are normally incident on the mica foil they are revealed only as small dots seen in the figure. Mica will not register the lighter spallation products (with charge below ~10; cf. Fig. 3.1b). The SSNTD method is elegantly simple, and excellently suited to recording events of low probability, such as multifragment fission, which are not easily detected by electronic counters. It also allows the corresponding cross-sections to be calculated. For example, in a 92 cm² area of the mica detectors scanned by Brandt and co-workers, ~1300 three-, four-, and five-pronged events were discovered in the above-mentioned 16.7 MeV/amu U + U experiment, giving a total cross-section for 3-or-more-pronged events equal to 4220 ± 580 mb (and that for 5-pronged events, 75 + 45 mb). The total reaction cross-section is theoretically computed to be 6089 mb at the energy concerned. The irradiation was carried out on the heavy-ion accelerator facility (UNILAC) at GSI, Darmstadt, Fed. Rep. Germany. (Photograph, courtesy P. Vater and E. U. Khan, University of Marburg.)*

fission tracks from the same event, which can be done if a thin layer of radioactive material is sandwiched between two sheets of plastic, it is possible to deduce the amount of energy carried away by the fission fragments, and thus to discriminate against the fission of actinides in the sample. It is believed that the fission of natural SHE (which had co-precipitated with the radiogenic lead) would result in a greater energy release than from the fission of actinides. The source of the tracks found by Malý and Walz[12] has yet to be established.

The presence in the early solar system of now-extinct superheavy elements could result in excess fossil fission tracks (above those expected from ^{238}U and ^{244}Pu fission) being present in certain locations. Bhandari *et al.*[13] found some long tracks in meteoritic pyroxenes, which they attributed to SHE fission. However, it now seems likely[14,15] that such tracks are due actually to Fe-group cosmic rays. More recently, it has been suggested that siderophile (metal-seeking) SHE were present in the metal of the lunar core[16] and of the iron meteorites[17] during the early history of the solar system. A search for fission tracks along the boundaries of silicate crystals in contact with meteoritic iron[18] has, however, failed to discover evidence for these postulated SHE. The observed track densities at the surface of silicate crystals adjoining the metal phase place upper limits on the SHE abundances in the distant past ($\sim 4 \times 10^9$ years ago) which fall short by several orders of magnitude[18,18a,18b] of the stipulated concentration required to drive the ancient lunar-core dynamo.[16]

Perelygin and co-workers[19,20] have examined meteoritic olivines for very long tracks produced by cosmic ray SHE. Although a very long residual track (etchable length, 365 μm) in an annealed olivine crystal has been reported by these authors,[21] which is tentatively assigned to a SHE nucleus ($Z \geq 110$), a more rigorous calibration of the high-Z response of these crystals is needed before this claim can be regarded as confirmed (see also reference 21a).

9.1.2   Other nuclear reactions

Track etch detectors are suitable for use with any type of reaction which results in products of sufficiently high specific ionization appropriate for a given detector.

Somogyi *et al.*[22] have used polycarbonate detectors to measure the angular distributions of the two energy groups resulting from the ^{27}Al(p, α)^{24}Mg reaction at various incident-proton energies. The two alpha-energy groups result from reactions leading to the ground and excited states of ^{24}Mg. These authors discriminated between α-particles of different energy by means of track diameter measurements, and obtained an energy resolution of 60 keV at 6 MeV. Balcázar-García and Durrani[23] were able to separate ^{3}He and ^{4}He particles of the same energy (ranging from 6.6 to 10.6 MeV) from each other, using the CA80-15 cellulose nitrate detectors etched from the "reverse" side (facing away from the beam) and measuring the diameters. The energy resolution for each particle was good ($\sim 1\%$, fwhm), but charge discrimination (based on the growth rate of the track diameter with etching time) was rather poor. The method could not distinguish between ^{3}He and ^{4}He of the same range (but different energies), though giving the detector a suitable annealing treatment before etching did help in this respect.

Another example of the measurement of α-tracks is the study of the $^{10}B(p, \alpha)^7Be$ reaction at very low proton energies.[24] This reaction may be an important one in stellar interiors, resulting in the low cosmic abundance of ^{10}B. Várnagy et al.[25] have studied total cross-sections and angular distributions of protons and alphas from the astrophysically interesting reactions $^6Li(d, p)^7Li$ and $^6Li(d, \alpha)^4He$, using cellulose nitrate track detectors.

Enge and co-workers[26,27] have utilized the particle identification properties of plastic detectors to study the spectrum of spallation products resulting from the bombardment of plastics with very fast nucleons and also to study[28] the fragmentation of energetic Ar and Ne ions upon entering plastic detectors. Khan and Durrani (1972; unpublished data) used a variety of detectors (several types of plastics, glass, fused silica, mica) and radiators (copper, gold, U foil, in addition to the materials of the detectors just mentioned), placed at different depths within large concrete blocks (up to $127 \times 76 \times 66 \, cm^3$) and exposed to 7 GeV/c protons and their secondary products, in order to study the charge spectrum of the spallation products as a function of the energy of the bombarding particles. Herz et al.[29] have also studied the spallation products resulting from the bombardment of gold targets with protons in the GeV energy range, using Lexan polycarbonate.

As the properties of plastic detectors are refined, it seems likely that they will find increasing use in many nuclear physics applications. The CR-39 plastic, with its ability to detect protons and other particles with $Z/\beta \gtrsim 10$ (where Z is the charge and β is the velocity relative to light),[30] is especially promising in this respect—though it must be remembered that, when it is desirable to *avoid* interference by such low-ionization particles, it is best *not* to use CR-39 but, rather, use a less sensitive plastic.* Fews and Henshaw[30b] have described techniques of high-resolution α-particle spectroscopy with CR-39, based on measuring several different track parameters simultaneously.

9.2 Elemental Distributions and Biological Applications

9.2.1 Elemental mapping

In Chapter 8 we described how, for the purposes of fission track dating, the uranium content of a material could be determined by irradiating a sample with thermal neutrons and counting the induced tracks from (n, fission) reactions on ^{235}U. This principle can be applied to the measurement of concentrations and distributions of a number of elements in a variety of materials.

---

* *Note added in proof.* Price and Tarlé[30a] have recently described a new, high-resolution cross-linked polymeric detector called CR-73 (made from a bisphenol-A, bisallyl carbonate monomer), which combines the excellent optical qualities of CR-39 with the lesser sensitivity characteristics of Tuffak bisphenol-A polycarbonate. A charge resolution as low as $\sigma_Z \approx 0.18e$ is claimed.

The only requirement is that some highly (or moderately) ionizing reaction product should issue from the element in question.

A schematic view of a typical experimental set-up is shown in Fig. 9.3. Particles (in this case neutrons) are incident upon a sample covered by a plastic foil. Reaction products from the sample enter the plastic (alternatively, in some applications, a mica detector may be used) which is subsequently etched and scanned. This method is appropriate to many "prompt" reactions (such as (n, fission) or (n, α) reactions). Where a comparatively long-lived radioisotope is produced by the particle bombardment, there is no necessity for the plastic to be in contact with the sample *during* irradiation (this relaxation may have advantages, e.g. avoidance of primary-particle tracks on the detector or of fading effects in a high-temperature environment such as a reactor core). The detector may then be placed in contact with the sample subsequently to irradiation.

If some standard material containing a known amount C_{sx} of the element x is irradiated along with the sample under investigation which contains an unknown amount of x, say C_{ux}, then this latter concentration can be easily determined for any small region of the sample. This is done by measuring the track density ρ_u in the area of plastic adjacent to the requisite region as well as the track density ρ_s in plastic placed next to the standard, and taking ratios. Thus:

$$\frac{\rho_u}{\rho_s} = \frac{C_{ux}}{C_{sx}}$$

(Here one assumes that the range (in g cm^{-2}) of the track-producing particle in the standard and in the sample is virtually the same.)

The spatial resolution of the elemental mapping will be limited primarily by the range, in the matrix material, of the particle detected by the plastic, as long as sufficient statistics can be achieved for the counted tracks. The resolution will amount to ~ 10 μm, for example, in the case of uranium in silicate minerals, which is very good for most practical purposes.

If an isotope of the element of interest is radioactive and emits α-particles, then, provided that its specific activity is high enough, it may be possible to map its distribution without the need for any irradiation.

In the case of elements requiring external irradiation, the requisite fluence of bombarding particles will depend on the concentration of the element of interest and on the value of the cross-section for some suitable reaction. A material of bulk density D, containing a weight-fraction C_x of element x, will, upon irradiation with a fluence F of particles, yield a track density ρ given by

$$\rho = \frac{1}{4} \frac{D C_x N_A}{A_x} \sigma F R$$

where A_x is the atomic weight of x, N_A is Avogadro's number, σ is the cross-section for the track-producing reaction, and R is the range of the

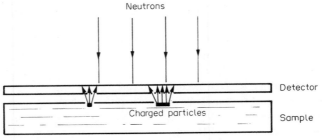

FIG. 9.3 *Experimental set-up for elemental mapping, using neutrons. The beam of neutrons produces charged-particle reactions (e.g. fission, or (n, α) interactions) in the sample. The charged particles, which have short ranges in the detecting medium, leave etchable tracks in the detector in the immediate vicinity of the target nucleus. This produces a readily-visualized replica, on the detector surface, of the spatial distribution of the relevant element (e.g. U in a rock, or B in boron glass) in the sample surface. The elemental concentration can be calculated from simple formulas (see text). Where a comparatively long-lived radioisotope is produced—decaying by α-emission, say—there is no need for the detector (e.g. a plastic) to be in contact with the sample during the irradiation. This can be advantageous, if the tracks of the projectile particle, or high temperature in, e.g., a reactor core, are to be avoided. An example of such an approach is the measurement of the distribution of lead in teeth, where teeth sections are exposed to high-energy (∼30 MeV) ^3He beams in a cyclotron, and the resulting distribution of the Po-206 (an α-emitter with a half-life of 8.8 days), produced in the reactions ^{206}Pb ($^3He, 3n$) ^{206}Po and ^{207}Pb ($^3He, 4n$) ^{206}Po, is recorded on LR 115 α-detectors, placed in close contact with the teeth sections, over the course of several weeks, following a period of post-irradiation cooling down of the radioactivity.[39a]*

emitted reaction product. (The factor $\frac{1}{4}$ comes from integrating the particles, over all emission angles, that are able to reach the "top" surface in a thick sample; see §4.2.3.) If R is replaced by the range in terms of mass per unit area d, such that $d = RD$, then

$$\rho = \frac{1}{4} \frac{C_x N_A}{A_x} \sigma F d \qquad (9.1)$$

It is assumed here that each reaction produces only one track-forming particle (cf. two fragments for induced fission, which would double the value of ρ in Eq. (9.1)), and that either the reaction is prompt or the plastic is in position long enough for essentially all of the induced radioactivity, if any, to have decayed and made its mark on the detector.

A number of elements have been studied using these techniques, including Li, B, Pb, Bi, Po, Th, U, Pu, and many others. It is not possible to describe here all the applications to which this method has been put. Instead, a few of the more interesting applications are described below.

Mapper *et al.*[31] have recently described a uranium fission autoradiography method of investigating "soft errors" in silicon memory devices. The cause of these soft errors is the alpha radioactivity from minute amounts of U and Th contaminants present in these devices. The passage of an α-particle through it can produce sufficient electronic charge to temporarily alter the contents of a bit location in the memory array area. Mapper *et al.*[31] spun a polyimide film on to the surfaces of the microchips and suitable component materials, and then irradiated these with high fluences of thermal neutrons. The film was then carefully etched, and the induced fission tracks (from ^{235}U) were observed under a microscope or on an image-analysis system. The technique is very sensitive, and amounts of U as low as 10^{-12} g, corresponding to an α-particle flux of 10^{-3} α cm^{-2} h^{-1} could be detected. Both the location and the amount of the U contaminant on the surface of the "random access memories" (RAMs) could, according to these authors, be correlated exactly with the precise details of the microcircuit, such as memory cells or bit-lines.

Where neutron irradiations are used in any elemental mapping, care must be taken, as emphasized in Chapter 8, to ensure a high degree of cleanliness in sample preparation to avoid the presence of fissionable impurities on the sample or detector surfaces.

Hashemi-Nezhad *et al.*[32] studied the diffusion of Pb in mica, a topic of considerable importance for theories of the origin of Po radiohaloes.[32a,b] Pb was introduced into the mica, which was then heated to various temperatures for different lengths of time. The mica was thereafter irradiated with ^3He or ^4He particles of energy 30 or 40 MeV, respectively, which produced α-active nuclei through reactions of the type Pb(He, n)Po. After the irradiation, the mica sheets were held in contact with cellulose nitrate, and the alpha tracks recorded in the plastic were etched to reveal the Pb distribution.

Track techniques can also be of value in metallography. Thus Rosenbaum and Armijo[33] have used the reaction $^{10}B(n, \alpha)^7Li$ to study the distribution of boron-rich particles in stainless steel by irradiating steel with thermal neutrons whilst in contact with alpha-sensitive plastic.

Many other examples of elemental mapping by track techniques exist in the literature, and the reader is referred to the book by Fleischer, Price and Walker[34] for further information.

9.2.2 Biological applications

The subject of biological applications of SSNTDs is a major field, and it is not possible to cover it in this chapter adequately (see the review paper by

Durrani[35] for a selective coverage, and also Chapter 7 for dosimetric applications). We shall confine ourselves to briefly describing just a few specific applications by way of illustration.

(a) INHALATION OF α-ACTIVE AEROSOLS

Investigations of radon emanation in dwellings, mines, and caves, which has an injurious effect on the inhabitants and mine-workers, etc., have been described in Chapter 7 (see §7.2), and its general coverage need not be repeated. Here we shall only deal, briefly, with the biological aspects of some α-emitting particulate materials.

Much interest has been taken, in recent years, in the radiological consequences of the inhalation of α-active particles (present, e.g., in tobacco smoke or in the atmosphere of uranium mines) and their deposition in the lung. A knowledge of the microdistribution of these particles is important for the purpose of estimating the dose delivered to the bronchial tissue. Henshaw et al.[36] sandwiched sections of bronchus between sheets of CR-39 for 3–4 months, and then etched the plastic to reveal α-particle tracks produced by radioactive particles lodged in the tissue. From the lengths of some of these tracks they were able to deduce that uranium in equilibrium with its daughter products was present in the lung. Gore et al.[37] have used neutron-induced autoradiography to examine the distribution of ^{235}U in the lungs of rats which were made to inhale an aerosol of UO_2. Sections of lung tissue were mounted onto sheets of Lexan, and an image of the section was produced by irradiating it with 4×10^9 α-particles cm^{-2} at an energy of <1 MeV. The lung tissue was sufficiently thick to stop these α's, and only in regions where the tissue was absent did the α-particles record tracks in the Lexan. The lung section was then irradiated with thermal neutrons, and induced fission tracks were produced from the ^{235}U. Thus the distribution of U was shown superimposed on an image of the lung tissue.[37] In later work, Gore and Jenner[38] replaced the Lexan with CR-39. Henshaw et al.[38a] have extended these studies to the monitoring of α-activity in the bloodstream by immersion (with deep freezing) of CR-39 detectors in fresh blood taken from smoking and non-smoking subjects.

(b) LEAD IN TEETH

Fremlin and co-workers have used SSNTD techniques to measure the spatial distribution of lead in human teeth,[39] and in particular to study the lead content as a function of the person's age.[39a] Tooth sections (~ 0.5 mm thick) were bombarded with 30 MeV ^{3}He ions from a cyclotron to produce a number of Po isotopes by reactions on lead. Of these, the most useful for SSNTD work was found to be the 8.8-day half-life α-emitter ^{206}Po produced by the reactions: $^{206}Pb(^3He, 3n)^{206}Po$ and $^{207}Pb(^3He, 4n)^{206}Po$.

The α-activity was then measured by placing the tooth sections in close contact with LR 115 cellulose nitrate foils for 8 weeks; these foils were subsequently etched. Absolute values of the Pb content were established by comparison with lead standards, included in the ^{3}He irradiation. From studying the spatial distribution of the α-emission from the teeth section these authors concluded that, while the Pb content of dentine and enamel showed hardly any variation with the age of the donor (from ~12 to 65 years), the Pb content of the "pulpal dentine" (a ~200–400 μm thick layer around the pulp chamber in the core of the tooth) showed a clear increase with age. It may be worth pointing out that the enamel Pb content gives an indication of the lead laid down at the time of tooth formation at a young age, while the pulpal dentine concentration of Pb is indicative of the cumulative exposure to lead throughout the owner's lifetime. There is, thus, a clear indication that lead enters the blood supply (through food and inhalation of car exhaust fumes and industrial pollution, etc.), and is progressively accumulated by parts of our bodies as a function of age. The same technique has been used to study the lead content and its distribution in human bones[39b]—e.g. by irradiating small sections of the tibia with 40 MeV ^{4}He ions, which yield the 138-day half-life α-emitter ^{210}Po from ^{208}Pb by the (α, 2n) reaction (see the review article[35] for further details).

(c) FILTRATION OF MALIGNANT CELLS BY MICROPOROUS FILMS

Collimated fission fragments, and more recently beams of heavy ions (e.g. ^{238}U of a few MeV/nucleon), have been used to produce uniform fine etched holes in plastics or mica sheets. These holes are, typically, a few micrometres in diameter (in plastics) or across (in the case of rhomboidal etch pits in mica; see Fig. 9.4). Commercially available filters (e.g. from Nucleopore Corporation, USA) have holes of specified size ranging from ~300 Å (30 nm) to ~10 μm. These filters have been used to separate cancer cells from blood,[34,39c] making use of the fact that cancer cells are both larger and more rigid than normal blood cells. These filters are finding increasing use as diagnostic tools in cancer studies and investigations of blood circulation disorders connected with heart diseases (e.g. by using a "blood-corpuscle rigidometer"). Thus Spohr and co-workers at the UNILAC heavy-ion accelerator facility at GSI, Darmstadt, in W. Germany, have developed techniques of producing single holes in plastic foils by defocusing the ion beam (to greatly reduce the flux) and interrupting the bombardment as soon as a single heavy ion penetrates the foil and is detected by a semiconductor placed behind it.[39d] These holes, when etched, have diameters of 3–5 μm. DeBlois and Bean [39e] have employed a single etched track (in a plastic partition separating two parts of an electrochemical cell) to count and size small particles in an electrolyte. This simple device can be used to count (and identify)

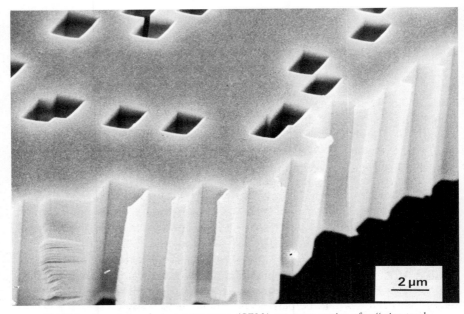

FIG. 9.4 *A scanning electron microscope (SEM) cut-away section of a "mica track microfilter". The rhomboidal holes (of side ~ 1.3 μm) result from etching a 20 μm thick mica foil after it had been irradiated with a normally incident parallel beam of 7.4 MeV/nucleon ^{238}U ions (~3 × 10^6 cm^{-2}) from the GSI heavy-ion accelerator. The etching had been carried out in 40 vol.% HF for 1 h at 22°C. The concept of a "molecular sieve" produced by the track-etch method, for use in biological and industrial applications, was first developed by R. L. Fleischer and co-workers in the 1960s. Plastic and mica filters, containing very uniform holes of sizes ~ 30 nm to 10 μm, are now commercially available; the etched holes in plastic are, of course, circular in shape. (Photograph, courtesy C. Riedel and R. Spohr, GSI, Darmstadt, W. Germany.)*

viruses and bacteria under controlled conditions. Micropore filters are being increasingly used nowadays in a number of industrial applications as well[39f,g] (ventilation studies; dust, smoke, and pollutant transport; etc.).

(d) MEASUREMENT OF α-EMITTERS IN THE ENVIRONMENT

As an example of many similar applications (other than radon measurements), just one may be cited here. Fremlin and Wilson[39h] have reported the measurement of the α-activity of the effluent from a nuclear-fuel reprocessing plant in England. The Sellafield (formerly known as Windscale) nuclear power and reprocessing plant releases a few thousand curies a year of α-emitting nuclides into the Irish Sea—about half are $^{241}Am(\tau_{1/2} = 433$ years) and the other half $^{239}Pu(\tau_{1/2} = 24.4 \times 10^3$ years), together with small amounts of ^{242}Cm and ^{238}Pu. Both Pu and Am are concentrated by edible shellfish, particularly mussels (*Myrtilus edulis*), to values of ~ 10 pCi of actinides per gram of tissue. The above authors have used the LR 115 cellulose nitrate detector to measure the transfer of plutonium and americium—and possibly uranium—through the mammalian gut into the blood-stream and hence into the liver and the skeleton. Mussels collected from the sea near the Sellafield effluent-discharge point were, after treatment, boiled, deshelled, homogenized, dried, and fed to rats. The rats were subsequently killed; their livers, which are expected to have the highest specific activity, were ashed (to further concentrate the activity); and the ashed material was compressed into thin discs. These were then placed between two sheets of LR 115 for 2 months. The total α-activity of the consumed mussels was found to be very low, ~ 2 nCi ($= 74$ Bq). Strippable LR 115 film (type II) was used so that measurements of the very low track densities could be made by a spark counter (see Chapter 7, §7.4.1). It was found that less than 0.1% of the α-activity in the mussels passed through the wall of the rat gut into its blood-stream and thence to its liver. The experiments thus indicated that human consumption of these effluent-contaminated mussels was quite safe on the basis of ICRP (International Commission on Radiological Protection) guidelines. (Fremlin and Wilson[39h] concluded that consumption of mussels up to a ton per person would be permissible under these guidelines!)

9.3 Extraterrestrial Samples

9.3.1 Lunar sample studies

One of the most exciting applications of the track-etch method during the last decade has been the investigation of lunar-surface processes and of ancient solar and galactic radiations through the examination of fossil tracks in lunar rocks and soils returned by the Apollo and Luna missions flown by the USA and the USSR, respectively.

Some of the features of the solar and galactic irradiation of meteorites and the lunar surface have been outlined in Chapter 8. Here we will describe some of the information about the history of the lunar surface that can be extracted from track studies. When galactic cosmic ray particles enter a rock, they are eventually brought to rest through ionization losses; they are also removed from the incident flux by nuclear interactions. For a simple, single-stage irradiation of a rock, the galactic (and solar) cosmic ray track density $\rho(X)$ falls monotonically with increasing depth X below the rock surface. The form of the track density profile, $\rho(X)$, depends on the energy spectrum of the galactic cosmic rays—and, at depths of <1 cm, on the energy spectrum of the solar flares. A number of measurements of track density versus depth for lunar rocks have been carried out (see, e.g., Walker and Yuhas,[40] Hutcheon *et al.*,[41] Blanford *et al.*,[42] and Durrani *et al.*[43]) in order to determine the energy spectra of the heavy cosmic rays and solar flare particles averaged over millions of years. In general, agreement seems to exist between these measurements and those made on contemporary cosmic ray fluxes.[40–44]

If detailed track density measurements are made on a rock with a well-determined cosmic ray exposure age, the rate of track production as a function of depth, $\dot{\rho}(X)$, may be found. Then, for any other rock having a single-stage exposure history, the track density at a given depth X_1 can be used to derive an exposure age T_{exp} from:

$$T_{exp} = \rho(X_1)/\dot{\rho}(X_1)$$

The initial calibration for this method must be made with the help of a rock for which a reliable cosmic ray exposure age has been determined by methods utilizing cosmic-ray-produced radionuclides or rare-gas isotopes (for reviews of these studies see Lal[45] and Crozaz[46]). Although simple in principle, such measurements are complicated by the fact that few lunar rocks appear to have had simple exposures to cosmic rays or solar flares.[47] Many of them have been tumbled, buried, re-exposed, or fragmented, and interpretation of the track data becomes more difficult as a consequence.

A vast amount of data has been collected on the track densities in lunar soil grains. Analysis of these results on lunar grains is even more difficult than for the rocks; nevertheless, some general and important features of lunar soil-deposition processes have emerged from these and other related studies.

Figures 9.5a,b illustrate, in a very simplified manner, some of the features of soil irradiation on the lunar surface.

If a slab of soil was deposited in one step, and then irradiated with solar flares and cosmic rays, a measurement of the average track density versus depth would look rather like Fig. 9.5a, which has essentially the same type of track density versus depth curve as would be expected for a lunar rock or meteorite. If, however, the soil was deposited as a series of discrete layers, each receiving a certain degree of irradiation before being covered by the next layer, the measured track density versus depth curve should exhibit a periodically varying form of the type shown in Fig. 9.5b. Measurements of track density versus depth within a soil layer have been carried out on cores of material obtained by driving tubes into the lunar soil to depths of a few metres (as was done in several Apollo and Luna missions) and then extracting them, with a portion of the lunar soil strata intact. Results of the kind shown in Fig. 9.5b are often obtained,[48–50] confirming this general pattern of soil deposition via discrete layers. Some soil layers showing evidence of a simpler irradiation history have also been found.[46]

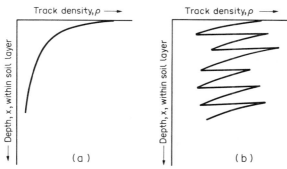

FIG. 9.5a,b *Two models of the variation of track density as a function of depth within the lunar soil. Model (a) assumes the deposition of a thick layer of soil in a single step, which then receives a single-stage irradiation by solar flares and cosmic rays. The track density distribution is similar to the one that would be found in a lunar rock or a meteorite, with a rapid attenuation of the softer components, with depth, at first, followed by a gentler fall, with increasing depth below the top surface, of the more energetic components of the primary flux. Model (b) assumes the deposition of soil as a series of discrete layers of varying thicknesses, each layer receiving a certain degree of irradiation before being covered by the next layer. Here the attenuation of the soft component with depth is highlighted and overlays the less steeply varying hard component. The track density distribution thus exhibits a periodically varying form of the type shown in (b). Lunar soil samples, scooped from the surface, or retrieved by drill-cores penetrating the top few (~3) metres of the lunar surface in Apollo and Luna missions, in fact usually display track density distributions of the form (b). The layers are believed to have been deposited on a millimetre to centimetre scale, by repeated micrometeorite impact. The whole, several metres deep, lunar regolith seems to have been churned on a scale of ~10^8–10^9 years by this "cosmic gardening" operation of meteorites and micrometeorites continually buffeting the surface of the moon.[49,49a,b,43] Track density gradients, observed on micrometre scale within many lunar grains, indicate that grains which are now several metres below the lunar surface must at one time have been at the very top of the moon's surface. Similarly, since the track distributions in soil grains at all levels in the lunar regolith bear a general resemblance to each other,[49,49a] the grains now at the top must formerly have been well below it.*

Superimposed on the above-mentioned patterns are observed the effects of small-scale stirring of the layers by micrometeorite impacts, which tend to transport some surface grains to greater depth and also to excavate deeper grains and redeposit them on the surface.[51]

Track densities observed in lunar soil grains and rock surfaces are often very high, ~10^8 to >10^{10} cm^{-2}, and track counting must be carried out with the aid of scanning or transmission electron microscopy.

Studies of lunar soil dynamics have achieved greater sophistication of late, with Maurette and co-workers,[52,53] for example, using a Monte Carlo type of computer code to generate histories of individual soil grains and to estimate the consequent radiation effects for comparison with experimental track density data.

9.3.2 *Meteorites*

Studies of cosmic ray tracks in meteorites are complicated by the unknown amount of ablation occurring when the meteorite penetrates the Earth's atmosphere. Indeed, cosmic ray track density measurements in meteorites have more often been used to determine this ablated thickness and hence the pre-atmospheric size of the meteorite[54-56] rather than to examine cosmic ray energy spectra. This method is based on the fact that attenuation of the cosmic ray flux over a distance dx is different at different depths x below the initial surface: close to the surface the flux attenuates more steeply than at greater depths where the surviving energy spectrum is harder. By assuming a form for the cosmic ray energy spectrum, and using a cosmic ray exposure age determined from studies of cosmogenic nuclides it is possible[57] to compute track production rate $\dot{\rho}(x, R_0)$ versus depth x for meteorites of any radius R_0. Then, by using a series of assumed ablation thicknesses for the meteorite, the measured track density profile can be plotted on a $\dot{\rho}(X, R_0)$ diagram. The assumed ablation thickness is adjusted until agreement with one of the theoretical curves is obtained. In this way, it is possible to determine both the ablated thickness and the pre-atmospheric radius of the meteorite (see also Fig. 8.10 and reference 58).

As was mentioned in Chapter 8, meteorites rich in rare gases usually also contain track-rich grains. Both of these features are believed to be the result of irradiation, by the solar wind and solar flares, of an ancient meteoritic soil layer (lunar and meteorite-parent-body soil layers are usually referred to as regoliths). Thus it is conjectured that surface grains were irradiated by solar flare particles which produced tracks; and by solar wind ions, which became implanted in grain surfaces, thus giving rise to high contents of various chemical species—most notably the rare gases. Track-rich grains in meteorites were first observed by Pellas *et al.*[59] and by Lal and Rajan.[60] A number of comparative studies of gas-rich meteorites and lunar soils have been carried out in order to seek clues to the nature of this primitive regolith on the meteorite parent bodies as well as to its interaction with solar radiations and the impinging micrometeorite complex.[61,62]

9.4 Track Detectors in Teaching

Because of their great basic simplicity, plastic track etch detectors offer advantages over many conventional detection systems for the demonstration of some nuclear phenomena in introductory courses at high schools and colleges, or even at polytechnics and universities. With the help of a constant-temperature bath, some simple chemicals, a few sheets of plastic, an optical microscope, an α-particle source, and perhaps a ^{252}Cf spontaneous fission source if available, a visual means of particle detection can be displayed to the students. The journal *Nuclear Tracks and Radiation Measurements* publishes an Educational Section which contains articles of an introductory nature designed for use by groups commencing track work, and particularly to help teachers in schools and colleges (see Bull,[63,64] and Enge[65]).

Some simple experiments that can be carried out include the measurement of α-particle source strengths and of α-particle ranges in air (or in simple solids, e.g. paper).

For the source strength determination, it is sufficient simply to place an alpha source next to a plastic foil for a measured time interval. Cellulose nitrate in the form known as LR 115 (manufactured by Kodak-Pathé) is ideal for this purpose as well as for many other simple experiments, since the etched tracks are observed as white holes against a dark-red background and are easily visible through low-power optical microscopes. Alternatively, various other forms of cellulose nitrate, as well as the polymer CR-39, can be used for these experiments. The number of tracks per field of view under the microscope may then be counted; and by calibrating this field of view the number of tracks per square centimetre may be obtained. Then, by measuring the total area of plastic covered by tracks, the source strength may be readily calculated (assuming 2π-geometry and a thin source, and $\sim 100\%$ revelation efficiency for α-particle tracks in plastic). See Table 4.1 for etchants and etching conditions.

Figure 9.6 shows a schematic diagram of the method for determining α-particle range in air. An alpha-emitting source (e.g. ^{241}Am) is covered (without physically touching it, to avoid damaging the source) with a metal, cardboard, or plastic foil containing a pinhole. A track-detecting plastic is placed at a distance d ($\simeq 2$–3 cm) from the source. After a sufficiently long irradiation time (e.g. ~ 100 second for a 5 μCi source) the detector is etched. The radius r of the region containing tracks is related to the range R_0 of the alpha particles in air through the equation

$$R_0 = (r^2 + d^2)^{1/2}$$

Some preliminary measurements would be needed to determine an appropriate exposure time, and also to select a value for d to ensure that the boundary of the track-bearing region was not determined by critical-angle considerations which would be the case if d were too small.

Experiments such as these can thus display some of the essential properties of alpha radiation with a minimum of complex apparatus. With slightly more

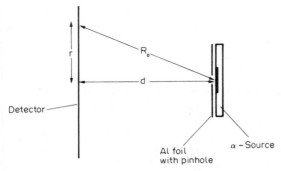

FIG. 9.6 *A simple experiment, useful in laboratory teaching, for determining the α-particle range in air. An α-emitting source is covered by a thin Al foil with a pinhole in it. A track-detecting plastic (e.g. a cellulose nitrate or CR-39) is placed at a distance of 2–3 cm from the source, with air intervening between the source and the detector. After a sufficiently long time of exposure (to produce a convenient track density, e.g. $\sim 10^5$ tracks cm^{-2}), the detector is removed and etched. The radius r of the detector area containing tracks is related to the range R_0 of the α-particles in air, and the source-detector separation d, by the simple equation: $R_0 = (r^2 + d^2)^{1/2}$. The same experiment can yield the source strength, based on the exposure time, solid angle (of a convenient central area), and the track detection efficiency of the plastic (taken as unity for medium-energy α's). Further simple SSNTD experiments, suitable for school or college laboratory teaching, have been described by Bull[63,64] and by Mold and Bull.[66]*

advanced facilities, a wider range of experiments could be contemplated. Experiments such as Rutherford scattering of alphas by heavy metal foils, and the detection of thermal neutrons by using ^{6}Li or boron converter-foils, are clearly amenable to demonstration using the track technique. A number of experiments are described in references 63–66.

9.5 Future Developments in Etched-Track Techniques and Their Applications

The scope of the track etch technique, and its successful applications, have expanded rapidly over the last two decades. There is no doubt that many developments lie in the future—although it is, of course, difficult to predict what these will be (for some speculations see the paper by Fleischer[67]).

One area in which there is great scope for improvement is our understanding of the formation of charged-particle tracks. On the theoretical side, there is a need for further calculations of the distribution of energy deposition on the microscopic scale around the paths of heavy charged particles. Such lines have already been pursued by Paretzke et al.[68] and by Fain et al.[69] On the experimental side, progress is already being made in arriving at a better understanding of track formation in both minerals and plastics. It seems that the thermalized ion-explosion spike[70] provides a most promising approach to the study of track formation in crystals, while O'Sullivan et al.[71] have made an important advance in the study of track formation in polymers. They find that the sensitivity of a polymer to the registration of particle tracks is closely related to its sensitivity to the formation of chain scissions under irradiation. This work suggests a means by which useful track-storing polymers may be identified, and also provides strong evidence that chain scission is of primary importance in the track formation process in such materials.

Another area of future development is the improvement of the sensitivity and uniformity of plastic track detectors. Effects of environmental factors (e.g. low temperature during irradiation,[72] which is of importance in high-altitude and in-space irradiation by cosmic rays) on the registration sensitivity of plastic track detectors also need careful and systematic investigations. The recent and dramatic addition[30] of CR-39 to the list of track-etch plastics suggests that the ultimate limit of sensitivity has by no means been achieved. Alpha-particles of tens of MeV in energy and protons of up to ~ 10 MeV are now detectable[72a] (as well as—via proton recoils—neutrons down to 150 keV). W. G. Cross (1985; private communication) and his co-workers have recently succeeded in revealing protons of energies as low as ~ 10 keV in CR-39 detectors by electrochemical etching along.[72b] As pointed out by Ahlen[73] and by Price,[74] since plastic track detectors respond largely to the abundant low-energy electrons produced along an ion path, they have potentially the highest charge resolution for particle identification attainable by any detector. At present their resolution is limited mainly by non-uniformities in the plastic.[75] Attempts are currently being made (see, e.g., reference 75a) to produce CR-39 plastic with a greater uniformity of response by controlling the temperature-time curing cycles and using various additives. Laser cutting of CR-39 detectors is now commonly employed—although local heating may alter the track-registration properties of CR-39 near to the cut edges.[75b]

An improved understanding of the annealing process in both polymers and minerals is also required. Gold et al.[76] have undertaken a theoretical study of the fundamentals of the annealing process in crystals. They believe that the use of the Arrhenius equation to describe the rate of track removal is incorrect.

Studies of annealing of minerals, and in particular of the extrapolation from the laboratory to geological time scales, are of particular importance in fission track dating. The effect of crystallographic anisotropies on the annealing properties of tracks (which may result from differences in the speed with which defects, etc., can diffuse in different directions during the "healing" process)[76a] also needs to be studied in greater depth. Other areas in the field of fission track dating requiring further work are those of thermal-neutron dosimetry in reactor irradiations and the associated problem of the ^{238}U fission decay constant.

The nature of the etching process is also an area worthy of further study. There is still considerable scope for the optimization of etching conditions to suit a given plastic. Furthermore, Enge and co-workers[77-79] have recently shown that, in addition to the simple etching process, diffusion of etchant ions into the bulk polymer also plays a rôle in the track development. This effect may be associated with the phenomenon of the etch induction time described, for example, by Ruddy *et al.*[80] Another recent discovery is that of Tarlé[81], and Price and O'Sullivan,[82] who found that the optical and etching properties of CR-39 can be greatly improved by incorporating additives such as dioctyl phthalate (DOP) into the polymer (see also reference 75a).

In the field of neutron dosimetry it is certain that further work will be devoted to the optimization of the response of CR-39 to fast neutrons. It is to be expected that electrochemical etching of this polymer will play an important part in the development of a practical dosimetry system. Interesting future work may follow from the suggestion by Price *et al.*[83] that the thin-film polymeric resists used in microelectronic circuitry may provide a means of neutron dosimetry in the ~ 10 keV energy region.

In radon dosimetry, an interesting recent development has been the use of "electrets", in conjunction with CR-39 detectors, to measure the concentrations of radon and thoron in air. An electret is a permanently charged (or polarized) dielectric, and is thus the electrostatic analogue of a permanent magnet.[84] Teflon "thermoelectrets" can hold an electrical charge for tens of years at ambient temperatures. Their use in specially constructed wire-mesh chambers[85] greatly enhances the collection efficiency of the (positively charged) decay products of radon or thoron (the electret face being kept in a negatively charged state). According to Kotrappa *et al.*[85], a Teflon electret can maintain an electric field inside the chamber equivalent to that provided by a battery of 2000–3000 V. The electrets are cheap; allow the apparatus to be used in the field without the need for electrical supplies; and are fairly easy to construct. (Basically, a high electric field is applied for a few hours, over certain temperature cycles, across a thin PTFE foil held between metal plate electrodes[86]; see also references 87, 88). They are thus bound to gain in popularity in radiation dosimetry, and in particular in the measurement of radon levels in houses, mines and the environment.

Another topic for potential future research is the chance discovery by Fisher *et al.*[89] that the resistance of a dielectric sample irradiated with heavy ions and then etched shoots up enormously. For example, irradiation with 7.5 MeV/nucleon xenon ions, followed by etching (such that the "porosity"— a texture parameter defined by the authors as track density × cross section of an individual etched track channel—had a value of 0.7) resulted in the increase of the resistance of the treated sample from 2Ω to $2 \times 10^{10}\Omega$. This effect has important potential uses in the subject area of breakdown of insulators at high electric field strengths. Changes in the electric conductivity of CR-39 (irradiated, for instance, with heavy ions or low-energy protons) would be worth further investigation.

M. A. Fadel and co-workers (private communication, 1984) have been investigating the effect on the electrical resistivity of phosphate glasses resulting from radiation damage caused by fission fragments emanating from a uranium radiator exposed to neutrons. The method offers the possibility of being exploited for personnel neutron dosimetry purposes. These authors have also studied the degradation of polymers exposed to fast neutrons and γ-rays, using the methods of physical chemistry; e.g. the mean molecular weight of cellulose nitrate and polycarbonate detectors (CA 80-15 and Makrofol E) is found to decrease with increasing neutron fluence and γ-ray dose[90,91]. They have also studied changes in the optical absorption properties of cellulose acetate foils as a result of exposure to mixed fast-neutron and γ-ray fields[92].

Effects of radiation damage caused in mineral crystals by natural fission and recoiling nuclei in α-decay as well as by heavy-ion bombardment in the laboratory, and the relationship of these effects to seemingly related phenomena of thermoluminescence, electron spin resonance and track formation in crystals and polymers, have been investigated by Durrani and co-workers.[93-97]

These studies, and others on changes in the physical, chemical and structural properties of plastic and mineral track detectors exposed to ionizing radiations (including charged particles), and both etched and unetched, will, it is hoped, lead to a better understanding of the track formation and revelation phenomena in dielectric materials.

Undoubtedly, the use of computer-based image-processing systems will spread widely, and the rate of acquisition and analysis of data be thus greatly accelerated. Tedious track-parameter measurements carried out by human scanners using eyepiece micrometers and graticules will then, it is hoped, be replaced by automatic means.

Among the new applications of track detectors, the work on charged-particle radiography[98] and on the imaging of a thermonuclear burn[99] looks particularly promising. Similarly, Stone and Ceglio[100] describe a "proton shadow camera", based on CR-39, for use as a diagnostic tool in laser fusion research. Other interesting developments include the ground-level monopole search described by Bartlett et al.,[101] and the large-scale monopole experiment in progress in Japan (Doke et al.[102]), where large stacks of CR-39 (total area ~ 1000 m^2) have been placed in a deep mine (at a depth of ~ 1 km) with the hope of detecting massive monopoles (of mass $\sim 10^{16}$ GeV/c^2) predicted by the "grand unified theories". Another promising field is that of earthquake predictions based on radon detection.[103] The most exciting applications of track detectors in the near future may well be the searches for exotic particles in accelerator experiments and in cosmic rays, as described by Price and co-workers.[74,75,104]

In the field of fossil track studies, the return of samples from Mars or from the surface of an asteroid or a comet would be of enormous interest.

Whatever the form of these future advances, it seems certain that solid state nuclear track detectors will find continuing, and ever-increasing, applications in science and technology for many years to come.

9.6 Epilogue*

The vitality, scope, and contemporary impact of the solid state nuclear track detection techniques are well exemplified by a seminal review paper given by P. B. Price[109] at the 13th International Conference on Solid State Nuclear Track Detectors, held at Rome in September 1985. This invited lecture reported, amongst others, the following recent advances.

1. The discovery of a new source of natural tracks in muscovite mica was reported in 1985: they are attributed to recoiling atoms (A1 and Si) struck by 8.8. MeV α-particles from the decay of ^{212}Po in the ^{232}Th natural-radioactivity series. These submicron tracks ($\sim 0.1-1$ μm long) have a radiation-damage rate and thermal stability similar to that expected from a slowly moving magnetic monopole in a bound state with a nucleus.

2. Arguments based on the etchability and morphological characteristics of tracks in mica, produced by slowly moving supermassive magnetic monopoles (of mass $> 10^{16}$ GeV/c^2) predicted by Grand Unified Theories, have led to new upper limits being placed on the flux of monopoles through the earth's crust as a result of experiments carried out by Price and co-workers.

3. Plastic track detectors (Rodyne polycarbonate and Cronar polyester) are being used, with great success, to discover new cases of recently reported novel, and extremely rare, modes of radioactive decay of heavy nuclides (with $Z \gtrsim 88$) involving monoenergetic heavy ions (e.g. the emission of ^{14}C from ^{222}Ra and ^{224}Ra, and of ^{24}Ne from ^{232}U, with branching ratios as low as $\sim 10^{-12}$ relative to α-decay).

4. Polymeric detectors—Rodyne, Cronar and CR-39 (DOP)—have been used to provide high charge-resolutions for the identification of heavy ions

---

* This note has been added in proof. As such it has necessarily had to be highly condensed and thus lacking in proper referencing and attributions, which may be found in reference 109. The Proceedings of the Rome Conference are to appear as Vol. 12 of *Nuclear Tracks* during 1986, and the reader is referred to them in order to learn of all the recent developments and advances reported at the conference, where a total of ~ 250 contributions were presented as oral papers, round-table discussions, and posters.

accelerated to relativistic energies: a charge resolution $\sigma_z < 0.25e$ is claimed for relativistic nuclei with Z/β from ~ 6 to ~ 120. These experiments have afforded valuable insights into, and discriminative criteria for, different track-forming models based on extended defects, K-shell excitation, Auger decay of inner-shell vacancies, total energy loss, and restricted energy loss, respectively.

5. Advances in automated scanning and measurements have opened up new physics applications requiring high statistics and resolution. Examples include searches for quark-nucleus complexes, "anomalons", extragalactic anti-matter, and ultraheavy cosmic ray nuclei in current and future space missions. In these searches, the high charge-resolution now available, the use of hybrid detector stacks, monitoring of registration temperature and time, control of environmental factors such as the provision of an inert (e.g. Ar) atmosphere during irradiation and "aging" prior to the etching of plastic detectors exposed in long-duration space missions, and, of course, the automatic scanning of the vast detector areas involved, all play an important part.

It is gratifying to note that solid state nuclear track detectors now occupy an essential, and sometimes a unique, place amongst the techniques and procedures deployed in scientific researches at the very frontiers of knowledge.

References

1. D. S. Burnett, R. C. Gatti, F. Plasil, P. B. Price, W. J. Swiatecki & S. G. Thompson (1964) Fission barrier of Thallium-201. *Phys. Rev.* **134B**, 952–63.
2. R. Brandt (1980) Overall review of the applications of SSNTD's in fission physics. *Nucl. Instrum. Meth.* **173**, 147–53.
3. P. Vater, H.-J. Becker, R. Brandt & H. Freiesleben (1977) Study of heavy-ion induced reactions on uranium with use of mica detectors. *Phys. Rev. Lett.* **39**, 594–8.
4. P. Vater, H.-J. Becker, R. Brandt & H. Freiesleben (1977) Multi-fragment decay reactions induced by heavy ions and studied with mica track detectors. *Nucl. Instrum. Meth.* **147**, 271–8.
5. P. A. Gottschalk, P. Vater, H.-J. Becker, R. Brandt, G. Grawert, G. Fiedler, R. Haag & T. Rautenberg (1979) Direct evidence for multiple sequential fission in the interaction of ^{208}Pb and ^{238}U with uranium. *Phys. Rev. Lett.* **42**, 1728–32.
6. H. A. Khan, P. Vater & R. Brandt (1980) Multiprong fission events produced by 1477 MeV ^{208}Pb ions in natural uranium. In: *Proc. 10th Int. Conf. Solid State Nucl. Track Detectors*, Lyon, and Suppl. 2, *Nuclear Tracks*. Pergamon, Oxford, pp. 915–20.
7. M. Debeauvais, S. Jokic & J. Tripier (1980) Etude a l'aide du Makrofol des interactions U et Pb sur U a des énergies supérieures a 7 MeV/uma. In: *Proc. 10th Int. Conf. Solid State Nucl. Track Detectors*, Lyon, and Suppl. 2, *Nucl. Tracks*. Pergamon, Oxford, pp. 927–32.
8. B. Grabež, Z. Todorović & R. Antanasijević (1980) The interaction of 300 MeV Ar ions with uranium studied by the use of a plastic track detector. In: *Proc. 10th Int. Conf. Solid State Nucl. Track Detectors*, Lyon, and Suppl. 2, *Nucl. Tracks*. Pergamon, Oxford, pp. 899–904.
9. M. Debeauvais, J. Ralarosy, J. Tripier & S. Jokic (1982) Fission of ^{238}U projectile by a ^{12}C target. In: *Proc. 11th Int. Conf. Solid State Nucl. Tracks Detectors*, Bristol, and Suppl. 3, *Nucl. Tracks*. Pergamon, Oxford, pp. 763–8.

10. R. Brandt (1978) Search for superheavy elements in nature and in heavy ion reactions. In: *Proc. Int. Symp. Superheavy Elements*, Lubbock (M. A. K. Lodhi, ed.). Pergamon, New York, pp. 103–17.
11. J. M. Nitschke (1978) Experimental prospect for the synthesis and detection of superheavy elements. In: *Proc. Int. Symp. Superheavy Elements*, Lubbock (M. A. K. Lodhi, ed.). Pergamon, New York, pp. 42–54.
12. J. Malý & D. R. Walz (1978) Long-range fission fragments from radiogenic lead. In: *Proc. Int. Symp. Superheavy Elements*, Lubbock (M. A. K. Lodhi, ed.). Pergamon, New York, pp. 216–26.
13. N. Bhandari, S. G. Bhat, D. Lal, G. Rajagopalan, A. S. Tamhane & V. S. Venkatavaradan (1971) Superheavy elements in extraterrestrial samples. *Nature* **230**, 219–24.
14. R. L. Fleischer & H. R. Hart Jr. (1973) Tracks from extinct radioactivity, ancient cosmic rays and calibration ions. *Nature* **242**, 104–5.
15. P. F. Green, R. K. Bull & S. A. Durrani (1978) The fission track records of the Estherville, Nakhla and Odessa meteorites. *Geochim. Cosmochim. Acta.* **42**, 1359–66.
16. S. K. Runcorn, L. M. Libby & W. F. Libby (1977) Primaeval melting of the moon. *Nature* **270**, 676–81.
17. L. M. Libby, W. F. Libby & S. K. Runcorn (1979) The possibility of superheavy elements in iron meteorites. *Nature* **278**, 613–17.
18. R. K. Bull (1979) A search for superheavy-element fission-tracks in iron meteorites. *Nature* **282**, 393–4.
18a. R. K. Bull & P. Mold (1982) Fission tracks in chondritic meteorites and the search for siderophile superheavy elements. In: *Proc. 11th Int. Conf. Solid State Nucl. Track Detectors*, Bristol, and Suppl. 3, *Nucl. Tracks*. Pergamon, Oxford, pp. 855–8.
18b. R. K. Bull & P. Mold (1982) A search for evidence of superheavy-element fission in chondritic metal. *Nature* **298**, 634–5.
19. V. P. Perelygin, S. G. Stetsenko, P. Pellas, D. Lhagvasuren, O. Otgonsuren & B. Jakupi (1977) Long-term averaged abundances of VVH cosmic ray nuclei from studies of olivines from Marjalahti meteorite. *Nucl. Track Detection* **1**, 199–205.
20. D. Lhagvasuren, O. Otgonsuren, V. P. Perelygin, S. G. Stetsenko, B. Jakupi, P. Pellas & C. Perron (1980) A technique for partial annealing of tracks in olivine to determine the relative abundances of galactic cosmic ray nuclei with $Z \geq 50$. In: *Proc. 10th Int. Conf. Solid State Nucl. Track Detectors*, Lyon, and Suppl. 2, *Nuclear Tracks*. Pergamon, Oxford, pp. 997–1002.
21. V. P. Perelygin & S. G. Stetsenko (1980) Search for the tracks of galactic cosmic nuclei with $Z \geq 110$ in meteorite olivines. *JETP Lett.* **32**, 608–10.
21a. R. Dersch, E. U. Khan, P. Vater, Shi-Lun Guo, E. M. Friedlander, V. Perelygin & R. Brandt (1986) On the track length of uranium ion in olivine crystals. *Nucl. Tracks* **11**, 67–72.
22. G. Somogyi, I. Hunyadi, E. Koltay & L. Zolnai (1977) On the detection of low-energy ^{4}He, ^{12}C, ^{14}N, ^{16}O ions in PC foils and its use in nuclear reaction measurements. *Nucl. Instrum. Meth.* **147**, 287–95.
23. M. Balcázar-García & S. A. Durrani (1977) ^{3}He and ^{4}He spectroscopy using plastic solid-state nuclear track detectors. *Nucl. Instrum. Meth.* **147**, 31–4.
24. J. Szabó, J. Csikai & M. Várnagy (1972) Low energy cross sections for ^{10}B(p, α)^{7}Be. *Nucl. Phys.* **A195**, 527–33.
25. M. Várnagy, J. Szabó & Z. T. Bódy (1980) Application of a cellulose nitrate detector for the simultaneous study of ^{6}Li(d, α)^{4}He and ^{6}Li(d, p)^{7}Li reactions. In: *Proc. 10th Int. Conf. Solid State Nucl. Track Detectors*, Lyon, and Suppl. 2, *Nucl. Tracks*. Pergamon, Oxford, pp. 881–4.
26. K. Grabisch, R. Beaujean, R. Scherzer & W. Enge (1977) Spallation products induced by energetic neutrons in plastic detector material. *Nucl. Instrum. Meth.* **147**, 263–6.
27. K. Grabisch, W. Enge, R. Beaujean & K. Fukui (1980) Detection of reaction fragments in cellulose nitrate produced by fast nucleons. In: *Proc. 10th Int. Conf. Solid State Nucl. Track Detectors*, Lyon, and Suppl. 2, *Nucl. Tracks*. Pergamon, Oxford, pp. 851–4.
28. R. Beaujean & W. Enge (1977) Fragmentation and isotope measurements on accelerated neon and argon particles of 280 MeV/nucleon. *Nucl. Track Detection* **1**, 19–25.

29. A. J. Herz, C. O'Ceallaigh, D. O'Sullivan & A. Thompson (1976) Production of nuclear fragments from the interactions of 24 GeV/c protons in a gold target. *Nuovo Cimento* **33A**, 487–92.

30. B. G. Cartwright, E. K. Shirk & P. B. Price (1978) A nuclear-track recording polymer of unique sensitivity and resolution. *Nucl. Instrum. Meth.* **153**, 457–60.

30a. P. B. Price & G. Tarlé (1985) CR-73: A new, high-resolution plastic track-etch detector. *Nucl. Instrum. Meth. Phys. Res.* **B6**, 513–6.

30b. A. P. Fews & D. L. Henshaw (1982) High resolution α-particle spectroscopy using CR-39 plastic track detector. In: *Proc. 11th Int. Conf. Solid State Nucl. Track Detectors*, Bristol, and Suppl. 3, *Nucl. Tracks*. Pergamon, Oxford, pp. 641–5.

31. D. Mapper, D. J. Bolus & J. H. Stephen (1982) The application of uranium fission track autoradiography as a method of investigating the problem of soft errors in VLSI silicon memory devices. In: *Proc. 11th Int. Conf. Solid State Nucl. Track Detectors*, Bristol, and Suppl. 3, *Nucl. Tracks*. Pergamon, Oxford, pp. 815–23.

32. S. R. Hashemi-Nezhad, J. H. Fremlin & S. A. Durrani (1979) Polonium haloes in mica. *Nature* **278**, 333–5.

32a. R. V. Gentry (1971) Radiohalos: some unique lead isotope ratios and unknown alpha radioactivity. *Science* **173**, 727–31.

32b. R. V. Gentry, T. A. Cahill, N. R. Fletcher, H. C. Kaufmann, L. R. Medsker, J. W. Nelson & R. G. Flocchini (1976) Evidence for primordial superheavy elements. *Phys. Rev. Lett.* **37**, 11–15.

33. H. S. Rosenbaum & J. S. Armijo (1967) Fission track etching as a metallographic tool. *J. Nucl. Mater.* **22**, 115–16.

34. R. L. Fleischer, P. B. Price & R. M. Walker (1975) *Nuclear Tracks in Solids: Principles and Applications*. University of California Press, Berkeley.

35. S. A. Durrani (1982) The use of solid-state nuclear track detectors in radiation dosimetry, medicine and biology. *Nucl. Tracks* **6**, 209–28.

36. D. L. Henshaw, A. P. Fews & D. J. Webster (1980) The microdistribution of α-active nuclei in bronchial tissue by autoradiography using CR-39 In: *Proc. 10th Int. Conf. Solid State Nucl. Track Detectors*, Lyon, and Suppl. 2, *Nucl. Tracks*. Pergamon, Oxford, pp. 649–54.

37. D. J. Gore, M. C. Thorne & R. H. Watts (1978) The visualisation of fissionable-radionuclides in rat lung using neutron induced autoradiography. *Phys. Med. Biol.* **23**, 149–53.

38. D. J. Gore & T. J. Jenner (1980) Alpha and fission fragment autoradiography with superimposed tissue images in CR-39 plastic. *Phys. Med. Biol.* **25**, 1095–1104.

38a. D. L. Henshaw, K. J. Heyward, J. P. Thomas, A. P. Fews, P. Gallerano & G. Sanzone (1984) Comparison of the α-activity in the blood of smokers and non-smokers. *Nucl. Tracks* **8**, 453–6.

39. J. H. Fremlin & M. I. Edmonds (1980) The determination of lead in human teeth. *Nucl. Instrum. Meth.* **173**, 211–15.

39a. T. Al-Naimi, M. I. Edmonds & J. H. Fremlin (1980) The distribution of lead in human teeth, using charged particle activation analysis. *Phys. Med. Biol.* **25**, 719–26.

39b. E. E. Laird (1983) Factors affecting *in-vivo* X-ray fluorescence measurements of lead in bone. PhD thesis, University of Birmingham.

39c. S. Seal (1964) A sieve for the isolation of cancer cells and other large cells from the blood. *Cancer* **17**, 549–68.

39d. H. G. Roggenkamp, H. Kiesewetter, R. Spohr, U. Dauer & L. C. Busch (1981) Production of single pore membranes for the measurement of red blood cell deformability. *Biomed. Technik* **26**, 167–9.

39e. R. W. DeBlois & C. P. Bean (1970) Counting and sizing submicron particles by the resistive pulse technique. *Rev. Sci. Instrum.* **41**, 909–16.

39f. G. Tress, M. Ellinger, E. U. Khan, H. A. Khan, P. Vater, R. Brandt & M. Kander (1982) Mica track microfilters applied in a cascade particle fractionator at an industrial plant. *Nucl. Tracks* **6**, 87–97.

39g. Shi-Lun Guo, G. Tress, P. Vater, E. U. Khan, R. Dersch, M. Plachky, R. Brandt & H. A. Khan (1986) New approach to increase the gas throughput through mica track microfilters by changing their pore structure. *Nucl. Tracks* **11**, 1–4.

39h. J. H. Fremlin & C. K. Wilson (1980) Alpha-emitters in the environment, II: Man-made activity. *Nucl. Instrum. Meth.* **173,** 201–4.
40. R. M. Walker & D. Yuhas (1973) Cosmic ray track production rates in lunar materials. In: *Proc. Fourth Lunar Sci. Conf.* Pergamon, New York, pp. 2379–89.
41. I. D. Hutcheon, D. MacDougall & P. B. Price (1974) Improved determination of the long term average Fe spectrum from 1–460 MeV/amu. In: *Proc. Fifth Lunar Sci. Conf.* Pergamon, New York, pp. 2561–76.
42. G. E. Blanford, R. M. Fruland & D. A. Morrison (1975) Long term differential energy spectrum for solar-flare iron-group particles. In: *Proc. Sixth Lunar Sci. Conf.* Pergamon, New York, pp. 3557–76.
43. S. A. Durrani, R. K. Bull & S. W. S. McKeever (1980) Solar-flare exposure and thermoluminescence of Luna 24 core material. *Phil. Trans. R. Soc. Lond.* **A297,** 41–50.
44. R. K. Bull, P. F. Green & S. A. Durrani (1978) Studies of the charge and energy spectra of the ancient VVH cosmic rays. In: *Proc. Ninth Lunar Sci. Conf.* Pergamon, New York, pp. 2415–31.
45. D. Lal (1972) Hard rock cosmic ray archaeology. *Space Sci. Rev.* **14,** 3–102.
46. G. Crozaz (1977) The irradiation history of the lunar soil. *Phys. Chem. Earth* **10,** 197–214.
47. R. M. Walker (1980) Nature of the fossil evidence: Moon and meteorites. In: *Proc. Conf. Ancient Sun.* Pergamon, Oxford, pp. 11–28.
48. G. Arrhenius, S. Liang, D. MacDougall, L. Wilkening, N. Bhandari, S. Bhat, D. Lal, G. Rajagopalan, A. S. Tamhane & V. S. Venkatavaradan (1971) The exposure history of the Apollo 12 regolith. In: *Proc. Second Lunar Sci. Conf.* Pergamon, New York, pp. 2583–98.
49. N. Bhandari, J. N. Goswami & D. Lal (1973) Surface irradiation and evolution of the lunar regolith. In: *Proc. Fourth Lunar Sci. Conf.* Pergamon, New York, pp. 2275–90.
49a. G. M. Comstock, A. O. Evwaraye, R. L. Fleischer & H. R. Hart, Jr. (1971) The particle track record of lunar soil. In: *Proc. Second Lunar Sci. Conf.* MIT Press, Cambridge, Mass., pp. 2569–82.
49b. G. Crozaz, R. Walker & D. Woolum (1971) Nuclear track studies of dynamic surface processes on the moon and the constancy of solar activity. In: *Proc. Second Lunar Sci. Conf.* MIT Press, Cambridge, Mass., pp. 2543–58.
50. R. L. Fleischer & H. R. Hart, Jr. (1973) Particle track record in Apollo 15 deep core from 54 to 80 cm depths. *Earth Planet. Sci. Lett.* **18,** 420–6.
51. S. A. Durrani, H. A. Khan, R. K. Bull, G. W. Dorling & J. H. Fremlin (1974) Charged-particle and micrometeorite impacts on the lunar surface. In: *Proc. Fifth Lunar Sci. Conf.* Pergamon, New York, pp. 2543–60.
52. J. P. Duraud, Y. Langevin, M. Maurette, G. M. Comstock & A. L. Burlingame (1975) The simulated depth history of dust grains in the lunar regolith. In: *Proc. Sixth Lunar Sci. Conf.* Pergamon, New York, pp. 2397–2415.
53. J. P. Bibring, J. Borg, A. M. Burlingame, Y. Langevin, M. Maurette & B. Vassent (1975) Solar-wind and solar-flare maturation of the lunar regolith. In: *Proc. Sixth Lunar Sci. Conf.* Pergamon, New York, pp. 3471–93.
54. P. B. Price, R. S. Rajan, & A. S. Tamhane (1967) On the preatmospheric size and maximum space erosion rate of the Patwar stony-iron meteorite. *J: Geophys. Res.* **72,** 1377–88.
55. R. K. Bull & S. A. Durrani (1976) Cosmic-ray tracks in the Shalka meteorite. *Earth Planet. Sci. Lett.* **32,** 35–9.
56. N. Bhandari, D. Lal, R. S. Rajan, J. R. Arnold, K. Marti & C. B. Moore (1980) Atmospheric ablation in meteorites: A study based on cosmic ray tracks and neon isotopes. *Nucl. Tracks* **4,** 213–62.
57. R. L. Fleischer, P. B. Price, R. M. Walker & M. Maurette (1967) Origins of fossil charged particle tracks in meteorites. *J. Geophys. Res.* **72,** 331–53.
58. S. A. Durrani (1981) Track record in meteorites. *Proc. Roy. Soc. Lond.* **A374,** 239–51.
59. P. Pellas, G. Poupeau, J. C. Lorin, H. Reeves & J. Audouze (1969) Primitive low-energy particle irradiation of meteoritic crystals. *Nature* **223,** 272–4.
60. D. Lal & R. S. Rajan (1969) Observations on space irradiation of individual crystals of gas-rich meteorites. *Nature* **223,** 269–71.
61. P. B. Price, I. D. Hutcheon, D. Braddy & D. MacDougall (1975) Track studies bearing on solar-system regoliths. In: *Sixth Lunar Sci. Conf.* Pergamon, New York, pp. 3449–69.

62. J. N. Goswami, I. D. Hutcheon & J. D. MacDougall (1976) Microcraters and solar flare tracks in crystals from carbonaceous chondrites and lunar breccias. In: *Proc. Seventh Lunar Sci. Conf*. Pergamon, New York, pp. 543–62.

63. R. K. Bull (1980) The use of solid state track detectors in teaching—I. *Nucl. Tracks* **4**, 59–65.

64. R. K. Bull (1980) The use of solid state track detectors in teaching—II. *Nucl. Tracks* **4**, 115–22.

65. W. Enge (1980) Introduction to plastic nuclear track detectors. *Nucl. Tracks* **4**, 283–308.

66. P. Mold & R. K. Bull (1980) Nuclear track detection in school. *The School Science Review* **62**, 262–71.

67. R. L. Fleischer (1979) Where do nuclear tracks lead? *Am. Sci.* **67**, 194–203.

68. H. G. Paretzke, G. Leuthold, G. Burger & W. Jacobi (1974) Approaches to physical track structure calculations. In: *Proc. 4th Symp. on Microdosimetry* (Verbania Pallanza, Italy, 1973) (J. Booz, H. G. Ebert, R. Eickel and A. Waker, eds.). EUR 5122 d-e-f. Commission of the European Communities, Luxembourg, pp. 123–40.

69. J. Fain, M. Monnin & M. Montret (1972) Spatial energy distribution around heavy ion paths. *Rad. Research* **57**, 379–89.

70. L. E. Seiberling, J. E. Griffith & T. A. Tombrello (1980) A thermalized ion-explosion model for high energy sputtering and track registration. *Radiat. Effects* **52**, 201–10.

71. D. O'Sullivan, P. B. Price, K. Kinoshita & C. G. Willson (1982) Predicting radiation sensitivity of polymers. *J. Electrochem. Soc.* **129**, 811–13.

72. D. O'Sullivan & A. Thompson (1980) The observation of a sensitivity dependence on temperature during registration in solid state nuclear track detectors. *Nucl. Tracks* **4**, 271–6.

72a. Y. V. Rao, M. P. Hagan & J. Blue (1982) Detection of 10-MeV protons, 70-MeV ³He ions and 52-MeV ⁴He ions in CR-39 track detector. *Nucl. Tracks* **6**, 119–24.

72b. W. G. Cross, A. Arneja & H. Ing (1985) The response of electrochemically-etched CR-39 to protons of 10 keV to 3 MeV. *Paper presented at the 13th Int. Conf. Solid State Nucl. Track Detectors*, Rome (and to be published in *Nucl. Tracks* **12**, 1986)

73. S. P. Ahlen (1980) Theoretical and experimental aspects of the energy loss of relativistic heavily ionizing particles. *Revs. Mod. Phys.* **52**, 121–73.

74. P. B. Price (1982) Applications of nuclear track-recording solids to high-energy phenomena. *Phil. Mag.* **45**, 331–46.

75. S. P. Ahlen, P. B. Price & G. Tarlé (1981) Track-recording solids. *Physics Today* **34** (No. 9), 32–9.

75a. D. L. Henshaw, N. Griffiths, O. A. L. Landen, S. P. Austin & A. A. Hopgood (1982) Track response of CR-39 manufactured in specifically controlled temperature-time curing profiles. In: *Proc. 11th Int. Conf. Solid State Nucl. Track Detectors*, Bristol, and Suppl. 3, *Nucl. Tracks*. Pergamon, Oxford, pp. 137–40.

75b. L. M. Kukreja, U. K. Chatterjee, D. D. Bhawalkar, A. M. Bhagwat & V. B. Joshi (1984) Studies on laser cutting of plastic track detector sheets and its effects on the track revelation properties. *Nucl. Tracks* **9**, 199–208.

76. R. Gold, J. H. Roberts & F. H. Ruddy (1981) Annealing phenomena in solid state track recorders. *Nucl. Tracks* **5**, 253–64.

76a. S. Watt, P. F. Green & S. A. Durrani (1984) Studies of annealing anisotropy of fission tracks in mineral apatite by track-in-track (TINT) length measurements. *Nucl. Tracks* **8**, 371–5.

77. G. Törber, W. Enge, R. Beaujean & G. Siegmon (1982) The diffusion-etch-model part I: proposal of a new two-phase track developing model. In: *Proc. 11th Int. Conf. Solid State Nucl. Track Detectors*, Bristol, and Suppl. 3, *Nucl. Tracks*. Pergamon, Oxford, pp. 307–10.

78. I. Milanowski, W. Enge, G. Sermund, R. Beaujean & G. Siegmon (1982) The diffusion-etch-model part II: First application for quasi-relativistic Fe-ion registration in Daicel cellulose nitrate. In: *Proc. 11th Int. Conf. Solid State Nucl. Track Detectors*, Bristol, and Suppl. 3, *Nucl. Tracks*. Pergamon, Oxford, pp. 311–14.

79. G. Braune & W. Enge (1982) Improvement of the α-particle registration applying the diffusion etch model. In: *Proc. 11th Int. Conf. Solid State Nucl. Track Detectors*, Bristol, and Suppl. 3, *Nucl. Tracks*. Pergamon, Oxford, pp. 315–18.

80. F. H. Ruddy, H. B. Knowles, S. C. Luckstead & G. E. Tripard (1977) Etch induction time

in cellulose nitrate: a new particle identification parameter. *Nucl. Instrum. Meth.* **147**, 25–40.

81. G. Tarlé (1981) Improvement of the etching properties of CR-39 plastic track detectors. In: *Proc. 17th Int. Cosmic Ray Conf.*, Paris, **8**, 74–77.

82. P. B. Price & D. O'Sullivan (1982) Improving the etching properties and tailoring the response of CR-39 plastic track detectors with dopants. In: *Proc. 11th Int. Conf. Solid State Nucl. Track Detectors*, Bristol, and Suppl. 3, *Nucl. Tracks*. Pergamon, Oxford, pp. 929–32.

83. P. B. Price, D. O'Sullivan & S. P. Ahlen (1982) Thin-film dosimetry for neutrons and alpha particles. In: *Proc. 11th Int. Conf. Solid State Nucl. Track Detectors*, Bristol, and Suppl. 3, *Nucl. Tracks*. Pergamon, Oxford, pp. 925–8.

84. G. M. Sessler (ed.) (1980) *Electrets*. Springer-Verlag, Berlin.

85. P. Kotrappa, S. K. Dua, P. C. Gupta & Y. S. Mayya (1981) Electret—a new tool for measuring concentrations of radon and thoron in air. *Health Phys.* **41**, 35–46.

86. G. Pretzsch, B. Dörschel & A. Leuschner (1983) Investigation of Teflon electret detectors for gamma dosimetry. *Radiat. Prot. Dosim.* **4**, 79–84.

87. S. R. Coover, J. A. Roseboro & J. E. Watson, Jr. (1977) Electret charger for pocket ion chamber dosimeters. *Health Phys.* **33**, 474–7.

88. P. Kotrappa, P. C. Gupta & S. K. Dua (1980) Design and performance of a Teflon electret dosimeter charger. *Health Phys.* **39**, 566–8.

89. B. E. Fischer, D. Albrecht & R. Spohr (1982) Preparation of superinsulating surfaces by the nuclear track technique. *Radiat. Effects* **65**, 143–4 (383–4).

90. M. A. Fadel, A. A. Abdulla & N. R. Adnan (1979) A method for measuring personnel neutron doses through induced changes in molecular weight of cellulose nitrate. *Nucl. Instrum. Meth.* **161**, 339–42.

91. M. A. Fadel, A. A. Abdalla & M. A. Hamied (1981) Degradation of polycarbonates with fast neutrons and gamma rays and its application in radiation dosimetry. *Nucl. Instrum. Meth.* **187**, 505–12.

92. M. A. Fadel & S. A. Kasim (1977) A devised light absorption method for measuring fast neutron fluences and gamma doses in mixed radiation fields using a cellulose acetate detector. *Nucl. Instrum. Meth.* **146**, 513–16.

93. S. A. Durrani, Y. M. Amin & J. M. Alves (1984) Studies of radiation damage in crystals using nuclear-track and thermoluminescence methods. *Nucl. Tracks* **8**, 79–84.

94. Y. M. Amin, R. K. Bull, P. F. Green & S. A. Durrani (1983) Effect of radiation damage on the TL properties of zircon crystals. *European PACT J.* **9**, 141–52.

95. J. M. Alves, J. E. Davies & S. A. Durrani (1983) Thermoluminescence of fluorapatites and other mineral apatites: High-temperature emission and the effects of radiation damage. *European PACT J.* **9**, 309–20.

96. S. R. Hashemi-Nezhad, R. K. Bull & S. A. Durrani (1985) Effects of electron and alpha-particle irradiations on the track-etching characteristics of biotite mica. *Radiat. Effects* **89**, 149–56.

97. E. A. Edmonds & S. A Durrani (1979) Relationships between thermoluminescence, radiation-induced electron spin resonance and track etchability of Lexan polycarbonate. *Nucl. Tracks* **3**, 3–11.

98. E. V. Benton, R. P. Henke, C. A. Tobias, W. R. Holley & J. Fabrikant (1980) Charged-particle radiography. In: *Proc. 10th Int. Conf. Solid State Nucl. Track Detectors*, Lyon, and Suppl. 2, *Nucl. Tracks*. Pergamon, Oxford, pp. 725–32.

99. N. M. Ceglio & E. V. Benton (1980) Imaging thermonuclear burn using solid state track detectors. In: *Proc. 10th Int. Conf. Solid State Nucl. Track Detectors*, Lyon, and Suppl. 2, *Nucl. Tracks*. Pergamon, Oxford, pp. 755–62.

100. G. F. Stone & N. M. Ceglio (1984) A proton shadow camera using CR-39 track detectors. *Nucl. Tracks* **8**, 605–8.

101. D. F. Bartlett, D. Soo, R. L. Fleischer, H. R. Hart, Jr., & A. Mogro-Campero (1981) Search for cosmic-ray-related magnetic monopoles at ground level. *Phys. Rev.* **D24**, 612–22.

102. T. Doke, T. Hayashi, I. Matsumi, M. Matsushita, H. Tawara, K. Kawagoe, K. Nagano, S. Nakamura, M. Nozaki, S. Orito & K. Ogura (1984) Search for massive magnetic

monopoles using plastic track detectors: Characteristics of CR-39 plastic for detecting monopoles. *Nucl. Tracks* **8**, 609–15.

103. R. L. Fleischer & A. Mogro-Campero (1982) Radon transport in the Earth: A tool for uranium exploration and earthquake prediction. In: *Proc. 11th Int. Conf. Solid State Nucl. Track Detectors*, Bristol, and Suppl. 3, *Nucl. Tracks*. Pergamon, Oxford, pp. 501–12.

104. P. B. Price (1982) Applications of plastic track detectors to atomic, nuclear, particle, and cosmic ray physics. In: *Proc. 11th Int. Conf. Solid State Nucl. Track Detectors*, Bristol, and Suppl. 3, *Nucl. Tracks*. Pergamon, Oxford, pp. 737–50.

105. P. Vater, E. U. Khan, R. Beckmann, P. A. Gottschalk & R. Brandt (1986) Sequential fission studies in the interaction of 9.03 MeV/n ^{238}U with ^{nat}U using mica track detectors. *Nucl. Tracks* **11**, 5–16.

106. H. A. Khan, I. E. Qureshi, W. Westmeier, R. Brandt & P. A. Gottschalk (1985) Elastic and inelastic scattering of uranium ions by holmium, gold and bismuth targets. *Phys. Rev.* **C32**, 1551–7.

107. E. U. Khan, P. Vater & R. Brandt (1985) Some salient features of U + U reaction of 9.0 MeV/N and 16.7 MeV/N using mica as a solid state nuclear track detector (SSNTD). *Paper presented at the 10th Int. Summer Coll. Phys. Contemp. Needs*, Nathiagali, Pakistan (June/July, 1985; and to be published in the Proceedings).

108. K. Jamil, F. R. Khan, H. A. Khan, R. Brandt & G. Kraft (1986) Interaction of 960 MeV/ nucleon ^{238}U ions with light target atoms of CR-39 polycarbonate. *Nucl. Instrum. Meth.* (in press).

109. P. B. Price & M. H. Salamon (1985) Advances in solid state nuclear track detectors. *Paper presented at the 13th Int. Conf. Solid State Nucl. Track Detectors*, Rome (and to be published in *Nucl. Tracks* **12**, 1986).

A Program to Calculate the Range and Energy-Loss Rate of Charged Particles in Stopping Media

In this appendix, a computer program (written by Henke and Benton[1]) is provided to calculate the range and energy-loss rate of a charged particle travelling through any given medium, as a function of four parameters: the atomic mass per electron (A/Z), and the mean ionization energy (I), of the *stopping material* (which may be the track detector); and the atomic number, and energy per nucleon, of the *incident ion*. The program is written in Basic-E language, but is easily adaptable for any other version of BASIC. (A FORTRAN version is also given by the authors[1].)

The numerical values listed in Table 1 are the necessary coefficients and should be stored consecutively, column by column, in a data file (called

TABLE 1. *Data held in File "NUM.DAT"*

64.03	−430.1	115.48	−30.142
−102.29	213.53	−24.598	−1.452
−30.77	−50.5	0.	18.371
132.98	8.56	18.664	3.4718
−84.08	4.	−31.2	−1.4307
20.19	3.	25.398	3.6916
0.94	4.	0. ·	−8.0155
0.	−9306.336	4.0556	1.
−108.57	12843.119	0.73736	16.5
327.06	−5558.937	−7.6604	2.
−367.09	752.005	0.	42.
199.14	−934.142	0.	3.
−51.96	1264.232	0.	38.
12.62	−528.805	0.	4.
57.52	74.192	0.	60.
−146.51	54.043	39.405	8.
131.1	−64.808	−33.802	96.
−42.03	11.139	−5.2315	10.
−7.05	18.797	−5.9898	131.
6.81	12.625	−68.538	12.
0.63	−13.396	71.303	156.
0.	8.309	−0.95873	
−139.4	−8.725	4.5233	
403.41	−99.661	23.603	

"NUM.DAT" in this program) as the read statements indicate. Table 2 lists the parameters of a number of detector materials of general interest; it is an expanded and modified version of a table given by Henke and Benton[2]. In the version of the program given below, I and A/Z are not required as data, but the atomic charge (Z), atomic mass (A), and the number of atoms per molecule have to be input for each component of the detector material. The values of I and A/Z are then calculated by the program. Further details of these calculations are given by Henke and Benton[2]. Finally, the parameters of the incident ion (energy (in MeV) per nucleon, atomic number, and atomic mass) are input. The program calculates the range of the ion in the given medium, together with either dE/dx or the restricted energy loss (REL), and prints out the computed values. The calculated ranges are given in microns.

TABLE 2. *Stopping parameters of various detector materials*

Material	I	A/Z	Density g cm^{-3}	Composition
Cellulose nitrate	81.1	1.939	1.4	$C_6H_8O_9N_2$
Cellulose acetate butyrate	63.4	1.835	1.23	$C_{20}H_{32}O_5$
Bisphenol-A polycarbonate (Lexan; Makrofol)	69.5	1.896	1.29	$C_{16}H_{14}O_3$
Polyallyldiglycol carbonate (CR-39)	70.2	1.877	1.32	$C_{12}H_{18}O_7$
Polyethylene terephthalate (Mylar)	73.2	1.915	0.93	$C_{19}H_{16}O_7$
Muscovite mica	128.3	2.011	2.8	$KAl_3Si_3O_{10}(OH)_2$
Phlogopite mica	129.4	2.009	2.7	$KMg_2Al_2Si_3O_{10}(OH)_2$
Albite (feldspar)	126.0	2.015	2.61	$NaAlSi_3O_8$
Anorthite (feldspar)	135.3	2.014	2.77	$CaAl_2Si_2O_8$
Olivine	158.9	2.050	3.35	$MgFeSiO_4$
Apatite (fluor-)	153.1	2.016	~3.2	$Ca_5(PO_4)_3F$
Zircon	209.5	2.128	~4.7	$ZrSiO_4$
Sphene	155.4	2.042	3.54	$CaTiSiO_5$
Enstatite	126.9	2.000	~3.2	$MgSiO_3$
Diopside	138.3	2.000	~3.3	$CaMgSi_2O_6$
Soda-lime glass	131.6	2.014	2.47	$67SiO_2:14Na_2O:14CaO$ $:5Al_2O_3$
Phosphate glass ("ZnP")	131.6	2.039	2.69	$34P_2O_5:7ZnO:4Al_2O_3$ $:4B_2O_3:SiO_2$

TABLE 3. *Sample output of range and energy-loss rate (dE/dx) values for* ^{40}Ar *ions in CR-39. The energy of the ion is expressed in MeV/amu; the calculated range is given in units of microns (μm), and dE/dx in units of MeV· cm² g⁻¹. Beta stands for β = v/c.*

Ar-40 in CR-39			
energy/nuc.	range	dedx	beta
1.00000E-02	.665577	8154.39	4.63451E-03
3.00000E-02	1.24287	12575.5	8.02708E-03
.1	2.5733	18613.2	1.46545E-02
.3	5.37454	23641	2.53783E-02
1	12.9907	29095.6	4.63082E-02

TABLE 4. *Sample output of range and restricted energy loss (REL) values for* ^{40}Ar *ions in CR-39. All the units are as given at the head of Table 3. A value of $W_0 = 350$ eV has been used for the upper limit on the δ-ray energy in this computation.*

Ar-40 in CR-39			
energy/nuc.	range	REL	beta
1.00000E-02	.665577	8154.39	4.63451E-03
3.00000E-02	1.24287	12575.5	8.02708E-03
.1	2.5733	18613.2	1.46545E-02
.3	5.37454	19856.4	2.53783E-02
1	12.9907	21358.9	4.63082E-02

(μm) (based on the density values shown in Table 2); and the units of dE/dx and REL are MeV. cm² g⁻¹.

Typical program printouts are given (for ^{40}Ar ions in CR-39) in Table 3 (range and dE/dx), and Table 4 (range and REL).

BASIC-E Version of Henke and Benton's Program[1] to Calculate the Range and Energy-Loss Rate of Charged Particles in Stopping Media

```
dim cjoin(7,4),order(3),crandi(4,4,3),zdifi(7),difi(7)
dim z(10),a(10),mu(10),percwt(10),indpot(10)
dim mbeta(50),cz(50),cpz(50),range(50),dedx(50)
dim crange(4,3),join(4),cdrde(4,3),enrg(50),ln1nrg(50)
dim bt(50)
a$ = "NUM.DAT"
file a$
for i = 1 to 4
```

```
for j = 1 to 7
read £ 1; cjoin(j,i)
next j
next i
for i = 1 to 3
read £ 1; order(i)
next i
for i = 1 to 3
for j = 1 to 4
for k = 1 to 4
read £ 1; crandi(k,j,i)
next k
next j
next i
for i = 1 to 7
read £ 1; zdifi(i),difi(i)
next i
input"name of material";name$
input"how many components in the material";nocomp
print"type charge,at. mass, and no of atoms of each element in material"
for i = 1 to nocomp
input z(i),a(i),mu(i)
next i
input"density of material = ";dens
rem
rem compute i and a/z for the material
sum = 0.0
for j = 1 to nocomp
percwt(j) = a(j)*mu(j)
sum = sum + percwt(j)
indpot(j) = 0.0
for k = 1 to 7
ab = z(j) − zdifi(k)
ab = abs(ab)
if ab < 0.1 then indpot(j) = difi(k)
next k
if indpot(j) > 1.0 then goto 10
if z(j) > 13. then goto 6
indpot(j) = 12.*z(j) + 7.
goto 10
6 indpot(j) = 9.76*z(j) + 58.8/z(j)^0.19
10 pot = 0.0
next j
```

```
aperz = 0.0
for j = 1 to nocomp
percwt(j) = percwt(j)/sum
pot = pot + z(j)*percwt(j)*log(indpot(j))/a(j)
aperz = aperz + z(j)*percwt(j)/a(j)
next j
aperz = 1.0/aperz
pot = exp(aperz*pot)
zpera = 1.0/aperz
print
print "Ioniz. pot", "A/Z", "Z/A"
print pot,aperz,zpera
rem
rem compute joining coefficients etc.
rem
pot3 = pot*0.001
for j = 1 to 4
join(j) = cjoin(1,j)
for k = 2 to 7
join(j) = cjoin(k,j) + pot3*join(j)
next k
next j
ln1pot = log(pot)*0.1
for j = 1 to 3
n = order(j)
for k = 1 to n
crange(k,j) = crandi(1,k,j)
for l = 2 to n
crange(k,j) = crandi(l,k,j) + crange(k,j)*ln1pot
next l
cdrde(k,j) = crange(k,j)*(n − k)*0.1
next k
next j
crext = (31.8 + 3.86*pot^0.625)*1.0e − 06
cprext = 137.0*crext
rem
rem read energy values and ion type
rem
print"number of energy values (max.50)"
input"if 50 then the program takes it as a plot";noenrg
if noenrg = 50 then goto 110
print"type energy values for which data is required"
for i = 1 to noenrg
```

```
input enrg(i)
next i
goto 39
110 input "type max and min energies for plotting";emax,emin
delx = (log(emax) − log(emin))/49
enrg(1) = log(emin)
for i = 2 to 50
enrg(i) = enrg(1) + (i−1)*delx
next i
for i = 1 to 50
enrg(i) = exp(enrg(i))
next i
39 input"type charge and at. mass of particle"; zion, aion
input"ion symbol";ion$
rem
rem compute energy termsn
rem
for i = 1 to noenrg
ln1nrg(i) = log(enrg(i))*0.1 + 0.000796811
erat = enrg(i)/931.141
beta = sqr(erat*(2.0 + erat))/(1.0 + erat)
bt(i) = beta
mbeta(i) = 938.59*beta*(1.0 + erat)^3.0
cx = 137.0*beta
x = cx/zion
if x < 0.2 then cz(i) = − 0.00006 + x*(0.05252 + 0.12847*x)
if(x > 0.2) and (x < 2.0) then cz(i) = − 0.00185 + x*(0.07355 + x*(0.07171 − x*
    0.02723))
if(x > 2.0) and (x < 3.0) then cz(i) = − 0.0793 + x*(0.3323 + x*(−0.1234 + x*
    0.0153))
if x > 3.0 then cz(i) = 0.22
if x < 0.2 then cpz(i) = 0.05252 + 0.25694*x
if(x > 0.2) and (x < 2.0) then cpz(i) = 0.07355 + x*(0.14342 − 0.08169*x)
if(x > 2.0) and (x < 3.0) then cpz(i) = 0.3323 + x*(−0.2468 + 0.0459*x)
if x > 3.0 then cpz(i) = 0.0
next i
rem
rem include ion dependence
rem
zion2 = zion*zion
axz = 5.0/3.0
apz = zion^axz
az = zion*apz
```

```
cb = crext*az
cbp = cprext*apz
mbyz2 = aion/(1.008*zion2)
print"do you want REL or dE/dx?"
print"type 0 for REL"
print"type 1 for dE/dx"
input itype
if itype then goto 30
input" what value of Wo";w0
w0 = w0*1.0E − 06
elim = w0/0.002176
rem compute proton ranges and dedx
rem
30 for j = 1 to noenrg
if enrg(j) < join(1) then rid = 1
if(enrg(j) > join(1)) and (enrg(j) < join(2)) then rid = 2
if enrg(j) > join(2) then rid = 3
if enrg(j) < join(3) then drdeid = 1
if(enrg(j) > join(3)) and (enrg(j) < join(4)) then drdeid = 2
if enrg(j) > join(4) then drdeid = 3
n = order(rid)
mp = order(drdeid)
m = mp − 1
prange = crange(1,rid)
for k = 2 to n
prange = crange(k,rid) + prange*ln1nrg(j)
next k
prange = exp(prange)
drdera = prange
pdrde = cdrde(1,drdeid)
for k = 2 to m
pdrde = cdrde(k,drdeid) + pdrde*ln1nrg(j)
next k
if (rid − drdeid) then goto 100 else goto 20
100 drdera = crange(1,drdeid)
for k = 2 to mp
drdera = crange(k,drdeid) + drdera*ln1nrg(j)
next k
drdera = exp(drdera)
20 pdrde = drdera*pdrde/(1.008*enrg(j))
rem
rem compute for ion
rem
```

```
range(j) = aperz*mbyz2*(prange + cb*cz(j))
range(j) = range(j)*10000.0/dens
print zpera,zion2,pdrde,cbp,cpz(j),mbeta(j)
dedx(j) = zpera*zion2/(pdrde + cbp*cpz(j)/mbeta(j))
if itype then goto 31
if enrg(j) < elim then goto 31
be2 = bt(j)*bt(j)
a2 = 130*bt(j)/(zion^0.6667)
zeff = zion*(1 − exp(−1*a2))
ze2 = zeff*zeff
subd = 0.15359*ze2/(aperz*be2)
wmax = 1.022*be2/(1−be2)
subd = subd*(log(wmax/w0 − be2))
dedx(j) = dedx(j) − subd
31 next j
rem
rem print out
rem
input"type char plus ctrl P";i2dum
print" ";ion$;"in";name$
if itype then goto 32 else goto 33
32 print"energy/nuc.","range","dedx","beta"
goto 34
33 print"energy/nuc.","range","REL","beta"
34 for j = 1 to noenrg
print enrg(j),range(j),dedx(j),bt(j)
next j
print
if noenrg < 49 then goto 111
input "filename for plot";b$
file b$
print £ 2;50
print"type 0 for de/dx vs energy/nuc"
print"type 1 for de/dx vs range"
print"type 2 for range vs energy"
input iplot
if iplot = 0 then goto 201
if iplot = 1 then goto 202
if iplot = 2 then goto 203
201 for i = 1 to 50
print £ 2;enrg(i),dedx(i)
next i
close 2
```

```
goto 205
202 for i = 1 to 50
print £ 2;range(i),dedx(i)
next i
goto 205
203 for i = 1 to 50
print £ 2;enrg(i),range(i)
next i
111 print
205 print
print"for another ion,type 0"
print"to terminate,type 1"
input nter
if nter then goto 38 else goto 39
38 print
stop
```

1. R. P. Henke & E. V. Benton (1968) Charged particle tracks in polymers: No. 5—A computer code for the computation of heavy ion range-energy relationships in any stopping material. *US Naval Radiological Defense Laboratory, San Francisco, Report* USNRDL-TR-67-122.
2. R. P. Henke & E. V. Benton (1966) Charged particle tracks in polymers: No. 3—Range and energy loss tables. *US Naval Radiological Defense Laboratory, San Francisco, Report* USNRDL-TR-1102.

Subject Index

(Page numbers in *italics* refer to terms used in figure captions and tables.)

Ablation   225, *227*, 260
Accelerated ions   27, *28*, (29), 267
Actinides   249, 257
Activation energy   37, 90, *97*, 98, 101, 105–111, 221
Activation foils/monitors   202, 203, *208*, 212–216
Active dosimeters   164, 165
Active species   96
Additives, to polymers   264
Adiabatic expansion   4
Adiabatic interaction   17
Age correction methods   111, 218 *et seq.*, 238
Age equation, fission track   200, 202, 216, 217, 229, 231
Age
   fission track   103
   effective   *104*
Age standards   *208*, 212, 216, 217
Aging   24, 129, 267
Albedo effect/principle   150, 152
Albedo neutrons   152
Albedo/threshold dosimeter   152, *153*
Allyl diglycol carbonate   *163*, 274
Alkali halides   25, 26, 42, 44
Alpha-active nuclei/particles   253, 254
Alpha activity   176, 253–257
Alpha detection   156, *162*, 163
Alpha emitters   *163–166*, 251–257, 261, *262*
Alpha-particle dosimetry   161 *et seq.*
Alpha-particle ranges   261, *262*
Alpha-recoil damage   87, 206, 209, 265
Alpha-recoil tracks (α-RTs) in dating/thermal history   109, 110, 206
Alpha-sensitive plastics   148, 154, 162, 171, (*252*), 253
Alpha tracks, spark counting   171
Alphagenic converter screen   148, 171
Alphagenic reactions   172
Americium   256, 257
Ammonia vapour track-location method   129
Ancient cosmic rays, *see* Cosmic rays
Analogue-to-digital converter (ADC)   190, *191*

Angled-polishing technique   *100*, 121, *123*
Angle of incidence   79, 80, 172
Angular distributions of nuclear reaction products   246, 249, 250
Anisotropic solids   83, *84*, *86*
Anisotropy of annealing   100, 111, 221, 263
Anisotropy of etching   (48), 83–87, 111, *138*, 206, 209, *220*, 221
Annealing (thermal)   11, 37, 89, 96–110, 140, 206, 210, 212, 217–*226*, 250, 263
Annealing correction   110, 218 *et seq.*
Annealing
   isochronal   101, *102*, 111, 219, *222*
   isothermal   101, *102*, *105*, 106, 111, 219, *222*
Anomalons   267
Antimatter   267
Antineutrino   5
Apatite   *50*, 101, *103*, 109, 111, 203–219, 228, 230, 234, 236, *276*
Apollo (and Luna) missions   (*35*), 167, 229, 257–*259*
Archaeology, application of FT dating to 236–239
Arrhenius plot   *97*, 98, 106, 108, *235*
Arrhenius equation   263
Asteroids   266
Atomic collisions   39
Atomic damage   34
Atomic defects   34
Atomic displacements   2, 19, 24, 39
Atomic planes   44
Au absorber   158
Au activation foil   202, *208*, 212, *213*, 216
Augite   99, *100*
Auger decay   267
Automatic evaluation   153, 169, 184, *188*–192, 265, 267
Autoradiography (*see also* Mapping)   (207, *208*), *252*, 254, (*255*)

Bacteria   256
Balloon flights   12, 56, 128, *224*
Basal plane of apatite crystal   101
Beryllium radiator   158
Bethe-Bloch formula   17

Biological effects 168
Biological applications of SSNTDs 1, 250, 253–257
Biotite mica 40, *50*, 110, 135
Bisphenol-A 48, 51, 250, *276*
Blood cells 255
Bloodstream 254
Blunted/rounded
  cone tip 61
  etch pits *62*
  tracks (in ECE) (178), 184
Bond rupture 24
Boron 148, 150, 161, *163*, 171, 250–253, 262
Bragg peak 99, 125, *137*
Bragg reflection *33*, 34
Branching ratios 266
Breakdown
  devices 173, *174*
  dielectric 178, 181, 265
  electrical 173, *178*
  spots 184
  "treeing" 179
Bremsstrahlung 13, 14
Bright-field microscopy *33*, *35*
de Broglie wavelength 14, 17
Broken molecular chains 10, 24, 37, 39, 96, 98
Bronzite 87
Bubble chamber 4–7, 11
Bubbles in ECE cell/on detector (*177*), 181
Bubbles, imperfections in glasses 237, *237*
Bulk-etch velocity, V_B 48 *et seq.*, 90, 99, *122*, 125, *133*, 139, 181, 184, 209
Butyrate *276*

CA 80-15 91, 148, *160*, *163*, 249, 265
Cadmium cover 150, 152, *153*
Cadmium cut-off 150
Cadmium dopant 10
Capacitance, electrical 169–174, *182*, *183*
Capacitor (thin film) 173, *174*
Calibration of SSNTDs 115, 124, 128, 129, 133
Carbon-14 dating 238
c-axis of apatite crystal 101
Cell damage, in human body 161
Cellulose acetate 88, *88*, 89, 156, 265, *276*
Cellulose nitrate (CN plastic) 13, 40, *50*, 89 *et seq.*, 118–120, 148, 152, *156*, 158, *163*–168, 171, 181, 249–257, 261, *262*, 265, *276*
^{252}Cf fission source/tracks *57*, 61, *84*, 90, 99, 101, *172*, *174*, *178*, 261

Chain scission in polymers 44, 263
Charge assignment 124, 125, (131)
Charge-coupled device (CCD) 187, *188*
Charged-particle interactions 11, 13 *et seq.*
Charged-particle parameters in stopping media 275–283
Charge registration threshold 161
Charge resolution 126, 127, 129, 133, (249, 250), 266, 267
Charge spectrum of cosmic rays, *see* Cosmic rays
Chemical etching 48 *et seq.*
Chemical reactivity 24, 39, 40, 44, 129
Chemical stage, in response to radiation 23
Cherenkov detector 6, 130
Cherenkov radiation 13, 130
Chlorapatite 230
Cleavage planes *57*, 83, *85*, 86
Clinopyroxene *50*, *100*, 102, *103*, 124
Closed-circuit television (CCTV) 87, *190*, 191
Closing (or closure) temperature 96, *97*, 103 *et seq.*, 200, 210, *211*, 234, *235*
Cloud chamber 4, *5*, 11
CN 85 148, *156*
CN plastic, *see* Cellulose nitrate
Cobalt, activation monitor 202, *208*, 212, *213*, 214, 216
Collimated neutron beams 160, 161
Collision time 17
Colloid layer of etch products 91
Comets 266
Conductivity
  electrical 38, 39, 169, *170*, (178), 265
  thermal 27, 39, 41
Cone angle/semi-cone angle 53 *et seq.*, 89, 121, *122*, 171
Cone length 51 *et seq.*, 89, 115, *116*, *120*–127, 133, 134, *140*, 181
Cone tip 54 *et seq.*, *178*, 179, 184
Confined track length 111, *138*, 139
Conical etching phase 60, *62*, *178*, 179, 210
Contact track density method 232–*234*, (249)
Conversion factor (neutron fluence to dose) 148, 149
Converter foil/screen 146–148, 150, 161, 171, 262
Cooling curve *104*, *105*
Cooling history 110, *211*, 234
Cooling rate 107, *108*, 210, *211*, *233*, 234, *235*
Cores, in lunar soil 258, *259*
Cosmic gardening *259*

Cosmic rays 1, 9, 11, 12, 50, 52, *56*, 93, 111, 121, 127–131, 167, 223–230, 249, 257–260, 266
 ancient 133, 136, 139, 140, 167, 223–228
 charge and mass resolution 127, 129
 charge spectrum 128, 130, 140, 167
 contemporary 140, (*227*), 258
 differential spectra *224*
 energy spectrum 223, *224*, 225, 257, 258, 260
 exposure age 225, 230, 258, 260
 Fe group 249
 flux *224*, 225, *227*, *259*, 260
 galactic 223, 228, 257
 heavy 11, 12, (90), 223, 224, 258
 high-energy (harder) *227*, 260
 induced fission by 228, *229*
 in-space irradiation 263
 low-energy (softer) *224*, 225, *227*, *259*
 in lunar rocks (minerals/crystals) 90, 100, 111, 136, 228, 230, 257–259
 in meteorites 11, 12, 133, 136, 139, 225–228, 230, 260
 physics 128, 131
 solar 225, 228, 257
 superheavy (SHE) 249
 tracks 50, *56*, 129, 130, *226*, 228
 tracks in plastic *56*
 ultraheavy 128, 129, 139, 267
 VH group *35*, (90), 100, 139, 140, *224*, 225, *226*, *227*
 VVH group 100, 140, *226*
Cosmogenic nuclides 260
Cosmos spacecraft 167
Coulomb force 16, *16*
Coulomb interactions 15
Coulomb potential 14
Counting efficiency *172*
CR-39 11, 13, *28*, 48–*50*, 55, 80, *81*, 89, 91, *92*, *103*, 127, 129, 135, 145, 152–156, 158, *159–163*, *166*, 167, 174–*180*, 181, 187, 246, 250, 254, 259–266, *276*, *277*
CR-39 (DOP) 264, 266
CR-73 250
Criteria for track formation 27–45
Critical angle 53, *55*, *56*, *58 et seq.*, 99, 145, 201, 209, 219, 261
Critical electric field 179
Cronar polyester 266
Crosslinking 24
Crystalline grains *35*
Crystallites *237*
Crystallization age of rock 210
Crystallographic anisotropy *138*, 263

Crystallographic effects 124
Crystallographic orientation 48, 100, 221
Crystals
 anisotropic annealing, *see* Anisotropy of annealing
 anisotropic etching, *see* Anisotropy of etching
 defects in 25, 34, *37*, 97
 etchants for *50*
 heavy-liquid separations for 203
 radiation damage in 23–26, 87, 89, 90, 265
 tracks formation in 10, 26, 263, 265
 tracks in *57*, *88*, *205*, *226*, *248*

Daicel/DaiCell 91, *156*, *163*
Damage trail (latent) 2, 10, 11, 33–*43*, 48, 51–64, 71, 80, 83, 96, *116*, 122, 132, 134, 139, *140*, 218, 221
Dark field microscopy *33*, *35*
Dating
 α-RT 109, (110)
 ^{39}Ar-^{40}Ar 232
 fission track (*see also* Fission track dating) 199–240 (Chapter 8)
 K-Ar 200, 210, 216, 220, 238
 radioactive 199, 200
 Rb-Sr 200, 210, 216
 thermoluminescent 237, 238
Decay modes, novel (of heavy nuclides) 266
Decay series/chains *163*, *164*, 200
Defects (along latent track) 25, 26, 98
 atomic 13, 34
 extended 34, *37*, 37, 97, 111, 267
 point 34, *37*, 97, 111
Degradation of plastics 89, 265
Delta rays 8, 19, 20, *29*–31, 40 *et. seq.*, *92*, 127, *277*
Dentine 255
Depth dose 160
Detection efficiency 51, 99, (129, 130, 201), 209, (221, 261)
Detection threshold 27, *28*, 29, 161, *162*
Detector minerals 233, (*233*), 234
Detector stack method, *see* Plastic-stack method
Developer 8, 10
Diameter of breakdown spots *174*
Diameter of ECE spots *177*, 181, 184
Diameter of etched tracks 38, 51 *et seq.*, 61–63, 79, 81, 90, 110, 126, 135, 160–162, 191, 218–*220*, 239
Diameter, evolution of 83

Diameter of sparked holes   170, 172
Diameter of unetched tracks   2, *3*, 10,
  (30, 33), 34, 38, 39
Dielectric constant   43, 192
Dielectrics   10, 26, 45, 178, 179, 181, 264,
  265
Diethylene glycol   49, 158
Differential annealing   225
Diffracted rays   *33*, 34
Diffraction contrast   33, *33*, (34), *35*
Diffusion, atomic   37
Diffusion of atomic defects   96, 98, 264
Diffusion coefficient   98, 100
Diffusion cloud chamber   5
Diffusion of electrons   24
Diffusion-etch model   91, 264
Diffusion rate   101
Digitizer, optical   168
Digitizing units/tablets   187, 189, 190
Digitrack   187, *188*, 189
Diopside   *276*
Dip angle   54, 61, *62*, *63*, 64 *et seq.*, 89, *134*
Dislocations (lattice)   10, 25, 83, 236
Displacements (atomic)   2, 10, 19, 34, 39
Displacement-energy threshold   19
Distant collisions   127
Dopant   10
Dose, absorbed   24, 144
Dose of charged particles   89
Dose, critical   40
Dose distribution   *20*, 145, 160, 161, 168
Dose distribution in phantom   145, 160,
  161
Dose distribution in radiotherapy   168
Dose equivalent   144, *149*–152, 158, 184
Dose, first-collision   158
Dose of ionizing radiation   144
Dose to population   163, 166
Dose response function   40
Dosimetry   89, 144–168, 172, 184, 192,
  254, 264
  alpha   145, 161–167, 245
  environmental   145, 148, 154, 163, 264
  heavy ion/high-LET radiations   145, 167
  neutron   145–161, 184, 203, 212, 216,
    245, 264
  personnel (*q.v.*)
  pion   168
  radon   163–167, 172, 264
  reactor   145, 161, (207)
  thermoluminescent   12, 153
Double-etch method   140
Drift chambers   6

Earthquake prediction   *166*, 167, 266
ECE, *see* Electrochemical etching

Efficiency of track revelation (*see also*
  Registration efficiency; Detection
  efficiency)   51, 64, 67, 69–72, 221, 262
Effective charge, Z_{eff} 15, 17, *20*, 30–32,
  51, *55*, 117, 118, 130, 131
Effective closing temperature   *97*, *108*, 109
Effective (electric) stress   179
Effective fission track age   *104*, *105*
Effective residual range   *123*
Effective threshold energy   *156*
Effective track-etch rate   131, *132*
Elastic collisions   154, *155*
Elastic strain (lattice)   10, 34, 42, *43*
Electret   264
Electrical conductivity, *see* Conductivity
Electrical shielding (in ECE)   *178*, *180*, 184
Electrochemical circuit diagram   *182*, *183*
Electrochemical etching (ECE)   152, 158,
  176–184, *190*–192, 263, 264
Electrochemical etching cell   (176), *177*,
  181, 255
Electrolytic cell/electrolyte   38, (*177*), 181
Electron avalanche   6
Electron-hole pair   25, 26, 42
Electron microscopy   2, 33, 34
Electron-phonon interaction   42
Electron-spin resonance   37, 39, 265
Electronic interactions   14, 33, *37*, 39, (41)
Electronic stopping power   *18*, 45
Electrostatic force/electrostatics   13, 178
Electrostatic stress   42, (179)
Elemental mapping   169, (171), 172,
  250–253
Elliptical track-opening   55, 58, *60*, *62*
Emulsion, nuclear/photographic   8–12,
  128, 152, 153, 192
Enamel, lead in   255
Energy loss   14 *et seq.*, 27, 29, 192, 275, 277
  electronic   15 *et seq.*, *18*
  radius-restricted (RREL)   32, 42
  restricted (REL)   27, *29*, 31, 32, 40–42,
    92, 114, 117, 192, 267, 276, *277*
  secondary-electron   19, (20), 31, 32, 40
  total   27, *28*, 29, (114, *117*), *162*, 267,
    275, *277*
Energy spectrometry   (91, *92*), *134*, 135,
  150, 158, *160*, 161, *178*, 181, 249, 250
Energy spectrum
  of α-particles   135, *178*, 181, 249, 250
  of cosmic rays, *see* Cosmic rays
  of neutrons   149, 158, *160*, 161
  of spallation products   250
Enstatite   *276*
Environmental factors in etching   24, 49,
  89–90, 93, 128, 129, 263
Environment factors during irradiation
  267

Environmental heating   96, 111
Environment, natural   163, 256
Environmental neutron spectrometry   158, *160*, 161
Epithermal neutrons   (148, 150), 213
Errors
    in cone-length measurement   124
    in FT dating   214, 239, 240
    in $f(t)$ factor   75
    in microscope measurements   126
    in RAM (random access memory)   253
Etchability threshold   32, 109, *178*
Etchable damage   38, 99, 221, (266)
Etchable range   99, *100*, *138*, 249
    maximum   137
    mean   110, 125, *125*
    total   122, *123*, 130, 136–139
Etchable tracks   2, 26 *et seq.*, 38 *et seq.*, 72 *et seq.*, 99, 116 *et seq.*, 136, *137*
Etchant concentration   49
Etchants   10, 49, *50*, 51, 93, 176, *177*, *178*, 181
    effect of concentration/temperature   90, 91, *92*, 93, 127, 181
Etched track
    cone length, *see* Cone length
    early studies   2
    electrochemical, *see* Electrochemical etching (ECE)
    environmental effects on, *see* Environmental factors in etching
    evolution of   *62*, 64, 80, (*81*, *82*)
Etched track geometry   51 *et seq.*
Etched track parameters   52 *et seq.*, *63*, 83 *et seq.*, 115 *et seq.*, 181, 189
Etched track phases   60, *62*
Etched track profile   62, 76, 77, 80, *81*, *82*, 89, 126, 135
Etched tracks in glass   *55*, 76, 236–239
Etched tracks in minerals   41, 54, *57*, 201 *et seq.*
Etched tracks in polymers   41, *56*, *81*, *82*, 89–93, 115 *et seq.*, 152 *et seq.*, 249–257, 263
Etch induction time   91, 139, 264
Etching conditions   62, 89, 98, 129, 162, *163*, 176, 181, 264
Etching efficiency (*see also* Detection efficiency; Registration efficiency)   64 *et seq.*, 147, 209
Etching method
    from both surfaces   *116*, (121), *134*, 135, *177*
    electrochemical, *see* Electrochemical etching (ECE)
    future developments   262 *et seq.*

multistage   132, *133*, 139, *140*
over-etching   *178*, 179, 184
pre-etching   158, *178*, *180*, 181, 184
Etching process   2, 93, 124, 127
Etching, prolonged, *see* Prolonged etching factor
Etch pit   1, 52 *et seq.*, *72–78*, *85*, 87, *88*, 134, 160, 173–178, *237*, 239
Etch pit evolution   64, 80, 191
Etch pit parameters   *63*, 64, 83
Etch pit profile   61, *62*, 76, 77, 89, 135
Etch products   90, 91, *92*, 93
Etch rate after annealing   100
Etch rate, bulk, *see* Bulk-etch velocity
Etch rate gradient   130, 131, *132*
Etch rate ratio, *see* $V (= V_T/V_B)$
Etch rate, track, *see* Track-etch velocity
Etch rate vs primary ionization   *118*, *119*
Etch rate vs residual range   *116*, *117*, *120*, *123*
Euler's equation/number   186, 187
Eutectic mixture   *50*, 204
Excitation, atomic/molecular   13, 23, 24, 41, 42
Excitation of electrons   20, 41
Excitation potential   17, (18)
Excitons   25
Exotic particles   266
Extended damage   *43*
Extended defects, *see* Defects
External detector   *66*, 70, 71, 146, 203, *208*, 209, 212, *237*, *246*, 251, *252*
External detector method   72, 207, *208*, 209, 247
External track source   64, 70, 71
Extraterrestrial samples   223, 230, 257–260

F centre   26
Fading (*see also* Annealing)   8, 10, 96 *et seq.*, 106, *108*, 129, 218, *220*, 251
Fast neutron dosimetry, *see* Dosimetry
Fast neutrons   148–161, 171, 173, 176, (181), 228, 229, (263), 264, 265
Fast neutrons in reactor irradiation   214, 216
Feldspars   *50*, 111, 124, 136, 203, 225, 232, *233*, 234, *237*, *276*
Fission barrier   127
Fission cross-sections   145–147, 150, *151*, 171, 201, 202, 214, 216, 217, 247
Fission decay constant, ^{238}U 105, 147, 200–202, 212, 214, *215*–217, *229*–234, 247, 264
Fission decay constant, ^{244}Pu 231, 233, 234

Fission events 64–68, 109, 145, 146, 201, 218, 247
Fission foils 150, 154, 161
Fission fragments 1, 2, *3*, 12, *36*, 41, 44, 61, 62, 65–76, 89, 90, 99, 101, 136, 145–147, 161, 171–*178*, 199, 201, 218, *233*, 247, 249, 252, 255, 265
Fission, heavy-ion 247, *248*
Fission, induced, *see* Induced fission
Fission-in-flight 247
Fission phenomena 145, 199, 245
Fission physics 247
Fission reactions 245–249
Fission threshold *151*
Fission track clock 96, 104, 218, 238
Fission track dating (FTD) 72, *86*, 87, 96, 101, 103 (105), 109, 146, 147, 173, 199–240, 245, 250, 263
  age equation 200, 202, 213, 216, 217, 218, 231
  in archaeology 109, 236–239
  contact track method 232, *233*, 234, (249)
  effects of annealing 96, 101 *et seq.*, 218–222
  errors in 214, 239, 240
  interpretation of 209–212
  of lunar samples and meteorites 223–236
  methods 206–209
  practical steps in 203–204
  using ²⁴⁴Pu 223, 225, 230–236
Fission tracks 2, *3*, 64–76, 90, 96–111, 145–152, *170*–174, 181, 184, 199–240, 245–249
Fission tracks in apatite *205*
Fission tracks in mica *3*, 38, *57*, *84*, 219, 247, *248*
Fission tracks, replicas *88*
Fission tracks, ²³⁷Np 150, *151*
Fission tracks, ²⁴⁴Pu 12, 223, 225, 230–236, 249
Fission tracks, ²³⁸U 12, 103–109, 199 *et seq.*, 249
Fitting constants 30, 115, 117, 118, *118*, 128
Flint 237
Flux depression 213
Flux distribution *213*
Flux gradient 212, 213
Flux monitors 202, *208*, 212–214, 216
Fogging 8, 152, 207
Fossil tracks 87, *88*, 110, 111, *205*–208, 218–*222*, 238, 239, 249
Fragmentation 168, 225, 228, 230
Free electrons 23, 44
Free radicals 23, 24

Fundamental electric strength of medium 179
Fused silica 247, 250
Fusion 158, 265
Fusion crust, *see* Meteorite
Future applications of tracks 245, 262, 265

G-value 44
Garnet 108, *108*, 109
Gases, solar wind, *see* Solar wind gases
Gas-rich meteorites 225, 260
Gemini spacecraft 167
Geological timeperiods/timescales 103, 109, 110, *205*, 206, *220*, *235*, 263
Geological uplift 208
Geometrical factor/geometry factor 105, 208, 209
Geometry, of track etching 51 *et seq.*
  2π & 4π detection/irradiation 67–76, (147), 172, 173, 201, 234, *248*, 261
Glasses
  age corrections 110, 218, 221, 238
  annealing experiments 219
  annealing temperature for fission tracks *103*
  in archaeology 237–239
  as detectors 70, 90, 96, 209, 229, 247, 250, 251
  effect of prolonged etching 72, 74, *75*, 207
  effect of radiation damage 89, 90, 265
  etchants *50*
  etched tracks in 26, *55*
  etching parameters 54, 61, 70, 72, 90, 93, 209, 218, 236
  etch rate 54, 72, 90, 93, 219
  man-made 236, 239
  natural (obsidians/tektites) *55*, *75*, *103*, 207, 219, *220*, 236–239
  phosphate *276*
  reference 202, 203, 212, *213*, 217, 229
  soda-lime *50*, 74–76, 90, *103*, *276*
Gold absorber/activation foil, *see* Au absorber/activation foil
Grand unifield theories 266
Grey-level 168, 185, 186, 187
Growth curve, of track density in cooling rocks *108*

Haloes, *see* Radiohaloes
Hardening of tracks 111
Hard-sphere scattering/collisions 15, 19
He radiator 158
Healing of tracks (*see also* Annealing) 11, 99, 206, 218, 221

Healing/repair process 97, 98, 264
Heavy cosmic rays, *see* Cosmic rays
Heavy-element impurities 228, 230
Heavy-ion accelerator (27, 128), 135, 140, *248*, 255, *256*
Heavy-ion induced fission 247, *248*
Heavy ions, *see* Ions
Heavy-liquid separations 203
Heavy projectiles 247
Helium isotopes 135, 249, 250, (253, 254)
High-energy physics 6
High-LET particles 145, 148, 161, 167, 168
High-resolution detector 250, 266, 267
"Hits"/many-hit process 40
Hole 25, 26, 44
Hole mobility 44
Hominids 236
Host tracks *138*
Hybrid detector systems 6, 130, 267
Hydrofluoric acid (HF) 35, 50, *50*, *55*, *57*, 74, 76, *84*, 87, *100*, 204, *237*, 247, *256*
Hydrogenous radiators 158
Hyperfragment; hyperon *9*
Hypersthene *226*

ICRP (International Commission on Radiological Protection) 144, 148, 150, *151*, 257
Identification, particle, *see* Particle identification
Image analysis systems 184–192, 253
Impact heating 211
Impact parameter *16*, 17
Imperfections (in crystals) 24
Implantation, of solar wind 255, 260
In-core measurements, in reactors 161
Induced fission 64, 70, 72, *75*, 97, 145–*151*, 201–206, 219–222, 228, *229*, 239, 247–253
Induction time, for etching 91, 139, 264
Instrumentation for SSNTDs 169–192, 265
Insulating crystals 24
Insulators 11, 42, 44, 265
Integrated circuits 181
Interactions of charged particles 13–21
Interference fringes (Moiré) 34, *36*
Intermediate-energy neutrons 148, 150, (152)
Internal calibration 129
Internal tracks 64–71, 72–76, 99, 101, 147, 201, 208, 218, 234
Interpretation of FT ages 209–212
Interstellar medium 128
Interstitials 8, 10, 26, 34, 39, 42, *43*, 96, 98

Intrinsic detector/reactions *65*, *66*, 70, (*134*), *155*, *156*
Intrinsic electric strength of dielectric 179
Intrinsic recoils 154
Intrinsic reaction products *134*
Intrinsic track density 155, *156*, 158
Intrinsic tracks *155*, 155, 158, (161), 176
Intrusive heating 211
Ion-explosion spike, model 42, *43*, 44, 45
  computer simulation of 44
  thermalized 45, 263
Ionization 10, 11, 13 *et seq.*, 18, 23 *et seq.*, 114 *et seq.*, 249
  chamber 168
  critical (J_c) 137
  peak 136
  -potential 19, 20, 30
  primary 17, 27, *28*, 30–32, 40, 44, 114, *116*–121, 130, 135–139
  rate 129, 136
  reaction products 250
  secondary 19, 31, 32, 40
  -threshold (11), 27, (*28*), 30, 31, 93, 136, 139, (162)
Ionizing particles 2, 4–6, 10, 12, 23, 27, 41, 42, 44, 129, 130, 173, 250
Ionizing radiation 11, 23, 25, 144, 145, 265
Ions (including heavy ions) 4, 5, *8–12*, 13 *et seq.*, 24, *28*, *29*, 31, 33, 39–45, 89–91, 99, 115 *et seq.*, *132*, 135, 137, 140, 161, 176, 247, 255, 260, 264, 266
Irradiation conditions 89, 90
Irradiation geometry (64 *et seq.*), *134*, *213*, (245), *246*
Irradiation history 258, *259*
Irradiation temperature 90, 129, 251, 263
Island of stability 247
Isochronal annealing 101, *102*, 219
Isochronal plateau method 220, 221, *222*
Isothermal annealing 101, *102*, *105*–107, 219
Isothermal heating *97*, *235*
Isothermal interval 105, *105*, 106, 107
Isothermal plateau method 221, *222*

J (primary ionization function), *see* Ionization, primary
Jogs 25

K (fitting constant in ionization equation) 30, 117, 118, *118*, 128
K-Ar dating/age 200, 210, 216, 220, 238
Knock-on 19, 39
KOH etchant *50*, *92*, 176, *180*, 204
Kr tracks 99

LR 115   80, *82*, 148, 152, *156*, 161, *162*,
    163, 171, 187, *252*, 255, 257, 261
LR 115 type I/type II film   171
L-R_T curve   (100), *123*
L-R data   125, *125*
L-R method   126, 131
L-R profile   118, *120*, 121, 128
L-R plot   *100*, 115, *117*, *120*, 121, *123*,
    124, 128
Laser cutting of CR-39   263
Lattice
    annealing of stress   98
    atoms   39, 41, 45
    of crystal   34, 39, 41, 96
    damage   41
    defects   25, 26, 96
    disruption   34, 42
    distortion   34, 98
    energy transfer to   42
    metallic   42
    planes   *33*, 34, 36
    relaxation   42, *43*
    sites   13
    strain   10, 34, 42, *43*
    temperature   41, 42
    vacancies   25, 26, 39, *43*, 98
    vibrations   25, 42
Latent damage trails, *see* Damage trail
Lead in teeth   254, 255
Lead, radiogenic   248, 249
Leitz T.A.S.   185
LET (*see also* High-LET particles)   145,
    148, 161, 167
LET threshold   168
Lexan polycarbonate   11, *28*, *29*, 32,
    48–51, *56*, 89–91, 127, 129, *132*, 140,
    161, 163, 207–209, 212, 250, 254, *276*
Lichtenberg figures   *180*
Light scattering   192
Lineal event density (LED)   32
Liquid-hydrogen bubble chamber   7
Lithium borate   148, 152
Lithium fluoride   1, 152, *153*
Long Duration Exposure Facility (LDEF)
    128, (267)
Loss of fission tracks, 100%   *103*
Loss of track density, 50%   *235*
Low-angle X-ray scattering   2, 34, *37*,
    97
Low-energy particles, identification   125
Luminescence centres   24
Lunar
    composition, rocks of   229
    core   249
    crystals/minerals   90, 100, 136, 140
    incident cosmic-ray spectrum   224

nucleon fluxes   229
radiation effects   260
regolith   *259*, 260
rocks/soil   90, 140, 211, 223, 228–232,
    257–260
    —exposure ages   230, 258
soil
    —dynamics   258, *259*, 260
    —grains   *259*, 259, 260
    —irradiation   258
surface   90, 111, 223, 229, 257–259
    —processes   257
thermal-neutron fission rate   229
Lunar samples   1
    cosmic-ray tracks in, *see* Cosmic rays
    fines   *35*
    fission track dating/age of   223, 228,
        230
    fossil-track annealing in   111, 140
    fossil tracks in   111, (136), 257
    ^{244}Pu tracks in   231
    soil grains   *35*, 258, 259
    spallation in   230
    studies   257

Magiscan   185, 187
Magnetic field
    galactic   223
    geo-   130, 224, *224*
    planetary   223
    solar   224, *224*
Major-axis, of track opening   54, 58, *58*,
    61, *63*, 85, 87, 126, 186
Makrofol   11, *103*, *119*, *120*, *163*, 168, 171,
    172, 265, *276*
Man-made glasses   236, 239
Mapping, of charged particles   160, 161,
    171, *252*
Mapping, of dose distribution   145, 168
Mapping, elemental   169, 171, (207), 208,
    238, 250–254, (255)
Mapping in U exploration   166, *166*, 167
Mapping, V_T variation along track   78
Marker   139, 140
Mars   266
Mason's equation   179, 181
Mass-resolution   126, 127, 129
Maximum etchable range, *see* Etchable
    range
Maximum permissible dose level   144
Measurement of track parameters, special
    techniques (*see also* SSNTD
    instrumentation)   87–89
Merrillite (Whitlockite)   *50*, *88*, 210, 228,
    232–236

Mesolithic cave settlements 238
Metamorphism 211
Meteorite 1, 11, 111, 136, 139, 223 *et seq.*, 230 *et seq.*, 257, 260
ablation 225, *227*, 260
atmospheric passage 225, *226, 227*, 260
chondritic 230, 232, 236
cooling rate *233*, 234, 236
cosmic-ray tracks in, *see* Cosmic rays
crystals and minerals 11, 12, *28*, 129, 136–140, 224, 225, 228–234, 249
exposure age 225, *227*, 230, 260
fission track dating/age 199, 223, 228, 230, 231 *et seq.*
fusion crust 225, *226, 227*
gas-rich/ion-implantation in 260
impact 111, 211, 225, *259*, 259
iron 249
parent body/progenitor 225, 260
preatmospheric size *227*, 260
^{244}Pu fission tracks in 223, 230 *et seq.*
Pu/U ratio 231–233, 236
regolith 260
replica of fossil tracks in *88*
shock-induced dislocations in 236
simulation 229
spallation tracks in 225, 230
track annealing in 140, 225, *226*
track production rate in *227*, 260
track retention age 231–234
track-rich grains in 225, 260
Mica 1, 2, *3*, *28*, 32, 38, 40, *50*, *57*, 83, *84*, 101, *103*, 109, 110, 146, 161, 203, 207, *208*, 229, 246–*256*, 266
biotite (*q.v.*)
muscovite (*q.v.*)
phlogopite *276*
synthetic 207, 246
Microchips 253
Microcomputers 185–187, *190*, 192
Microdensitometry 184, 192
Microelectronic circuitry 264
Microfilter/Micropore 255, *256*, 256
Micrometeorites *259*, 259, 260
Microprocessors 185, 187–191
Microscope, *see* Optical microscope
Microscope, resolution limit 38, 87, 126
Microtome 77
Miller index 83
Minerals (*see also* Crystals, Lunar, Meteorite, and under individual mineral names)
anisotropic etching in 83–87
annealing/annealing corrections 96–111, 218–222, 263

closing/closure temperatures 103 *et seq.*, 210, *235*
etchants for *50*
etched tracks in *28*, *57*, *205*, *206*
etch rate dependence 93
fission track dating 199–240
radiation damage 23 *et seq.*, 90, 265
separations 203, 204
as track detectors 2, 11, 136, *233*, 233
track formation in 11, 26 *et seq.*, 263
track parameters in 54
track retention characteristics *103*
Minor-axis, of track opening 54, 59, *60*, 63, (86), 87, 186
Modulation, of cosmic rays 224, *224*
Moiré fringes 34, *36*
Molarity/molality 49
Molarity of etchant 49, 50, (*50, 81, 82*), 91, 92, (159)
Molecular chains 10, 13, 24, 39, 96, 98
Molecular chain scission 44, 263
Molecular excitation 23, 24
Molecular fragments 37, 96
Molecular sieve 256
Molecular weight 24, 265
Molybdenum trioxide 34, *36*
Monopoles, magnetic 265, 266
Monte Carlo calculations 161, 260
Multidetector devices (*see also* Plastic-stack method) *166*
Multifragment fission *248*
Multipronged events 247
Multi-sample ECE cell *177*, 181
Multistage etch procedure 132, *133*, 139, *140*
Multi-wire proportional counter 6
Muscovite mica *3*, 32, 40, *50*, *57*, 83, *84*, 207, *208*, 246, 266, *276*
Mussels (*Myrtilus edulis*) 257
Mylar *155*, 169, *170*, *172*, 248, *276*

(n, α) cross-sections *156, 157*, 171
(n, α) reactions 148, 152–156, 158, *160*, 161, *163*, 171, 176, 251–253, (262)
aOH etchant 48–51, *81, 82*, 90–93, *159*, *172*, 175–*177*, 204, *226*
Natural tracks, sources of 266
Negative pion (π⁻) dosimetry 168
Neptunium (^{237}Np) 150–154
Neutron activation 238
Neutron dose equivalent (*see also* Dose equivalent) *149*, 149, *151*, 152, 158, 184
Neutron dosimetry (*see also* Dosimetry) 145–161, 212–216
environmental 145, 148, 154, 161

Neutron dosimetry (*continued*)
  personnel   (144, 145), 146, 148,
    (151–153), 154
  in FT dating   147, 202, 203, 212–214,
    216, 264
Neutron energy spectrum, *see* Energy
  spectrum of neutrons
Neutron fluence/flux   145–*151*, (154), 161,
  172, 201–203, 207, *208*, 212–214, 216,
  217, 240, (251), 253, 265
Neutron recoils   152–154, 158–161, 173,
  176, 181, 207, *208*, 263
Normality   49
Nuclear
  collisions   14, 15, (18)
  emulsion, *see* Emulsion
  physics, application of particle
    identification to   131–136
  physics, application of track methods to
    1, 121, 169, 245–250
  reaction products   145
  stopping   14
  tracks, *see* Tracks, Etched track,
    SSNTDs
Nucleosynthesis   128
Numerical aperture (of microscope
  objective)   87, 126

Obsidian   236, *237*–239
Oklo prehistoric fission reactor   223
Olivine   *50*, *103*, 124, 139, 224, 225, 228,
  233–235, 249, *276*
Omnicon (Pattern Analysis System)   185
Optical absorption methods   184
Optical densitometry   152, 184, 192, 265
Optical microscope/Microscope   2, 33, 38,
  87, 89, 121, 126, (129), 169, 171, *172*,
  176, 184, 187–191, 204, *237*, 253, 261
Optical qualities of CR-39   250, 264
Optical spectroscopy   238
Optomax III   186
Orbital electrons   13, 15, 17, (18)
Orthopyroxene   *50*, *103*, *226*, 235
Oxygen effects on
  etching properties of plastics   90, 129
  radiation damage   24, 90
  track image in emulsion   8

Partial annealing of tracks   111, 140, *222*
Particle identification   6, 8, 10, 91, *92*,
  114–140, 245, 250, 263, 266
  applications of techniques   127–140,
    245, 250
Particle trajectory   13, 32, 44, 115, 125,
  127, 168
Passive detector   128, 247

Passive dosimeters   161, 164, 165
Peak-to-plateau dose ratio (for π^-)   168
Peking man (*Sinanthropus*)   238
PENS (Passive Environmental Neutron
  Spectrometer)   161
Permittivity of free space   15, 16, 43
Personnel dosimetry   12, 144, 146, 148,
  154, 169, 172
Phantom, dose distribution in   145, 160, 161
Phase transformations   99
Phosphate glasses   265, *276*
Photoelectron   8, 10
Photo-fission   153
Photographic emulsion, *see* Emulsion
Photographic film   130
Photographic map   208
Photo-neutron reactions   149
Photo-oxidation   24, (90, 129)
Physical/Physico-chemical stage of
  response to radiation   23
Pion (π^-) dosimetry   168
Pixel (picture element)   186, 187, 189
Plastics, *see* Polymers
Plastic-stack method   56, 115, *116*, 129,
  130, (158, 161)
Plateau age corrections
  isochronal   111, 220, 221, *222*
  isothermal   111, 221, *222*
Plate-out effects   166
Platinum electrodes (ECE cell)   176, *177*
Plutonium (^{238}Pu, ^{239}Pu, ^{244}Pu)   12, 150,
  223, 225, 230–234, 249, 252, 256, 257
Point defects, *see* Defects
Poisson statistics   239
Polarization   17, 30
Polishing (*see also* Angled polishing)
  (139), 204, 208, (212), 221, (*226*)
Polonium   *163*, *164*, *180*, *252*–255, 266
Polyallylalcohol (PAA)   49
Polycarbonate (*see also* CR-39, Lexan, and
  Makrofol)   28, *29*, 38, 41, 48, *50*, 89,
  132, 152, *156*, 161, 163, 181, *208*, 250,
  266, *276*
Polyethylene radiator   152, *153*, 158
Polygonal etch figures   83, *84*, 255, *256*
Polyimide film   253
Polymers/plastics (*see also* Bisphenol-A,
  CA 80-15, Cellulose acetate, Cellulose
  nitrate (CN), CR-39, Daicel, LR 115,
  Makrofol, Polycarbonate)
  alpha-sensitive (*q.v.*)
  annealing of   37, 96, 263
  annealing temperature (for FT)   *103*
  biological applications   253–257
  broken molecular chains/molecular
    fragments (*q.v.*)
  chain scission   44, 263

Polymers/plastics (*continued*)
cosmic ray tracks in 56, 129, 130
crosslinking and bond rupture in 24
curing process 127, 263
dosimetry with 145 *et seq.*, 264
ECE/ECE spots in 176–184, *180*
elemental mapping with 250–253
energy discrimination/spectrometry with 91, *134*, 135, 150, 158–161, *178*, 181, 249, 250
environmental factors 24, 49, 89–93, (96)
etchants for 49, *50*, 51
etch-cone lengths in *116*, *120*, 131
etching of 48, 49, 91–93
G-value for 44
intrinsic tracks in 154–161
irradiation temperature effect 90, 129, 263
molecular chains (*q.v.*)
molecular weight 24, 265
non-uniformities in 127, 263
optical properties *103*, 207, 264, 265
particle identification with 115 *et seq.*, 250, 263
pyrolysis 39
radiation damage/effects in 23, 24, (44)
refractive index 121
replica *88*
registration threshold (11), *28*, (29), 161, *162*
scintillator-filled etch-pit counting in 173–176
spark counting with 169–173
stopping parameters of *276*
stress ratio 43, 44
in teaching 260–262
tracks in 56, *159*, *160*
track detection with 96, *133*–136, 154–161, 192, 246–253, *262*, 263, 266, 267
track formation in 26, 44, 263
track parameters 54, 61, 62
track profile technique 89, (126), 135
track profiles in *81*, *82*
track registration efficiency 11, 165, 184, 263
as track-storing solids 26, 263
V_B, V_T variation in 48, 89–92, 130–*132*
Polymeric detectors 250, 266
Polymeric resist 264
Pooley mechanism 25
Population method 206, 221
Pore radius 38, *38*, (255, 256)
Potential energy, in radon dosimetry 164, 165
Pre-annealing 89, (90), 98, (206), 246

Pre-etching 158, *178*, 179, *180*, 184
Pre-irradiation 89, 90
Pre-sparking 171, *172*
Primary ionization (*see also* Ionization) 17, *28*, 30, 114, 117
Profile, etched track, *see* Etched track profile
Projected length 58, 60, 61, *63*, 64, 82, 121, *122*, 126
Projectile ions/particles 246, 247, *252*
Prolonged etching 162
Prolonged etching factor, $f(t)$ 72–76, (111), 206, 209, 219
Prompt reactions 251, 252
Proton detection 7, (*28*), 145, 154, 155, 158, 161, 250, 263
Proton irradiations 13, 90, (160, 168), 228, 229, 249, 250, 265
Proton radiator 152, (158)
Proton recoils *155*, 158, *159*, 161, 263
Proton-sensitive plastics *155*, (263)
Proton shadow camera 265
Protons in cosmic rays 223, *224*, *229*, 230
PTFE foil 264
Pyroxene (*see also* Clinopyroxene, Orthopyroxene) *50*, 224–*226*, 228, 232–234, 249

Q-value *156*, 161
Quality factor 144, 148, *149*, 152
Quantimet 184, 185, 187
Quantitative fission track retention 96, (104)
Quantum mechanics 17
Quartz 50, 102, 161, 203, 247
Quark-nucleus complexes 267
Quaternary fission 247, *248*

Radiation damage (*see also* Damage trail) 1, (2), 10, 23–26, (33), 34, 37, (38), 42, (*43*), 87, 89, 90, (99), *100*, *117*, 127, (161, 179, *180*), 206, 218, (260), 265, 266
Radiation dosimetry, *see* Dosimetry
Radiation effects 24, 89, 260
Radiation exposure 129, 166
Radiation field 111, 144
Radiator foils 152, 154, 158, 250, 265
Radicals 24, 39, 90
Radioactive dating (*see also* Fission track dating and Dating) 199, 200
Radioactive series, natural 163, *163*, 164, *164*, 200, 266
Radiogenic (163, 164), 248, 249

Radiography (*see also* Mapping)   265
Radiohaloes   253
Radiolytic action   41
Radiotherapy   145, 158, 160, 168
Radium determination   176
Radius (*see also* Diameter)
   of unetched track   33, 38
   of etched track   38, *38*
Radius-restricted energy loss (RREL)   32, 42
Radon detection   *166*, *180*
Radon dosimetry   163–167, 172, 264
Radon emanation/emission   163, 166
Radon measurements   161–167, 264
Random Access Memory (RAM)   *188*, 189, *191*, 253
Range, *see* Etchable range
Range of alpha particles   (252), 261, *262*
Range of charged particles   275–277, *277*
Range–energy relationships   20, 21, *116–119*, (*120*, *123*), 130, 135, 191, 192, 275–277
Range of fission fragments   64–76, 99, 105, (110, 111), 145–147, (*178*), 201–203, 208, (219), *233* (236, *248*, 251, 252)
Range of ions   276, *277*
Range of pion (π^-)   168
Range of recoils   155, *155*
Range, residual (*see also* L-R plot)   76, *81*, *82*, *100*, 114–*117*, *120*–126, 130, *133*–140
Rare-gas atoms   225, 258, 260
Rare-earth elements   236
RBE (Relative biological effectiveness)   144, 148, *149*
Rb-Sr dating   200, 210, 216
Reactive species   40
Reactor dosimetry, *see* Dosimetry
Reactor irradiation   70, 161, (201, 206), 207, *208*, 212–214, 251
Recoils (*see also* Neutron recoils, Intrinsic tracks)   39, *153*, 154, *155*, 158, *159*, 161, 176, 181, 265, 266
   spallation   230
Registration/detection efficiency   51, 64, 69, 71, 90, 99, 105, 129, (130), 165, (*172*), 201, (207), 209, 228, 263
Registration/etchability threshold   (11), 27–29, (31, 32), 93, (100), 109, (114), 136, *137*, 139, (*159*), 161, 162, *162*, (*163*), 246, (263)
Reference age   232
Reference glass   202, 203, 212, *213*, 217, 229
Regolith   *259*, 260
Reichert-Jung IBAS   185

Relative biological effectiveness, *see* RBE
Relativistic correction term, δ 30
Relativistic energies/nuclei   267
Removed layer thickness   *52*, 63, 72–76, 78–86, 121, *122*, 126, *163*, (208)
Repair of damage (*see also* Annealing, Healing)   *97*, 98, 264
Replicas, of tracks   78, *88*, 89, 126, *170*, 171, (*252*)
Reprocessing plant   148, 256
Residual range, *see* Range, residual
Response functions/curves (V_T vs J)   92, 118, *119*, (*120*), 135
Response of neutrons (R_T)   149, 150
Resolution
   charge (*q.v.*)
   energy   249, (250)
Resolution of microscope   38, 87, 126
Restricted energy loss, *see* Energy loss
Revelation efficiency   201, 206, 221, 261
Rodyne polycarbonate   266
Rounded cone tip   61, (*62*, 184)
Rutherford scattering   *3*, 14, 15, 19, 262

Saturated vapour pressure   4
Scanning electron microscope, *see* SEM
Scanning of plastic sheets (for cosmic-ray tracks, etc.)   129, 130, 267
Scattering
   hard-sphere   15, 19
   light (*see also* Optical densitometry)   192
   low-angle X-ray   2, 34, *37*, 97
   Rutherford   *3*, 14, 15, 19, 262
Scintillation counter   192
Scintillation, track counting method   173–176
Seasoning of tracks   111
Secondary electrons (*see also* Delta rays)   19–21, 31, 40, 167
   model   40
Secondary interaction/particles   *7*, 228
Secondary reheating episode   *211*
Secondary standards   202, 212
Secular (radioactive) equilibrium   163–165
Seitz model   25
SEM (Scanning Electron Microscopy)   78, *84*, 87, *88*, 89, 126, 191, *256*, 259
Semiconductors   2, 24, 26, 44
Sensitivity (tracks/ECE spot; or per neutron)   (146), *160*, 184, 202, 217
Sequential fission   247, *248*
Shape discrimination   185
SHE, *see* Superheavy elements
Shock wave/dislocations   111, 236

Shortening of tracks (*see also* Annealing, Fading) (90), 99–*102*, (105), 110, 111, (*180*, 184), 212, 218–221, (*222*, 249)
Sievert (*see also* Dose equivalent) 144
Silver halide crystals 8, 10
Silver speck 8
Skylab 128, 167
Slip planes 111
Soft errors (in silicon memory devices) 249
Soil, lunar, *see* Lunar
Solar cosmic rays, *see* Cosmic rays
Solar flare tracks *35*, 90, (*224*), 225, 228, 257–260
Solar wind gases 90, (*224*), 225, 228, 260
Solid angle 64, *66*, 71, 262
Solid-state damage 127
Sources
    crystals (in contact method) 232–234
    external *65*, *66*, 70, 71, (72)
    internal 64–70, 72, (110), 147, (201)
    intermediate 70
    thick 64, *65*, 67–70, 74, (147)
    thin *66*, 69, 71, 261
Space dosimetry (128), 167
Space Shuttle 128
Spallation 225, (229), 230, *248*, 250
    products *248*, 250
Spark chamber 4, 6, 11
Spark counter 169–173, 184, 257
Sphene *50*, 87, *103*, 203, 206, 209, 210, 218, 221, *235*, 238, *276*
Spinel 27
Spontaneous fission of actinides 249
Spontaneous fission of ^{244}Pu 223, 230–234
Spontaneous fission of SHE 247–249
Spontaneous fission (of ^{238}U) 103, 147, 148, 200–203, *214*–216, 219, 223, *229*–231, 248
Sputtering 39, 45
SSNTDs (*see also* Track, Etched track)
    instrumentation for 169–192
    introduction to 1, 2, 10–12
Stainless steel, boron content 253
    electrode in ECE cell 176
Standards (*see also* Secondary standards) 251
    age *208*, 212, 216, 217
    glasses 202, 203, 212, *213*, 217, 229
Statistical tests 240
Statistics, Poisson 239
Stellar interiors 250
Stirring of etchant solution 51, 127
Stopping medium 13, 14, 17–19, 23, 30, (32, 39), 275, 277

Stopping parameters (275), *276*
Stopping power 21
    electronic 14, (17), *18*, 45
    nuclear 14
Storage of natural tracks 111
Strange particle *7*
Stratigraphy 217, (258, *259*)
Streamer formation 6
Stress ratio 43, 44
Superheavy elements (SHE) 135, 136, 247, 249
Surface-barrier detectors 247
Surface opening/etch-pit opening 51–63, 76–89, (*121*), *122*, (126), *134*, (135)

Target-detector arrangements *246*
Target-projectile arrangements *134*, 247
Target theory (in dose response) 40
Teaching experiments 260–*262*
Teflon 152, 204, 264
Tektites (*see also* Glasses, natural) *55*, *75*, *103*, 207, *220*, *235*, 238
TEM (Transmission Electron Microscopy) 2, *3*, 26, 33–36, 259
Temperature
    annealing (*see also* Annealing) 97–*103*
    closing, *see* Closing/closure temperature
    etching 90–93, 127, 181
    excursion 110, *211*, (*212*)
    irradiation 90, 129, 251, 263
    registration 267
Ternary fission 247, *248*
Thermal conductivity 27, 39, 41
Thermal excursions 110, (*211*), 212
Thermal fading of tracks (*see also* Fading, Annealing) 96 *et seq.*
Thermal history (*see also* Cooling history) 110, (111), 210–212, (*222*), 230
Thermal neutrons 145–148, 150, *153*, 171, 176, 201, 202, (*212*–)214, 229, (*252*), 253
Thermal-spike model 41, 42
Thermalized ion-explosion spike 45, 263
Thermoluminescence 12, 37, 129, 152, *153*, 237, 238, 265
Thermonuclear burn 265
Thermoset plastic 158
Thick-source geometry, *see* Sources, thick
Thin-film capacitor 173, *174*
Thin source geometry, *see* Sources, thin
Thoron (^{220}Rn) 163, 164, *164*, *166*, (167), 264
Three-pronged tracks 158, *160*, *248*
Threshold dosimeter *153*
Threshold energy (151), *156*, 158, 160, (162, 163)

Threshold for track registration, *see*
   Registration/etchability threshold
Through holes   129, 161, 169–*172*
Tissue-equivalent   160
Total energy loss, *see* Energy loss
Total etchable range, *see* Etchable range
Topological techniques   186, (187)
Tracer   167
Track(s) (*see also* SSNTDs, Etched tracks,
   Fission track dating, Damage trail)
   annealing   37; 96–111 (Chapter 5)
   annealing correction methods   218–*222*
   annealing temperatures for fission—
      101–*103*
   applications to nuclear physics
      131–136, 245–250
   charged-particle—, nature of   23–26
   confined—   111, *138*, 139
   cosmic ray—   128–131, 136–140,
      223–230
   contact—density method   232–236
   in dating   199–240 (Chapter 8)
   density gradients   (258), *259*
   density reduction   99
   detectors, further applications of
      245–262
   diameter   54, 79, 85
   in dosimetry   144–168
   early studies   1–*3*
   environmental effects on—etching   89–93
   etchability/etchable—   27, 31, 34–44,
      52, 90, *123*, 266
   etch parameters   *52*, *53*, *58*, *78*, 90, *122*
   etch pit, *see* Etch pit
   etch polymers   127
   etch rate function $V(R)$ *81*, *82*
   etch(ing) velocity (*see also* V_T)   48 et seq.
      constant   52–61
      varying   76–83
   etched—radius   *38*
   etching   48–93 (Chapter 4)
   etching efficiencies   64–76
      for external source of tracks   70–72
      for internal tracks   64–70
   etching geometry   51–87
      in anisotropic solids   83–87
      with constant V_T 52–64
      methodology   48, 135
      parameters   *78*, 87, 126
      with varying V_T 76–83
   etching recipes   49–51
   fading   96–111 (Chapter 5)
   fission—   3, 64–76, *103–108*, 145–152,
      170–174, 199 et seq., 245–249
   fission—age
      equation   200–203

interpretation of   209–212
practical steps in obtaining   203–209
^{244}Pu—equation   231–232; 233–234
fission—annealing temperatures, typical
   101–*103*
fission—dating   199–240 (Chapter 8)
   in archaeology   236–239
   errors in   239–240
   of lunar samples and meteorites
      223–230
fission—retention (*see also*—retention)
   effective   104
   quantitative   96
formation
   criteria   27
   in crystals/minerals   31, 263
   in dielectrics   265
   mechanisms   11, 23–33, *37*
   models   23, 27, 29–33, 41–44, 263,
      267
      critical appraisal of   39–45
   in plastics/polymers   2, 31, 41, 44, 263
fossil—   111, 218–*222*, 239
future applications   262–267
geometry formulae (for constant V_T)   *63*
host—   *138*
image-enhancement   169
-in-cleavage (TINCLE)   *138*, 139
-in-track (TINT)   133, 137, *138*
introduction to   1–*3*, 10–12
latent—(*see also* Damage trail)   129
length   *52*
length distributions   *138*
loss (%)   *97*, *103*, *235*
major and minor axes   54, *58–61*, *63*
nature of charged-particle—   23–26
non-retention interval   232–234
opening   51, 54, *58–63*
parameters   *52*, *53*, *58*, *78*, *122*
   determination of R and V_T 61–64
   special techniques for measurement of
      87–89
in particle identification   114–140
   (Chapter 6)
pit, *see* Etch pit
pre-annealing effect on etched—
   98–101
production rate   *227*, *258*, *260*
profiles   *62*, *77*, *81*, *82*
profile technique (TPT)   89, 126, 135,
   138
projected length   60–61, *63*, 82, 121, *122*
^{244}Pu fission   223, 230–236
   age equation   231, 233–234
   in very ancient (meteoritic/lunar)
      samples   230–236

Track(s) (*continued*)
in radiation dosimetry 145–168
radius 33, *38*
recoil— 39
α- 39, 109–110, 266
intrinsic 154, *155*
neutron 152–154, 158–161
spallation 230
recorders in particle identification
114–140 (Chapter 6)
reduction 101
replicas 78, *88*
retention 96, 104, 231–*235*
retention age 232, 234
retention characteristics of detectors
103
retention temperatures *103, 104, 235*
-rich grains 225, 260
seasoning 111
shapes (*see also*—profiles) 54
shortening 99, 212, 218–*222*
size 38
size correction method 218–*220*
size reduction 96
solar flare— *35*, 225, 228, 258, 260
source, internal and external 64–72
spallation recoil— 230
storage at elevated temperatures 111
storing materials 26–27
structure of unetched— 33, 39
surface opening 54, *58–63*
in teaching 260–262
thermal annealing of—, *see*—annealing
thermal fading of—, *see*—fading
threshold for—etchability/storage 26, 27
threshold for—formation 26, 27
tip 54, 59, 61, 179
triple-α— 158, *160*
unetched—radius 33, 38
^{238}U— 199–240 (Chapter 8)
width of etched— 38, 85
width of unetched— 34
Track-in-Cleavage (TINCLE) 111, 135,
138, 139, 225
Track-in-Track (TINT) 110, 111, 133,
137, *138*, 225
Transient heating/temperature 110, (211)
Transmission electron microscope, *see* TEM
Trapping centres 8
Treeing *178*, 179, 181, 184
Triple sequential fission *248*
Triple tracks, *see* Three-pronged tracks
Troposphere 167
Tuffak polycarbonate 250
Two-stage etching 139, (*140*)
Type B film (LR 115, CN 85) 148

^{235}U 64, 145–148, 150–152, 163, 172,
201, 212, *229*, 250, 253, 254
^{238}U 12, 64, 101, 103, 105, 109, 139, 146,
147, *151*, 163, 164, 199–201, 212, 214,
215, 223, *229*, 231–234, 236, 246–*248*,
255, *256*
U-bearing contaminations 207
U-bearing crystals 103
U-bearing glasses (*see also* Reference glass)
228
U-foil *3*, 146–148, 229
U-poor/U-rich phases 225, 228, 230,
232–236
U-rich crystal/material 213, 228
Ultraheavy cosmic rays, *see* Cosmic rays
Ultrasonic agitation 127
Ultraviolet light 2, 8, 10, 24, 89, 90, 127,
132, *133*, 140, *163*
Uranium content/concentration 167, 171,
(199), 201–209, 212, 239, 240, 247,
252, 253
Uranium exploration *166*, 167
Uranium ore 165
Uranium-poor/rich, crystals/phases, *see*
U-poor, etc.
Uranium radiator 265

$V(=V_T/V_B)$ *53–63*, 78–86, *92*, 184, (209)
V_B 48, 51–*92*, *122*, 125, (184), 209
V_T 27, 29, 48, 51–*92*, 114–128, 130–*132*,
162, *178*, 179, 209
$V_{T_{eff}}$ 131, *132*
$V_T - J$ curve 118, 136
V_T (J) function 31, *119*, 128, 130, (135)
$V_T - J$ plot *118, 119*
$V_T - R$ curve/profile 117, 118, *119*
V_T (R) function 115, 116, 121, 126
Vacancies
inner-shell 267
lattice (*q.v.*)
Van Allen belts *224*
Varley mechanism 25, 42
VH group of cosmic rays, *see* Cosmic rays
Videoplan 189
Video cassette recorder (VCR) 191
Video position analyser (VPA) 189, *190*,
191
Vidimet 186, 187
VIDS 189
Viruses 256
Voids 34
VVH group of cosmic rays, *see* Cosmic
rays

W 37
W$_0$ *29*, 31, 32, 92, *277*
Wearer-dose *151*, 154
Whitlockite, *see* Merrillite
Wilson cloud chamber *4*, 5
Working level (WL) 165

Xenon
  fission product 230–232
  heavy ions 247
  retention 232

X-ray diffraction 34
X-ray fluorescence 238
X-ray scattering 34, *37*, 97

Young's Modulus 43

Z_{eff}, *see* Effective charge
Zircon *50*, *103*, 203, 204, 206, 209–*211*,
  235, 238, *276*

Titles in the International Series in Natural Philosophy

Vol.   1. Davydov—Quantum Mechanics (2nd Edition)
Vol.   2. Fokker—Time and Space, Weight and Inertia
Vol.   3. Kaplan—Interstellar Gas Dynamics
Vol.   4. Abrikosov, Gor'kov and Dzyaloshinskii—Quantum Field Theoretical Methods in Statistical Physics
Vol.   5. Okun'—Weak Interaction of Elementary Particles
Vol.   6. Shklovskii—Physics of the Solar Corona
Vol.   7. Akhiezer *et al.*—Collective Oscillations in a Plasma
Vol.   8. Kirzhnits—Field Theoretical Methods in Many-body Systems
Vol.   9. Klimontovich—The Statistical Theory of Nonequilibrium Processes in a Plasma
Vol. 10. Kurth—Introduction to Stellar Statistics
Vol. 11. Chalmers—Atmospheric Electricity (2nd Edition)
Vol. 12. Renner—Current Algebras and their Applications
Vol. 13. Fain and Khanin—Quantum Electronics, Volume 1—Basic Amplifiers and Oscillators
Vol. 14. Fain and Khanin—Quantum Electronics, Volume 2—Maser Theory
Vol. 15. March—Liquid Metals
Vol. 16. Hori—Spectral Properties of Disordered Chains and Lattices
Vol. 17. Saint James, Thomas and Sarma—Type II Superconductivity
Vol. 18. Margenau and Kestner—Theory of Intermolecular Forces (2nd Edition)
Vol. 19. Jancel—Foundations of Classical and Quantum Statistical Mechanics
Vol. 20. Takahashi—An Introduction to Field Quantization
Vol. 21. Yvon—Correlations and Entropy in Classical Statistical Mechanics
Vol. 22. Penrose—Foundations of Statistical Mechanics
Vol. 23. Visconti—Quantum Field Theory, Volume 1
Vol. 24. Furth—Fundamental Principles of Theoretical Physics

Vol. 25. Zheleznyakov—Radioemission of the Sun and Planets
Vol. 26. Grindlay—An Introduction to the Phenomenological Theory of Ferroelectricity
Vol. 27. Unger—Introduction to Quantum Electronics
Vol. 28. Koga—Introduction to Kinetic Theory: Stochastic Processes in Gaseous Systems
Vol. 29. Galasiewicz—Superconductivity and Quantum Fluids
Vol. 30. Constantinescu and Magyari—Problems in Quantum Mechanics
Vol. 31. Kotkin and Serbo—Collection of Problems in Classical Mechanics
Vol. 32. Panchev—Random Functions and Turbulence
Vol. 33. Taipe—Theory of Experiments in Paramagnetic Resonance
Vol. 34. ter Harr—Elements of Hamiltonian Mechanics (2nd Edition)
Vol. 35. Clarke and Grainger—Polarized Light and Optical Measurement
Vol. 36. Haug—Theoretical Solid State Physics, Volume 1
Vol. 37. Jordan and Beer—The Expanding Earth
Vol. 38. Todorov—Analytical Properties of Feynman Diagrams in Quantum Field Theory
Vol. 39. Sitenko—Lectures in Scattering Theory
Vol. 40. Sobel'man—Introduction to the Theory of Atomic Spectra
Vol. 41. Armstrong and Nicholls—Emission, Absorption and Transfer of Radiation in Heated Atmospheres
Vol. 42. Brush—Kinetic Theory, Volume 3
Vol. 43. Bogolyubov—A Method for Studying Model Hamiltonians
Vol. 44. Tsytovich—An Introduction to the Theory of Plasma Turbulence
Vol. 45. Pathria—Statistical Mechanics
Vol. 46. Haug—Theoretical Solid State Physics, Volume 2
Vol. 47. Nieto—The Titius-Bode Law of Planetary Distances: Its History and Theory
Vol. 48. Wagner—Introduction to the Theory of Magnetism
Vol. 49. Irvine—Nuclear Structure Theory
Vol. 50. Strohmeier—Variable Stars
Vol. 51. Batten—Binary and Multiple Systems of Stars
Vol. 52. Rousseau and Mathieu—Problems in Optics
Vol. 53. Bowler—Nuclear Physics
Vol. 54. Pomraning—The Equations of Radiation Hydrodynamics
Vol. 55. Belinfante—A Survey of Hidden Variables Theories
Vol. 56. Scheibe—The Logical Analysis of Quantum Mechanics
Vol. 57. Robinson—Macroscopic Electromagnetism
Vol. 58. Gombas and Kisdi—Wave Mechanics and its Applications
Vol. 59. Kaplan and Tsytovich—Plasma Astrophysics
Vol. 60. Kovacs and Zsoldos—Dislocations and Plastic Deformation
Vol. 61. Auvray and Fourier—Problems in Electronics
Vol. 62. Mathieu—Optics
Vol. 63. Atwater—Introduction to General Relativity
Vol. 64. Muller—Quantum Mechanics: A Physical World Picture
Vol. 65. Bilenky—Introduction to Feynman Diagrams

Vol. 66. Vodar and Romand—Some Aspects of Vacuum Ultraviolet
Radiation Physics
Vol. 67. Willett—Gas Lasers: Population Inversion Mechanisms
Vol. 68. Akhiezer *et al.*—Plasma Electrodynamics, Volume 1—Linear Theory
Vol. 69. Glasby—The Nebular Variables
Vol. 70. Bialynicki-Birula—Quantum Electrodynamics
Vol. 71. Karpman—Non-linear Waves in Dispersive Media
Vol. 72. Cracknell—Magnetism in Crystalline Materials
Vol. 73. Pathria—The Theory of Relativity
Vol. 74. Sitenko and Tartakovskii—Lectures on the Theory of the Nucleus
Vol. 75. Belinfante—Measurement and Time Reversal in Objective
Quantum Theory
Vol. 76. Sobolev—Light Scattering in Planetary Atmospheres
Vol. 77. Novakovic—The Pseudo-spin Method in Magnetism
and Ferroelectricity
Vol. 78. Novozhilov—Introduction to Elementary Particle Theory
Vol. 79. Busch and Schade—Lectures on Solid State Physics
Vol. 80. Akhiezer *et al.*—Plasma Electrodynamics, Volume 2
Vol. 81. Soloviev—Theory of Complex Nuclei
Vol. 82. Taylor—Mechanics: Classical and Quantum
Vol. 83. Srinivasan and Parthasathy—Some Statistical Applications in
X-Ray Crystallography
Vol. 84. Rogers—A Short Course in Cloud Physics
Vol. 85. Ainsworth—Mechanisms of Speech Recognition
Vol. 86. Bowler—Gravitation and Relativity
Vol. 87. Klinger—Problems of Linear Electron (Polaron) Transport Theory
in Semiconductors
Vol. 88. Weiland and Wilhelmson—Coherent Non-Linear Interaction
of Waves in Plasmas
Vol. 89. Pacholczyk—Radio Galaxies
Vol. 90. Elgaroy—Solar Noise Storms
Vol. 91. Heine—Group Theory in Quantum Mechanics
Vol. 92. Ter Haar—Lectures on Selected Topics in Statistical Mechanics
Vol. 93. Bass and Fuks—Wave Scattering from Statistically
Rough Surfaces
Vol. 94. Cherrington—Gaseous Electronics and Gas Lasers
Vol. 95. Sahade and Wood—Interacting Binary Stars
Vol. 96. Rogers—A Short Course in Cloud Physics (2nd Edition)
Vol. 97. Reddish—Stellar Formation
Vol. 98. Patashinskii and Pokrovskii—Fluctuation Theory of Phase Transitions
Vol. 99. Ginzburg—Theoretical Physics and Astrophysics
Vol. 100. Constantinescu—Distributions and their Applications in Physics
Vol. 101. Gurzadyan—Flare Stars
Vol. 102. Lominadze—Cyclotron Waves in Plasma
Vol. 103. Alkemade—Metal Vapours in Flames
Vol. 104. Akhiezer and Peletminskii—Methods of Statistical Physics

Vol. 105. Klimontovich—Kinetic Theory of Non-Ideal Gases and Non-Ideal Plasmas

Vol. 106. Ginzburg—Waynflete Lectures on Physics

Vol. 107. Sitenko—Fluctuations & Nonlinear Interactions of Wave in Plasma

Vol. 108. Sinai—Theory of Phase Transitions: Rigorous Results

Vol. 109. Davydov—Biology and Quantum Mechanics

Vol. 110. Demianski—Relativistic Astrophysics

Vol. 111. Durrani and Bull—Solid State Nuclear Track Detection